Reason and Less

Reason and Less

Pursuing Food, Sex, and Politics

Vinod Goel

The MIT Press
Cambridge, Massachusetts
London, England

The MIT Press would like to thank the anonymous peer reviewers who provided comments on drafts of this book. The generous work of academic experts is essential for establishing the authority and quality of our publications. We acknowledge with gratitude the contributions of these otherwise uncredited readers.

This book was set in Stone Serif and Stone Sans by Westchester Publishing Services.

Library of Congress Cataloging-in-Publication Data

Names: Goel, Vinod, author.
Title: Reason and less : pursuing food, sex, and politics / Vinod Goel.
Description: Cambridge, Massachusetts : The MIT Press, [2022] |
    Includes bibliographical references and index.
Identifiers: LCCN 2021017752 | ISBN 9780262045476 (paperback)
Subjects: LCSH: Decision making. | Reasoning. | Logic. | Cognitive neuroscience.
Classification: LCC BF448 .G64 2022 | DDC 153.4/3—dc23
LC record available at https://lccn.loc.gov/2021017752

152941459

To the memory of my mother and father

For Kalpna, Amit, and Natasha,
who taught me there is often less
to life than reason

Contents

Preface

After more than 20 years of studying the neural basis of rationality, it dawned on me that there was very little consequential human behavior that I could explain. Nothing I have learned about rationality was relevant to understanding my teenage daughter. Nothing I have learned about rationality is relevant to explaining the behavior of my MAGA (Make America Great Again) Florida friends and neighbors who profess an unshakable faith in American exceptionalism (which I accept and have benefited from) but then deny and ridicule the sciences of vaccines and climate change emerging from exceptional American institutions. Nothing I have learned about rationality seems particularly relevant to explaining certain views of my ultraliberal friends and colleagues, such as gender being just a social construct, despite scientific evidence to the contrary. Nothing I have learned about rationality is relevant to explaining why intelligent, powerful men engage in sexual indiscretion, even assault, at great personal risk and harm to others. Nothing I have learned seems particularly relevant to explaining why I overindulge in chocolate cake and pizza, despite being overweight. Based on the standard models of reasoning, the only explanatory tools available are appeals to "heuristics," some form of "motivated reasoning," poor education, or perhaps cognitive deficiency. Such explanations may apply in specific individual cases, but they cannot account for all or even much of human behavior. I have come to believe that we are making a fundamental mistake in bringing only the tools of rationality to explain human behavior.

My main message is that, while we *are* rational animals, explaining real-world human behavior just in terms of reasoning does not get us very far. We have to recognize that nonreasoning systems also affect actual behavior. We need to look beyond (or below) reason to *noncognitive* factors to fully account for human behavior. Much human behavior that does not conform to our expectations of rationality is not irrational but rather *arational*, by which I mean that it is not reason based. Some nonreasoning systems are initiating and/or modulating the behavior.

The goal of this book is to undertake a commonsense reconsideration and recalibration of theories of human behavior. Human behavior needs to be explained in terms of the workings of autonomic systems, instinctive systems, associative systems, and reasoning systems. Each of these systems has been extensively studied. How these systems communicate and interact to account for human behavior is rarely considered. I sketch out a proposal that I call *tethered rationality*, in which human behavior is a *blended response* incorporating inputs from each of these systems. The challenges are to provide empirical data for the blended response hypothesis, show how the tethering is supported by the neurophysiology, propose a common currency that would allow these systems to communicate and interact, and provide a control structure for the overall system. Meeting these challenges takes us on a fascinating journey through psychology (cognitive, behavioral, developmental, and evolutionary), neuroscience, philosophy, ethology, economics, and political science, among other disciplines.

One key insight that holds the model together is that *feelings*—generated in old, widely conserved brain stem structures—are evolution's solution to initiating and selecting all behaviors and provide the common currency for the four different systems to interact. Reason is as much about feelings as is lust and the taste of chocolate cake. All systems contribute to behavior and the overall control structure is one that maximizes pleasure and minimizes displeasure. Such an account drives human behavior back into the biology, where it belongs, and provides a richer set of tools to understand how we pursue food, sex, and politics.

Models not only explain behavior but also have consequences for changing it. The model of tethered rationality is no exception. For those engaged in changing behaviors—such as sexism, racism, cheating, or even climate change denial—tethered rationality may have the unwelcome message that such behaviors cannot be easily changed by changing beliefs through a few days of "sensitivity training." This is not to say that they cannot be changed at all, but rather that more drastic measures will be required, the nature of which will depend on the specific behavior in question. Having an accurate model of human behavior is the first step in this endeavor.

Utopia, Ontario, Canada

May 2021

Acknowledgments

I would like to thank the following colleagues from diverse areas of expertise for reading one or more chapters of the manuscript and offering valuable feedback: Ellen Bialystok, Christopher von Bülow, Ron Chrisley, Hugo Critchley, Wim De Neys, Shira Elqayam, Larry Fiddick, Jordan Grafman, Sam Gilbert, Ira Novak, Magda Osman, Jerome Prado, Steven Sloman, and Hongkui Zeng. Among friends and students, I'm grateful to Mark Hewitt, Ron Bean, Adam Burnett, and Claire Quenneville, for reading and commenting on multiple chapters. I thank Sophie Goss and Jenna Zorik for reading all chapters multiple times and forcing me to articulate the story more clearly. Finally, I am grateful to my American MAGA friends and neighbors for many months of discussions that provided insight into their underlying thought processes and confirmed what I had learned from my teenage daughter: there is indeed often less to life than reason.

1 The Rational Animal

Man is the only animal capable of reasoning, though many others possess the faculty of memory and instruction in common with him.
—Aristotle

There's a logical explanation for everything, often mistaken for the reason it happened.
—Robert Breault

To ask questions about the role of reason in human affairs is, in the broadest sense, to ask questions about our place in the universe. What is the nature of man? Who and what are we? We have struggled with such questions for as long as we have been able to think about such things. Are we reasoning animals? Are we only reasoning animals? Is reason necessary? Is it sufficient? What ever happened to the "animal passions"? Have socialization and culture—constructions of the reasoning mind—allowed us to rise above them (like Katharine Hepburn's character in the film *The African Queen* advocated [Huston, 1951]: "Nature, Mr. Allnutt, is what we were put on this world to rise above"), or do we need an account of human nature that reconciles the two? The reader will guess from the title of the volume that I make the case for the latter.

1 Food, Sex, Politics, and the Rational Animal

To proceed on this track, investigators would need to accept one grand but empirically robust premise—that higher aspects of the human mind are still strongly linked to the basic neuropsychological processes of "lower" animal minds.

—Jaak Panksepp

Much of life is about pursuing food, sex, and politics. Any adequate theory of human behavior must be able to explain these pursuits.

By far the most popular academic accounts of human behavior place the *rational* mind front and center (Cassirer, 1944; Durkheim, [1895] 2014; Simon, 1955). Humans bring the tools of reason to bear on these problems. Reason sets us apart from other animals. It allows us to successfully pursue not only food, sex, and politics but also art, science, and technology. This model is often referred to as the standard cognitive or social science reasoning model of human behavior (Tooby & Cosmides, 1995). After more than 20 years of trying to understand human decisions and choices just through the lens of reason, I have become skeptical of the explanatory scope of this standard model.

I'm convinced that reason is an integral part of who and what we are. I'm also convinced that, on its own, it is inadequate to explain much, if not most, real-world human behavior. It is only half the story. We do not have to look very far to understand what is missing. There is a commonsense model of behavior, embedded in the Western-Christian intellectual tradition, that recognizes not only reason but also "animal passions" (often characterized as the four Fs: feeding, fornicating, fighting, and fleeing) as determinants of human behavior. Our choices and decisions are a function of both. Not only is this much more intuitive, but we will see that the data demand such a model.

Despite common sense and data, such a model no longer gets serious consideration in large segments of modern society, including much of academia. I worry that the main reason is that many people, some academics included, hold variations on the meritless belief that "humans no longer need to rely on instinct to survive, not when we have education, technology, and social norms" (Pomeroy, 2011). The goal of this book is to push back against this widespread misconception, and articulate a commonsense model of human nature, called *tethered rationality*, that preserves the basic intuitive insight of the Western-Christian model—that both reasoning and nonreasoning systems are in play in human behavior—and can be discharged without divine intervention.

The "animal passions," or nonreasoning behaviors in technical parlance, include autonomic behaviors, instinctive behaviors, and associative learning behaviors. These behaviors and their underlying mechanisms have been studied extensively over the past hundred years. They differ not only from reasoned behaviors but also from each other. They are hierarchically organized in terms of appearance on the evolutionary tree, are integrated, and are widely available across species, including humans. Humans also exhibit reasoning or rational behavior, which (I will argue) is unique to us. However, it does not supplant the evolutionarily older behaviors. Reason evolved on top of them, but it does not "float" untethered above them; it is tightly integrated with both bottom-up and top-down connections. This means that human behavior is a blended function of all these systems, not just reason (or any other individual system). Humans have a reasoning mind, but it is tethered to and modulated by evolutionarily older associative, instinctive, and autonomic minds.

I begin this chapter by introducing five examples of real-world decisions that are widely thought to be explained by reason. Before we can consider whether these examples are actually explained by models of reasoning, we need to introduce the notion of reason and rationality. This is initially done informally. With this preliminary understanding of reasoning in hand, I then evaluate each example to see if it can be explained just in terms of reason. I conclude that four of the five examples cannot be so explained. Satisfactory explanations for these require the introduction of evolutionarily older nonreasoning systems. A roadmap is then provided to foreshadow the argument for the model of tethered rationality and guide the reader through the subsequent chapters.

Examples of Reasoning in the Real World

Let's begin by considering five real-world examples of reasoning and decision-making scenarios.

The first example is climate change, the ultimate existential issue of our time. The best science we have agrees that human activity is contributing to rising temperatures, which will reshape planetary weather patterns and geography and have detrimental, even catastrophic, effects on all life on earth. The scientific models could be wrong by either overestimating or underestimating the changes that will occur, but they provide the best information we currently have. Most governments and citizens accept the science and are willing to take some (limited) steps to mitigate the impact of human activity. However, the forty-fifth president of the United States, a number of US senators, and 40% of the American public believe that "man-made global warming is the greatest hoax ever perpetrated on the American people" (Revkin, 2003). They claim, without evidence, that the scientific models are incorrect. Even among the other half of Americans who do accept the science, there is considerable reluctance to undertake full remedial measures. This example illustrates two separate issues: that many people simply deny the science, without evidence to the contrary, and others seem to accept the science but fail to act on it. There seems to be a lack of rationality in both cases.

The second example involves weight management. Last year, I went to my doctor's office for my annual checkup. After I stepped on the scale, my doctor advised me to lose 30 pounds. I agreed but complained that my busy schedule did not allow time to eat healthy meals and exercise regularly. My doctor replied, "What fits your busy schedule better, eating healthy and exercising one hour a day or being dead 24 hours a day?" Many of us have been in this situation, but few of us actually manage to follow our doctor's advice. Notice that we do not question the doctor's judgment. There seems to be considerable evidence linking obesity with the onset of various diseases (e.g., diabetes and heart disease) and premature mortality. Most of us do not have a death wish. Given that we want to live a long, healthy life, and given that we accept that obesity will impair and even shorten our lives, the rational, reasonable thing to do would be to lose weight. So, why don't many of us comply with our doctor's advice?

For our third example we turn to sex. In December 2006, John Edwards, a handsome, charismatic lawyer and politician, announced his candidacy for the 2008 Democratic nomination for president of the United States. He was among the frontrunners, along with Barack Obama and Hillary Clinton, for the nomination. In March 2007, it was revealed that his wife, Elizabeth, was suffering from stage IV breast cancer. Shortly thereafter, it came to light that he was having an affair with one of his campaign workers. In what world was this a rational choice? He was running for the highest office in the world, in a country that contains some of the most socially conservative, prudish, judgmental, evangelical voters. He must have known that if there was any hint of infidelity—even in the best of circumstances—his campaign was over. His circumstances were such that his wife was dying of cancer and receiving enormous emotional and moral support from the public. Any hint of infidelity in such circumstances would be suicidal. Evidence of the affair emerged in early 2008 and ended his candidacy overnight. How do we explain his choices?

The fourth example concerns healthcare, a topic that often comes up in discussions with my American friends. The conversations often take the following form:

Me: Given your very high premiums and the large deductible in your private healthcare plan, why don't you support overhauling your healthcare system into a universal Canadian/European-type system whereby everyone can receive good equivalent healthcare at a lesser cost?

My American friend: Affordable healthcare would certainly be a great benefit to me. However, you see that guy over there? Yes, that one. He doesn't work. He doesn't pay taxes. He is a freeloader. If we had universal healthcare, he would get the same healthcare that I do, but he doesn't *deserve* it. Therefore, I cannot support a universal system. (Another interesting response is the admission that, "yes, that would probably be better than what we have," followed by passing shame and a disappointed sigh, "but that would be socialism.")

My friend is willing to forgo a benefit for himself just so that someone "undeserving" does not receive an equivalent benefit. Again, it is hard to see the rationality in this choice.

For the fifth example we turn to a drug warning issued to doctors and patients by the UK Committee on Safety of Medicines in 1995. The warning stated that the third generation of birth control pills doubled (i.e., increased by 100%) the risk of life-threatening blood clots in the legs and/or lungs. Unsurprisingly, this caused great anxiety among women and resulted in a sharp increase in unwanted pregnancies and abortions in subsequent years. A closer examination of the study showed that for every 7,000 women who took the second-generation pill, one developed thrombosis. By contrast, for every 7,000 women who took the third-generation pill, two developed thrombosis. So, while the relative risk did increase by 100% as advertised, the absolute risk was an increase of 1 in 7,000 women (Gigerenzer, 2015). This hardly seems to warrant the panic that ensued, so how can we explain it?

These are five (very different) examples of everyday, real-world decisions or choices. Other examples will be introduced throughout the book. Even though I have not yet formally introduced the idea of "rationality," I'm confident most readers will agree that each example illustrates a choice that seems less than fully rational. I will not go so far as to say that they are *irrational*. In the cases of examples one through four, I will argue that they are *arational*—that is, they involve noncognitive factors.

The most popular academic models that we have for explaining these behaviors are the cognitive reasoning and decision-making models, buttressed by distinctions between analytic and "heuristic" reasoning, such as the "fast and slow" thinking model popularized by Daniel Kahneman (2012), or by notions of motivated reasoning (Kunda, 1990) or even sloppy reasoning (Pennycook & Rand, 2019). Such models will be introduced and considered in chapters 7 and 13. They provide satisfactory explanations for a number of phenomena, including example five, but lack the requisite machinery to deal convincingly with examples one through four, which are the ones of interest in this book.

To make sure we are all on the same page, I offer an initial introduction to the notion of rationality and decision-making and then return to address the preceding examples.

What Is Rationality?

Man is widely considered to be the "reasoning" or rational animal. But what does this mean? To invoke reason or rationality is to say that human behaviors or actions are explained by postulating beliefs and desires and a principle of *coherence* that guides our pursuit of the latter in the context of the former. By coherence I mean roughly "making sense." Coherence is a relationship that holds between thoughts, propositions, or sentences. In the first instance, it is a basic, primitive, intuitive notion, though it can be considerably enhanced with education. For example, if I believe that all Americans are intelligent, and all Fox News viewers are American, then it would be coherent or reasonable for me to infer that all Fox News viewers are intelligent. Given the same beliefs, it would not be coherent to infer that no Fox News viewers are intelligent. This example illustrates a particularly extreme case of coherence found in deductive arguments, referred to as *validity*, where the truth of the given information (or beliefs) is sufficient to guarantee the truth of the conclusion, but it is worth noting that validity does not evaluate the veracity of the premises that all Americans are intelligent and all Fox News viewers are Americans; it merely determines whether a conclusion follows from or is entailed by them. We can consider validity as coherence in the narrow sense of the term and additionally have a broader sense of the term, corresponding to *soundness* in logic, that also takes into consideration the veracity of the premises. In this broader use of the term, we would step back and evaluate (and either accept or reject) the truth of the premises before drawing the inference.

On a recent trip to New Delhi, India, one afternoon I observed Indian fruit bats dangling from tree branches like so many brown and black cloth sacks. Based on this observation, I formulated the belief that Indian fruit bats spend the afternoon dangling from tree branches. This is a plausible or coherent inference based on my observations, but notice that it lacks the certainty of the preceding inference about Fox News viewers. Further observations (or consultation with bat experts) might reveal that this behavior is a peculiar habit of fruit bats in this particular region of India. In this case, I would have to modify my belief for it to be consistent with the facts in the world. Absent additional information, it is coherent for me to believe that Indian fruit bats spend afternoons dangling from tree branches. Given the same evidence, it would be incoherent for me to conclude that Indian fruit bats do not spend the afternoon dangling from tree branches or spend the afternoons diving for crayfish in shallow rivers.

Coherence relations between premise and conclusion are disrupted by inconsistency, indeterminacy, or irrelevance. Inconsistency is illustrated where the conclusion "No Fox News viewers are intelligent" is drawn from the beliefs that "All Americans are intelligent" and "All Fox News viewers are Americans." An example of indeterminacy occurs if I tell you "Mary is taller than George, and Mary is taller than Michael" and ask you the height relationship between George and Michael. The premises do not provide sufficient information to draw any inferences about the relative heights of George and Michael. An example of failure of coherence through irrelevance would occur if, given the belief that all Americans are intelligent and the belief that Indian fruit bats spend afternoons dangling from tree branches, I conclude that global warming is caused by human activity. In this case, the issue of coherency does not even arise, because the three propositions are unrelated.

From Rationality to Decision Theory

Reasoning is about maintaining coherence in belief networks. Life is about actions. Reason mediates action by determining choices consistent with specific goals, given specific beliefs. Choice selection is studied by decision theory. We get from reasoning to decision-making by overlaying some model of human goals on top of the model of rationality. These models are usually based on maximizing self-interest. A historically popular one is the *Homo economicus* model. In this account, man is intrinsically a self-interested utility maximizer as a consumer and a self-interested profit maximizer as a producer.[1] These become the goals of the individual. Rational actions are those that are expected to advance goals in light of beliefs.

I will illustrate this standard model of decision-making with the controversial US decision to invade Iraq in March 2003. While I have no privileged access to the particulars of the decision-making process, its overall *form* would be something like that depicted in figure 1.1. It would begin with a goal or desire that needs to be achieved, such as securing the Iraqi oil leases. This goal would be explored or expanded via subgoals. One subgoal option might be negotiation. Another might be to take the oil by force if certain conditions can be met, such as: assurance of success, clean surgical intervention and withdrawal, that the value of the oil leases be greater than the cost of the invasion, and that Iraq be able to pay for its own reconstruction costs. In this example, these conditions are believed to be met

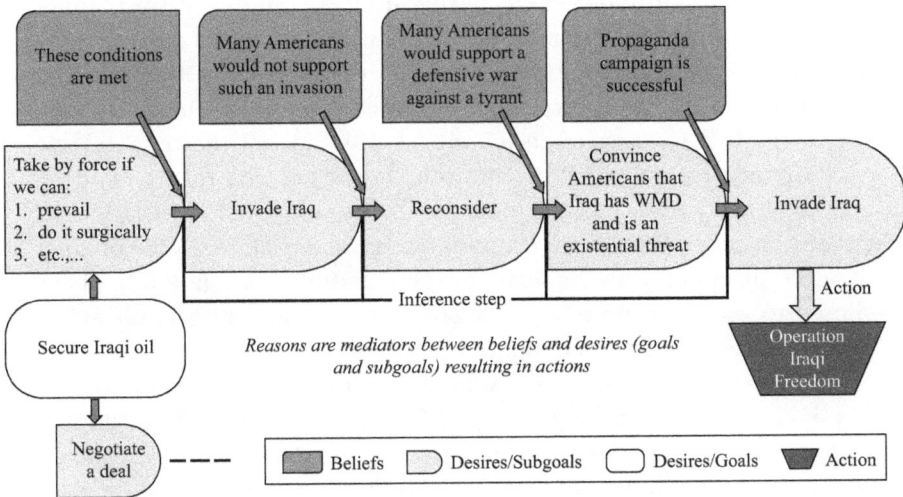

Figure 1.1
An example of the rational mind at work using a hypothetical reconstruction of the US decision to invade Iraq in 2003. Each subgoal follows coherently from the preceding goal or subgoal plus beliefs, eventually resulting in an action. The integration of goals and subgoals plus beliefs via the coherence relation is the nexus of the reasoning step.

(and negotiation is not considered feasible or cost-effective), leading to the subgoal of invading Iraq. However, there are accompanying beliefs that suggest most Americans (and the world community) will not support an unprovoked invasion, even if it means access to cheap oil. This results in another subgoal to pause and reconsider. There are accompanying beliefs that most Americans (and the world community) would support a defensive war against a tyrant. This generates another subgoal of launching a campaign to vilify Saddam Hussein and convince Americans that Iraq has weapons of mass destruction that are an imminent threat to the United States (which has more weapons of mass destruction than all other countries combined) and its allies. It is determined that the propaganda campaign is successful and there is sufficient support within the country for the invasion. Given all this, the rational decision is to invade Iraq; each step follows coherently from the previous goal or subgoal plus beliefs.

However, this model is an oversimplification. It assumes that the beliefs or information at hand are complete and certain. But how certain are we that Iraq can repay its own reconstruction costs? 100%? 10%? 73%? Are there any constraints on the desire to take the oil by force? If the financial cost of the war equals or exceeds the benefits of the oil, do we still want to pursue this desire? In real-world situations, information is always incomplete and uncertain, and even the relative utility of different desires cannot be confidently ascertained and ordered. These complications transform the problem of inferential coherence from the realm of logic to the realm of probability theory (see figure 1.2). Coherency is then determined by applying the probability calculus to the model. The rational choice is the one with the highest utility value. One consequence of this shift is that the criterion of coherence morphs to an optimality criterion. However, for our current purposes, these complications are not material. It is still coherent to select the option with the highest expected utility (see figure 1.2). I have chosen to use the concept of coherence rather than utility as central to rational decision-making throughout the book.

This example is offered as a simplified illustration of the machinery of standard decision-making models. There are two points worth noting. First, the postinvasion justification (when no weapons of mass destruction were found)—that American lives and resources were expended so the Iraqi people could benefit from regime change and democracy—is irrational because it violates the basic tenets of maximizing self-interest. Second, I'm not claiming that this rational model is sufficient to explain the invasion of Iraq. On the contrary, I'm certain that a number of nonrational factors considered in this book were significant factors in making the decision.

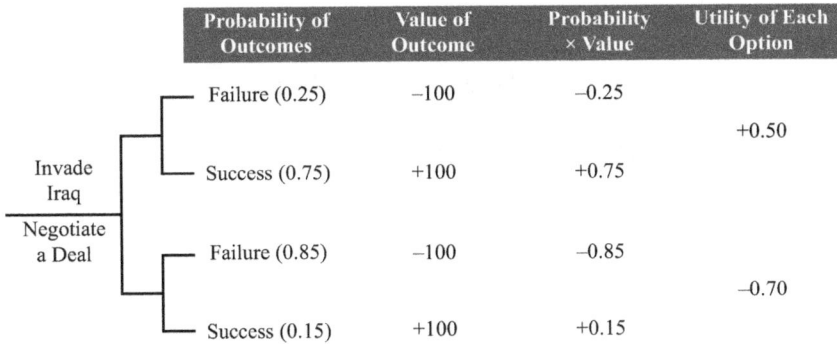

Probability of Outcomes	Value of Outcome	Probability × Value	Utility of Each Option
Failure (0.25)	−100	−0.25	
			+0.50
Success (0.75)	+100	+0.75	
Failure (0.85)	−100	−0.85	
			−0.70
Success (0.15)	+100	+0.15	

Invade Iraq

Negotiate a Deal

Figure 1.2
Simple decision tree and utility function. One might model the decision to invade or make a deal as follows. The chances of a successful invasion are 0.75, while the chances of failure are 0.25. The chances of a successful negotiation are 0.15, while the chances of failure are 0.85. The value assigned to both the successful invasion and successful negotiation is +100. The value assigned to a failed invasion and failed negotiation is −100. Based on these values, the utility of invasion is +0.5 and the utility of negotiation is −0.7 (utility=Σ(probability$_{outcome}$×value$_{outcome}$)). Notice that decision theory provides no guidelines for assigning probabilities of outcomes and the value of the outcomes, but once these numbers are (magically) assigned, simple probability theory allows us to coherently calculate expected utility. The rational choice is the one with the highest expected utility.

More generally, I'm claiming that such standard models of rationality cannot adequately account for much of human behavior, including the invasion of Iraq and four of the five examples introduced earlier. Understanding this claim requires reviewing each example more closely, beginning with global warming.

Rationality in the Real World: Global Warming Example

The basic questions around climate change are "Is the earth warming?" and "Is human industrial activity contributing to it?" Most scientists answer "yes" to both questions (The National Academy of Sciences & The Royal Society, 2020). Many members of the public agree, but at least 40% of Americans vehemently disagree. The same data are available to all. We are all rational, so why the discrepancy in opinion? Let us consider the argument and the various sources of dissent to see rationality working, failing, and being irrelevant.

The argument climate scientists make for man-made climate change is summarized as follows by the National Aeronautics and Space Administration

(2020): data indicate global temperatures have been steadily rising since the 1800s (the start of the Industrial Revolution), resulting in melting of the polar ice caps and rising sea levels. There can be many natural sources for temperature increases, such as variation in solar activity, volcanic activity, and even slight shifts in Earth's trajectory around the sun, and these have indeed resulted in past climatic changes. But the timescale and "fingerprint" of the changes we are currently experiencing are not consistent with any of these natural causes. Examination of ice cores from Antarctica reveals that carbon dioxide levels have been relatively stable throughout the past 800,000 years but have shot up dramatically over the past hundred years. When we incorporate the data about excess introduction of carbon dioxide into the atmosphere as a result of human fossil fuel activity and disruption of the natural carbon-oxygen cycles, the projected greenhouse effect is very similar to what we are actually experiencing. Therefore, it is reasonable to believe that human activity (such as carbon dioxide emissions) is a large causal factor in global warming.

This conclusion is plausible, perhaps even compelling, but not certain. One can probe, question, and doubt. Let's examine some possible "reasons" for rejecting the argument offered by nonbelievers by reviewing a question-and-answer session on climate change, held in June 2010 at the University of New South Wales, called "The Sceptics" (2010). It was moderated by Jenny Brockie and featured climate scientist Professor Stephen Schneider from Stanford University and some ardent skeptics from the Australian general public. The first skeptic questioned by the moderator was Tania.

Moderator: Tania, do you believe in man-made climate change?

Tania: Man-made? Not at all.

Moderator: Why?

Tania: No one has proven to me that it's man-made at all. What I say is it's a big hysteria just for money. The only reason you're getting grant money is because of climate change. The planet is warming is the only reason you're getting grant money. If we didn't have this hysteria there would be no grants. There would be no people making money at all.

In this case, the argument for climate change is not actually in play. Tania's objection does not consider the relation between the evidence and the conclusion. Tania is attributing a disingenuous or malicious motive to climate scientists and is offering an ad hominem response. Scientists are simply lying to pad their pockets with grant money. This objection is a case of disagreeing with a conclusion but for reasons that have nothing to do with the coherence of the argument. Many real-world disagreements fall into this category. A similar technique can be used to endorse arguments that are offered by friends and people that one admires. The argument itself does not matter. Coherence relations between evidence and conclusion are not in play. Therefore, such objections (or endorsements) do not belong to the realm of the rational. One might think that educating Tania about the individualistic and competitive nature of science and scientific grant funding may dissuade her from her misconception, but as we will see in chapter 13, it probably will not.

More valid reasons to reject the climate change argument would be to question the data and/or measurement techniques. Another skeptic, John, voiced the concern that he had read that 89% of the thermometers were placed too close to artificial heat sources, such as buildings, and this was artificially inflating temperatures. The accuracy of the methods of calculating temperature changes from thousands of years ago using tree rings data was also questioned. If these concerns are correct, whatever coherency the initial argument had would need to be reevaluated. Schneider acknowledged the challenges of accurate historical measurements, corrected John's belief about 89% of thermometers being placed near heat sources, and explained some of the techniques scientists use to ensure accuracy of the data (e.g., pulling out from the record those temperature readings affected by urban heat sources and covarying population growth with temperature increase).

Case, another skeptic, had just returned from a trip to Alaska and raised two issues regarding glacial melt. On his trip, he had learned that in 1750 Glacier Bay was completely occupied by a glacier. By 1860, half of it had melted. This melting occurred prior to any significant human industrial activity, so how can we assume that the melting of glaciers is proof of global warming? Furthermore, the Alaskan glaciers that originate at low altitudes are indeed receding, but those that originate at higher altitudes are actually advancing. Both these observations seem inconsistent with the global warming models. If so, the models need to be revisited to make them cohere with the data.

Professor Schneider replied that it is not correct to say that human activity was inconsequential prior to the 1800s. We have been involved in large-scale agricultural land clearing for thousands of years, and this has had an impact on CO_2 accumulation, albeit on a much smaller scale than present industrial activity. Schneider explained that if we average across all glaciers around the world and the rates of melt, the data show accelerated rates of melting in the twentieth century relative to prior centuries. With respect to glaciers actually building up and advancing at higher altitudes, that is exactly what the models predict. If you begin with a very cold temperature, say −10°C, and you warm it up by 5°C, to −5°C, the warmer atmosphere will hold more moisture, resulting in increased snowfall and ice buildup, until it warms up past 0°C and starts to melt. This has been observed not only in Alaska but also in Antarctica and Greenland. All this is consistent with expectations of the theory. So, the apparent contradiction was based on some incorrect information in the belief network of the dissenter. Once this misinformation is corrected, the inconsistency should disappear and coherence emerge.

The final skeptic we will consider is Ian. He raised the following objection: "I understand that carbon dioxide that man produces is 3% of what nature produces. How can small changes to our production of CO_2 impact upon something as large as the Earth? It seems absurd." Schneider responded to Ian by briefly explaining the annual carbon cycle, whereby carbon dioxide is taken up from the atmosphere by vegetation during photosynthesis in the spring and summer growth seasons and released back into the atmosphere in the autumn and winter when the leaves fall and decompose. The amount of carbon involved here is much greater than that generated by human activity, but critically, the cycle is in balance. Burning of fossil fuels by humans disrupts the balance of the cycle by adding CO_2.

Ian: Sorry to butt in on this. Look, you're not answering the question. I said that we produce approximately 3% of natural production. You haven't really addressed that. You've given us some prevaricative answer.

Prof. Schneider: I mean perhaps you do not understand the answer. What I said is the amount of carbon dioxide coming from the atmosphere goes in and out and it's larger than what we inject. But it's in balance.

Ian: It's 3% carbon dioxide of the total production of carbon dioxide. It's still a small percentage. If we reduce our carbon dioxide by 50% and send ourselves back to the Stone Age we've made very little difference. Could you answer that question? I did understand what you said perfectly.

Prof. Schneider: Let me give you an example. If you have a bathtub, you can turn it on so you are getting a gallon coming in a minute, right? Now the drain is opened up to the point where a gallon is going out in a minute. So, there's a flow in and there's a flow out. That's an analogy to the fact that there is a very large flow of carbon dioxide naturally going into the system in the summertime and coming out in the winter. Much larger than the 3%, I agree with that. However, it's in balance. The amounts are the same, so when you add the 3%, it's 3% this year and next year and next year. . . . And it accumulates. So, if all of a sudden, I go to the bathtub and I make the one gallon into 1.2 gallons and I don't change the drain size in the bottom, the water in the bathtub is going to rise [and overflow].

In this particular case, the dissenter does not assail the motives of climate change scientists or question the measurement techniques, nor is he giving indications of harboring erroneous beliefs, but he nonetheless simply refuses to accept the coherency of Schneider's argument. Most of us can readily see the rationale in the bathtub analogy. Even if the human contribution of CO_2 is a small fraction of the naturally occurring amount, as long as it is in addition to the natural input/output cycle, such that the input becomes greater than the output, we can readily understand that an overflow will eventually occur (figure 1.3). But this skeptic simply fails to understand or acknowledge the coherency of the argument. If this is a genuine failure of coherence (rather than a contrived stance), it is not clear what more can be said to convince the dissenter. Simple coherence relations are primitive intuitive notions. Either you "see it" or you don't. Everyone with normal cognitive capacity should be able to "see" that the bathtub will overflow.

These exchanges between Schneider and the skeptics illustrate various sources of disagreement in real-world arguments, including assigning

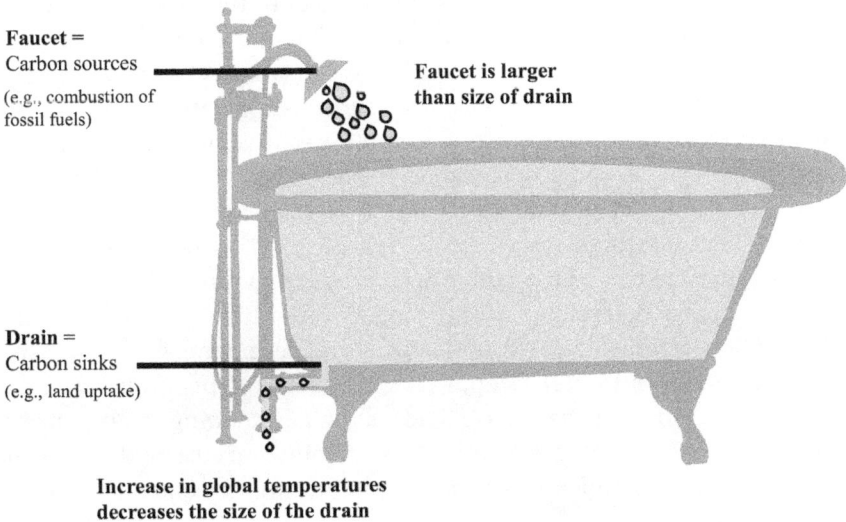

Figure 1.3
The carbon bathtub analogy. If more water is dripping into the bathtub than is leaving via the drain, no matter how small the difference, the coherent conclusion is that the bathtub will eventually overflow. This is an example of a basic, intuitive coherency judgment. If one fails to acknowledge it (in good faith), it is not clear what more can be said to change one's mind. This would constitute a cognitive failure in detecting coherency.

disingenuous motives to the individual putting forward the argument and therefore simply not believing it; questioning measurement techniques and the accuracy of data; having false beliefs about data or misunderstanding parts of the argument; and failure of coherency judgments. Professor Schneider did provide evidence and reason-based answers regarding measurement techniques and corrected false beliefs among participants. But at the end of this exchange, only one individual changed their mind from "sitting on the fence" to accepting the reality of climate change. *The other 20 or so skeptics were equally as skeptical at the end of the session as at the beginning.* The reasoned responses provided by Schneider had no impact on their beliefs. The skeptics did not question the evidence and arguments he presented. They did not offer corrections or additional evidence to the contrary. They simply refused to change their beliefs. This is not rational. How can this be explained?

The most ubiquitous explanation for failures of reasoning is the prominence of "heuristics" over analytical reasoning (Evans & Over, 1996; Kahneman, 2003; Sloman, 1996). We will see in chapter 7 that heuristics come in several different flavors and can play useful roles in theories of reasoning, but they are not particularly relevant to explaining the types of examples under consideration here. They are part of the machinery of reasoning and are sensitive to coherence relations. The heuristic explanation is often combined with the "sloppy reasoning" and "motivated reasoning" explanations.

The "sloppy reasoning" explanation is exactly what it sounds like (Pennycook & Rand, 2019). While coherence itself is a basic, intuitive notion, determining coherence between data and theory need not be a trivial matter. We often need to call on the formal apparatus of logic and mathematics to guide coherence determinations in complex cases, highlighting the value of education, training, and effort in honing and developing basic, intuitive coherency judgments. In the sloppy reasoning account, one would say that the audience did not have sufficient education and training to understand the argument. As we will see in later chapters, when we take up this issue in earnest, this may be true in individual cases but cannot explain the overall phenomenon.

A third explanation is that the skeptics have a vested interest in some status quo and were engaging in motivated reasoning (Kunda, 1990). Whereas ideal reasoning involves going from data to conclusions in a disinterested manner, motivated reasoning is guided by a preexisting goal or desire (e.g., continuing to burn fossil fuels) that serves to filter the data in order to support the preferred conclusion (man-made climate change is a hoax). This explanation also falls short. In fact, motivated reasoning is genuine reasoning and is part of scientific reasoning.

Indeed, science is rife with motivated reasoning. But scientific disagreements usually involve issues of metatheoretical frameworks, such as technical methodological differences having to do with study design and analysis (e.g., confounding variables, underpowered studies, appropriate statistical techniques) or assigning different weightings to existing beliefs, theories, or data points that favor one's preferred theory. For example, climate scientists may disagree on the relative roles of solar radiation and atmospheric aerosol concentrations versus greenhouse gases in causing global warming (Hansen & Lacis, 1990), the most accurate method of reconstructing preindustrial global temperatures (Holland, 2007), or the numerical values that should be assigned to some of the assumptions built into the computational models (Lindzen, 1994). These judgments will undoubtedly be affected by one's pre-existing theoretical commitments. But even in such cases data are collected, vetted, and interpreted to maximize overall coherence with existing knowledge and only then added to the knowledge base. Incorrect beliefs are revised or discarded. This is how the reasoning mind works. Why wasn't this the case among Schneider's audience? This issue will be revisited in chapter 13, once we have described the machinery necessary for tethered rationality. We will see that neither heuristics, motivated reasoning, nor sloppy reasoning can explain Schneider's inability to change minds among his audience. We require an explanation involving nonreasoning systems.

It was noted earlier that there were two issues involved in the climate change example: (1) accepting the scientific conclusions and (2) acting on them. The preceding discussion dealt with some of the challenges involved in getting people to accept the science. Getting people to *act* on the science raises a different set of issues. Societal participation in actions to combat climate change constitutes what economists refer to as a "tragedy of the commons" dilemma (Hardin, 1968). The dilemma is that as an individual you receive a higher benefit from not cooperating (using excess energy, continuing to pollute) than from cooperating, irrespective of what other members do, but if everyone cooperates, everyone is better off. These are nontrivial problems, but as we will see, there are some known solutions. I will take up this issue in chapter 9 and argue that a model that recognizes a blended response, incorporating both reasoning and nonreasoning systems, takes us further than just a reason-based model in understanding this failure to act.

We now turn to the four other examples from the introduction and see that the standard rational model fares no better on three of the four.

Rationality in the Real World: Other Examples

In the weight management example, it would seem more advantageous for me to eat less and exercise more rather than risking poor health outcomes. One complicating factor is that the reward for a long and healthy life is in the future, while modified eating and exercise habits need to be implemented in the present. In the decision-making literature, these types of situations are often framed and analyzed as temporal discounting problems (Frederick, Loewenstein, & O'Donoghue, 2002; Reuben, Sapienza, & Zingales, 2010). In this account, we assign a value to a present utility (or profit) and a value to future utilities (or profits). Distant utilities or profits are always discounted (after all, "a bird in the hand is worth two in the bush"). So, for example, if I am giving away money and give you the choice of receiving $10 today or $12 next week, most people would opt for the $10 today, for obvious reasons: it can be spent or invested immediately, serving to maximize utility or profit. By accepting the $10 today, you reduce your chances of receiving nothing in case you do not see me next week, I change my mind, or some other reason. However, if the choice is between $10 today and $100 next week, many people will bypass the $10 today and wait for the $100 next week, calculating that it is more beneficial to delay gratification and take the risk associated with waiting for the larger sum. Where monetary rewards are concerned, this type of explanation often makes sense. Present and future values of monetary sums can be quickly and accurately calculated, given the rate of inflation, interest rates, and other factors. Where individuals diverge in terms of the future value they will trade for the current value, we can explain this in terms of the shape of personal preference or discounting functions and cognitive differences in ability to carry out temporal discounting calculations.

How does this type of explanation fare with my overweight problem? The problem can certainly be formulated as a temporal discounting problem. It could be argued that I do not have the cognitive ability to carry out the temporal discounting calculations or that I have a "skewed preference function." I think this formulation is ultimately unsatisfactory. Even if I don't have the cognitive ability to do the temporal discounting calculations, others making the same choices I do will, so this cannot be the correct general explanation. This leaves the "skewed preference function" explanation, which is just to say that I didn't make the expected, rational choice.

I overeat chocolate cake and pizza because they *taste good*. The decision theorist may want to associate this craving with the skewed discounting function. That is fine, but it just begs the question. A more satisfactory answer requires a description of the systems driving the craving and how there can be individual differences, which can then explain the different discounting functions. Furthermore, the debilitating health consequences of overeating behaviors are not only distant but abstract and do not have any immediate *feelings*—pleasurable or not—like the taste of chocolate cake, associated with them (until they are actually realized). I will argue that without feelings the consequences cannot even enter into the temporal discounting function to actually impact my decision making. These issues are taken up in considerable detail in chapters 11 and 12.

Consider the third example, where sexual gratification jeopardizes the prized goal of the presidency of the United States. This also can be cast as a temporal discounting problem, and we can postulate that perhaps Edwards was not smart enough to do the calculation. But this is simply not convincing. Furthermore, there are some interesting differences between this problem and the one involving weight management. The goal in this scenario is not an abstract commodity in some distant future. Someone running for the presidency of the United States must *taste* it, *feel* it, *crave* it, every living day they are engaged in the pursuit. So there is an affective component associated with both the immediate sexual gratification and the path to the presidency. If questioned, I do not imagine that Edwards would find greater utility in a current transitory sexual encounter than in the future prize of the presidency. So why did he choose the former and jeopardize the latter? It is possible that he thought he could get away with it. If he had good reasons to believe so, we might consider the choice rational, but as an experienced politician, he should have known better. It is possible that he may have deluded himself into believing this, but then the question becomes, what is the source of the delusion? I think a better explanation for his behavior is offered by the old joke attributed to Mae West: "God gave man two heads, but only enough blood to use one at a time." Jokes aside, any convincing explanation of Edwards's behavior requires an acknowledgment of rationality tethered to evolutionarily older biological systems. This type of behavior is also discussed in chapters 11 and 12.

The fourth example involves the reasoning of my American friend with respect to healthcare. He is willing to incur personal cost or forgo personal benefits just so that guy over there—that one—who doesn't pay taxes (i.e., is a freeloader) doesn't get any benefits. According to Kane (2012), some of the relevant facts about healthcare costs are as follows. In 2012, Americans spent on average $8,233 per person per year on healthcare, or 17.6% of GDP. Countries that have a universal single-payer system spent much less for equivalent or better healthcare. Canadians and Germans spent $4,400 per person, or 12% of GDP, over the same period, for equivalent or better healthcare.[2] The French and the Japanese spent even less. It is not rational to pay $8,000 rather than $4,000 per person per year for equivalent or inferior healthcare.

This situation is not unlike one I encountered several years ago when my children were teenagers. They were squabbling and fighting over the TV. Failing to restore peace and quiet with simple requests and threats, I offered each of them five dollars if they would stop fighting. My son turned down the offer, stating that his little sister "did not *deserve* five dollars." This was a real choice made in real time, but it makes no sense in terms of maximizing utility. Our theories of rationality cannot account for it. From the perspective of rationality, my son should have been concerned about the fact that he is getting five dollars, irrespective of whether his sister was getting one dollar, five dollars, or five hundred dollars! But his sister's behavior had outraged his sense of justice, and he was determined to punish her (by withholding the five dollars from her) *even at the expense of losing five dollars himself*. When I remind him of this today, he realizes that it was a stupid decision. It would have been more advantageous to take the five dollars. Similarly, many Americans are willing to incur costs or forgo benefits just so someone they feel is not deserving doesn't get any benefits. How do we account for this behavior?

When it is laid out in economic terms, a universal single-payer healthcare system also has a tragedy of the commons component, but as already noted, there are solutions. This is again a situation where evolutionarily older, nonreasoning systems (and their highly affective manifestations) are short-circuiting the rational decision-making process. These systems are considered in greater detail in chapters 9 and 13.

Now let's consider the fifth and final oral contraceptive example. Such problems are discussed in chapter 7. I will agree with Gerd Gigerenzer that the distinction between relative risk and absolute risk, and our preference for natural frequencies over conditional probabilities, go a long way in explaining the poor decision-making in this example. Such explanations implicate issues internal to the reasoning mind. What I have to say in this book does not impact the work on these types of problems. Conversely, this research does not address the concerns that I'm raising.

These examples (excluding the oral contraceptive problem) have three common features that I would like to highlight: (1) the problem or decision seems to lie within the realm of rationality; (2) seemingly "irrational" choices are being made; and (3) there are underlying nonreasoning mechanisms such as autonomic, instinctive, and associative systems modulating the behavior. In subsequent chapters, we will encounter other examples, but all will share these three features.

In discussing these examples, I have emphasized reasoning and rationality because these are by far the most popular models in the academic literature for construing and analyzing such scenarios. At this stage, it is worth pointing out a second, very different academic account of human behavior offered by sociobiologists and evolutionary psychologists, most recently popularized by Steven Pinker (1997). It emphasizes the continuity between human and nonhuman animals and postulates similar mechanisms to explain the behavior of both. In this type of model, human behavior, like the behavior of nonhuman animals, is not a function of reason but rather a function of a large collection of instincts, which in this literature are referred to as "modules." The most popular version of this model goes by the name of "massive modularity" and states that any particular situation that we encounter will trigger one or more instincts or modules, resulting in a particular choice or behavior. There is not much role for rationality in this model (Cosmides & Tooby, 1994a, 1994b). Some proponents argue that rationality may even be an illusion. This is very much a minority position. I address it in chapter 9.

Given that I'm questioning the explanatory scope of the reasoning models and appealing to evolutionarily older nonreasoning systems (including instinctive systems), the reader may be thinking that I will be advocating a massive modularity type model. This is not the case. I believe that evolutionary psychologists provide a critical insight that needs to be incorporated into the solution. But despite my appeal to nonreasoning systems, I am confident that we are not simply steered by them. We do have the ability to reason and make choices. As an illustration, consider the following example in which I used my rational mind to modulate my (nonrational) eating behavior.

A few years ago, I was attending a conference on reasoning sponsored by the Parmenides Foundation, held on the isle of Elba. The host organization was taking very good care of us, offering food and drink on every possible occasion. After several days of this, I was satiated and determined to limit my food intake for the sake of my health. After a particularly interesting talk on the neurobiology of addiction, I began conversing with the presenter. (I was fascinated by the claim that addictive behavior is not a choice.) It was lunchtime so we were all walking toward the beach, where lunch would be served. I said to my colleague, "I will come with you so we can continue our conversation, but I'm satiated so I won't eat anything." We sat down and continued talking about his presentation. The waiter brought menus. I thought, I will not order anything, but I'll just look at the menu. Upon examining the menu, the pizza looked very appetizing. I said to myself, "Well, you know what, I will order the pizza but I will only eat half of it." The pizza arrived; I cut it in half and ate the first half. I then ate the other half. I'm overweight and suffer from the typical consequences. I do not need an extra slice of pizza. Why did I eat it? I derived such pleasure from it that I could not help myself.

The next day, at lunchtime, my wife was with me. We walked to the beach where lunch was being served, and I said to her, "The pizza is very good, but don't order your own. I will order one and we will split it." I ordered the pizza. When it arrived, I cut it in half and ate my half. I then looked for the other half, but my wife had already eaten it. So, reluctantly, I did without. I used my reasoning abilities to put myself in a situation where I would not be confronted with the temptation of eating the other half of the pizza, and I was thus able to control my food intake.

The point of this anecdotal story is to highlight that the rational mind is able to exert some control over behaviors through various strategies to avoid being totally at the mercy of deep-seated evolutionarily older mechanisms. To prevent overeating, I placed myself in a situation where food was not readily available. Some of the questions we will need to explore are: To what extent is this possible? What is the nature of the interaction between reasoning and nonreasoning systems? What is the common language used for communication across different systems? How do we account for individual differences in behavior? Who is in charge of the tethered mind?

The idea that reason alone is not sufficient to account for human behavior is being voiced by an increasing number of researchers, particularly in the social, economic, and political sciences (Kahan, 2016; Oliver & Wood, 2018; Young, 2019). These researchers frequently contrast reason with "heuristics," "emotions," "gut feelings," and "unconscious processes." They are trying to account for their intuitions and data, but this vocabulary lacks substantive conceptual machinery to allow them to say what they want and need to say. (The first two terms are part of the reasoning mind, and the latter two are undefined or unhelpful.) It is not their fault. They are not in the business of developing the models to explain behavior but rather applying machinery developed by cognitive scientists to their respective problem domains. The cognitive sciences have come up short. I will argue that tethered rationality provides a much richer repertoire of conceptual machinery to explain their intuitions and data.

Finally, models of human behavior are not only necessary for explaining political, economic, social, and moral behaviors but also have consequences for changing these behaviors. Tethered rationality is no exception.

Some readers may be disappointed in its implications. It suggests that many behaviors—such as racism, sexual harassment, cheating, adherence to false beliefs despite counter evidence, and overeating resulting in obesity—however unacceptable, are often driven by early maturing autonomic, instinctive, and associative neural systems and cannot be easily changed simply by changing beliefs. Even behaviors based on reasoned social constructs can become deeply entrenched once neural systems mature. Attempts at belief revision through a weekend of "sensitivity training" will be ineffective. This does not mean that such unacceptable behaviors cannot be changed at all, but it does mean we will need to understand the underlying biology of each specific behavior and apply behavior-specific remedies. Even then, there may be limits.

Organization of the Book

This volume is organized into six parts. Part I introduces the rational animal and the enigma of rationality. It is accepted that we *are* the rational animal but our rationality is not disembodied. It is tightly tethered to evolutionarily older autonomic, instinctive, and associative systems. Before we can tell the story of the tethered mind we need to have a common understanding of each of these systems. Part II is devoted to characterizing autonomic, instinctual, associative, and reasoning behaviors and systems. Each is characterized in terms of the following five dimensions: (1) function of the behavior, (2) tightness of causal coupling between stimulus and response, (3) origin of behavior, (4) underlying mechanisms, and (5) brain structures involved. The behaviors are found to differ along these five dimensions and are accordingly assigned to different systems or "kinds of minds." Considerable effort is made to explain what behaviors each type of mind can and cannot explain.

The characterizations of each type of mind are reasonably standard. The autonomic mind (chapter 3) is characterized in the manner found in most biology textbooks. The instinctive mind (chapter 4) draws on the models of Konrad Lorenz, Nikolaas Tinbergen, and other ethologists. The characterization of the associative mind (chapter 5) follows that of B. F. Skinner and other behaviorists, enriched by the insights of William James. The exposition of the reasoning mind (chapter 6) draws on the ideas of twentieth-century philosophers and cognitive scientists, including Ernst Cassirer, Donald Davidson, John Searle, Noam Chomsky, Herbert Simon, Allen Newell, Jerry Fodor, and Zenon Pylyshyn. I will always use the terms *autonomic, instinctive, associative,* and *reasoning* in the manner specified in the corresponding chapters. What is new in this book is my assertion that *all* these systems are in play (to various degrees) in *all* human behaviors; that is, human behavior is a blended response. This will be a blatant truism for many general readers. It will be less obvious to many of my colleagues. It was expunged from us during graduate school.

Once an understanding of the kinds of minds associated with each type of behavior has been established, part III reviews the theoretical frameworks of reasoning, built from the conceptual machinery of the cognitive mind. Chapter 7 considers models of formal reasoning, particularly various dual mechanism accounts of reasoning resulting in the widely accepted distinction between heuristic and analytical systems. I flag a number of sources of confusion in this literature, but by and large, the literature is not relevant to the types of issues of interest in this volume, so it is set aside. The reader not encumbered with the belief that heuristics explain the examples raised earlier could bypass chapter 7.

Chapter 8 reviews the literature on conceptual coherence (inductive reasoning). It begins by raising a number of issues largely of concern to philosophers and cognitive scientists, and then shifts to considering real-world problems from the realm of science (Galileo's arguments about motion) and politics (first impeachment of Donald Trump). Once we reach the latter we are confronted with a whole set of issues that the cognitive science literature cannot address. Any satisfactory explanation requires an appeal to the engagement of nonrational systems.

Part IV begins the development of the positive account of the tethered mind. Chapter 9 provides the behavioral data for the "blended response" hypothesis. It begins by considering instincts in their modern reincarnation as "modules," from the work of Leda Cosmides and John Tooby. While I reject the massive modularity model, I find value in their insights regarding the role of instincts in human behavior, specifically their explanation of behavioral data from a famous reasoning task in terms of "cheater detection" instincts, rather than coherence relations. An exploration of the related concepts of self-interest maximization, fairness, trust, cheating, and punishment suggests that they are reasonable candidates for instincts, albeit all but the first may be specific to humans. Given the types of decision-making examples of interest (e.g., climate change, universal healthcare), I turn to the work of a small but influential group of economists and mathematical biologists, such as Ernst Fehr and Martin Nowak, who explain human choices on such problems as interactions of the instincts noted above. A careful examination of the data from this literature shows that these instinctual systems are modulated by reasoning systems and that a full account of the data requires postulating a blended response involving both instincts and reason;

that is, a model of tethered rationality. The appendix to chapter 9 reiterates the distinction between reason and instincts and discusses the conceptual pitfalls in trying to account for human behavior just in terms of "modules" or instincts. The reader more interested in the positive account of tethered rationality could bypass this appendix without sacrificing continuity.

Chapter 10 turns to comparative neuroanatomy for the neural underpinnings of tethered rationality. The challenge here is to show that the hierarchy of evolved behaviors (autonomic, instinctive, associative, reasoning) is mapped onto a hierarchy of evolved brain structures. The interconnections between brain structures supporting tethering are readily apparent at the level of anatomy and physiology. A further challenge is to illustrate the differences in brains of organisms that can reason and those that cannot.

Once we have a story of hierarchically organized behaviors, underwritten by hierarchically organized (but interconnected) brain structures, it is necessary to account for how these various systems contribute to behavioral responses. The tethered rationality model allows each of the four systems to generate responses to any environmental perturbation, but the organism is restricted to a single behavioral response at a time. This requires some global integration function that takes input from each of the systems and generates a blended response. Chapter 11 advances the speculative conjecture that what is common across each system is *feelings*. Feelings are generated in old, widely conserved brain stem structures, and are evolution's solution to initiating and selecting behaviors. Reason is as much about feelings as lust and the taste of chocolate cake. Feelings provide the common currency that allows communication across systems and the calculation of an overall blended response. This controversial solution has the additional benefit of bridging the divide between the cognitive and noncognitive and driving reason back into the biology, where it belongs. The works of neuroscientists such as Jaak Panksepp, Kent Berridge, and Morten Kringelbach play a central role in putting together some of these ideas.

Chapter 12 considers the control structure for tethered rationality. Who is in charge of the tethered mind? Reason is not the CEO. In fact, I will conclude that there is no CEO. All four systems affect the resulting behavior in the currency of feelings. The system is set up to maximize pleasure and minimize displeasure. The model is illustrated with several examples and, in particular, my difficulty in losing weight. At this point, I return to complete the explanations of some of the other examples introduced throughout the book.

Part V takes up the question of why it is so hard to change certain beliefs, ranging from climate change being a hoax, to the MMR vaccine causing autism, to gender identity being a socially constructed choice independent of biology, to "Democrats are evil people, they hate America," among others, despite evidence to the contrary. Insofar as reason is viewed as untethered to the biology (as in mind/body dualism or the cognitive science computer program/hardware metaphor), it should have an unfettered ability to update and revise beliefs (perhaps constrained only by time and memory resources). This is not the case. Driving reason back into the biology provides some answers.

Chapter 13 provides one answer to this puzzle by applying the tethered rationality model: reason is only one component of the system. The other systems reason is tethered to may prevent belief revision or belief revision may not be sufficient for behavioral change. The introduction of the in-group/out-group instinct allows us to complete the explanations for climate change denial, the impeachment debate, and some Americans' aversion to universal healthcare. The other constraint on belief revision is neural maturation. This phenomenon is independent of the tethering of reason and largely comes into play where large-scale global belief systems, known as worldviews, need to be revised late in life. In chapter 14 I propose that with the maturation of the association cortex in adulthood there may not be sufficient neuronal resources left for large-scale architectural neural reorganization, making global belief revision challenging.

Part VI briefly considers some of the consequences of the tethered mind and concludes the volume. Different models of human behavior come with different control structures and have different social and legal consequences. One consequence of the tethered rationality model is that changing certain deeply seated behaviors (however socially unacceptable) is not a matter of just changing beliefs. To consider remedies beyond belief revision, we need to understand the biological underpinnings of the specific behaviors. This sensitive topic warrants a separate volume. However, in chapter 15 I very briefly consider some concerns and consequences of the tethered mind, and conclude the volume by offering a few closing thoughts to colleagues and the general reader.

Let us begin the journey by stepping back and reconceptualizing the problem of rationality and human behavior in a broader context.

2   The Enigma of Rationality: Fallen Angel or Risen Ape?

Plac'd on this isthmus of a middle state,
A being darkly wise, and rudely great:
With too much knowledge for the skeptic side,
With too much weakness for the Stoic's pride,
He hangs between; in doubt to act, or rest;
In doubt to deem himself a god, or beast;
In doubt his mind or body to prefer;
Born but to die, and reas'ning but to err;
Alike in ignorance, his reason such,
Whether he thinks too little, or too much:
Chaos of thought and passion, all confus'd;
Still by himself abus'd or disabus'd;
Created half to rise, and half to fall;
Great lord of all things, yet a prey to all;
Sole judge of truth, in endless error hurl'd:
The glory, jest, and riddle of the world!
—Alexander Pope (*Essay on Man*, Epistle 2)

There is an enigma associated with the reasoning mind: human behavior ranges from the "four Fs"—feeding, fornicating, fighting, and fleeing—which we share with all animals to behaviors such as the development of art, science, mathematics, and sending a man to the moon, which no other animal is capable of. How are these abilities, unique to us, explained and reconciled with the four Fs? Is there something special about us? Is the difference qualitative or merely quantitative? Are we risen apes or fallen angels?

In this chapter, I want to consider three very different answers to these difficult questions. The first is provided by the Western-Christian intellectual tradition and eloquently articulated by Alexander Pope: reason belongs to the ethereal realm while "animal passions" have more earthly origins. Man uniquely straddles these two worlds. A second, very different answer is provided by Darwin in the theory of evolution: reason is not all that special; clear precursors of it can be found in bees, bats, and baboons. The third answer we will consider is provided by cognitive science: reason-based cognitive abilities are unique to humans.

I argue that the cognitive science view is the most reasonable. However, it comes with an unfortunate flaw: the "animal passions" have been abandoned. In the cognitive account, volitional human behavior is explained *exclusively* in terms of reason. Common sense tells us that this is only half the story. We need to reclaim the insights of Alexander Pope and his contemporaries about the reality and importance of "animal passions" and use the insights of the modern theory of evolution and cognitive science to weave together a picture of human behavior that is more complete, and can be discharged without invoking God's grace. This is what the model of tethered rationality sets out to do.

Rationality as God's Grace

To Alexander Pope (and his contemporaries) it was clear that there was a vast qualitative gulf between our intellectual abilities and those of other animals but not between our "passions" and the "passions" of other creatures. In this chapter, I use the term *passion* as a generic reference to all behaviors not based on reason. Pope's formulation of the problem, and the answer he offered, was one that runs deeply through Western-Christian intellectual tradition at least up to the Enlightenment and even beyond. It is known as the Great Chain of Being (Pope, *Essay on Man*, Epistle 1):

> Vast chain of being, which from God began,
> Natures ethereal, human, angel, man,
> Beast, bird, fish, insect! what no eye can see,
> No glass can reach! from infinite to thee,
> . . . .
> Where, one step broken, the great scale's destroy'd:
> From nature's chain whatever link you strike,
> Tenth or ten thousandth, breaks the chain alike.

The idea dates back to at least Plato and Aristotle and became the great organizing and classification principle of the Western-Christian worldview (Lovejoy, 2011; Tillyard, 2011). The world is conceived as a hierarchical structure (ladder/pyramid/concentric rings/chain) based on intrinsic value that begins with the most basic elements at the bottom and rises to the highest perfection, God, at the top. Minerals and inanimate objects occupy the lowest rungs or links. Plants can grow and reproduce, so they occupy a higher stratum. Animals are additionally animated, so they occupy a still higher stratum. But they are still limited, because they are comprised only of physical appetites and sensory organs. At the top of the pyramid sits God, embodied with divine powers of omniscience, omnipresence, and omnipotence, along with reason, love, and imagination. Below God are the angels. They are pure spirits, lacking the divine powers but possessing reason, love, and imagination. Man straddles the divide between the ethereal world of angels and the base physical world of animals. He is subject to the base passions such as pain, hunger, and sexual desire but also shares in the spiritual attributes of reason, love, and imagination.

The Great Chain of Being model served not only as a religious organizing principle but also as a social, political, and economic organizing principle that permeated every aspect of society. It gave order and meaning to the world. There are several notable features about the model. First, it is a linear, static, essentialist model with fixed boundaries. Each level is defined by necessary and sufficient conditions. The levels are hierarchically organized, and there can be no movement between levels (without disrupting God's plan and introducing confusion into the world). Second, man is unique in having a dual nature; that is, in straddling the realm of being, characterized by reason, and the realm of becoming, characterized by animal passions. Reason is explained as a function of our divine soul and accounts for our ability to create art, science, and mathematics and to go to the moon. The passions, or the four Fs, which we share with other animals have a more earthly origin. Third, there is a tension between reason and the passions; they often initiate different actions. While God did give us free will to choose between the different actions generated by the two systems, he is most pleased when we reject the passions and embrace reason (and hence God himself).

The Western-Christian model is valuable because it explicitly recognizes that human behavior must be explained by both reason and the four Fs. What is problematic is that it leaves us with a dilemma: if we restrict ourselves to the laws of the natural sciences, we can explain only the animal passions. To explain reason, we need to reach beyond the natural sciences to God's grace. Both options are unpalatable. A model that recognizes the full range of human behaviors and is committed to an explanation consistent with the natural sciences is needed.

Rationality and Darwin: Nothing Special

The Darwinian revolution blew up the linear, static, essentialist world of the Great Chain of Being and replaced it with a world where mutability, change, and transformation are the organizing principles of life. Interestingly, the hierarchical organization was largely preserved. The basis of the revolution was the convergence of three preexisting ideas: variation, inheritance, and survival of the fittest.

The ideas of individual variance in traits and inheritance of traits from parent to child were previously well established from livestock breeding. A number of intellectuals, such as Maupertuis, Buffon, and Diderot, discussed how these processes could lead to gradual change within species and even across species (Lovejoy, 1968). Wallace ([1871] 2009) and Darwin ([1859] 1995) independently took up these ideas, combined them with a third idea—"survival of the fittest"—from Malthus's (1798) economic theory, and applied the resulting framework to all species, including man.

The basic outline of the Darwin-Wallace evolutionary story is as follows. There are naturally occurring variations or individual differences in traits in members of all species. Certain traits may give the individual an advantage in reaching for fruit, fleeing from predators, attracting mates, defending territory, and so on. Individuals with these advantageous traits will thrive better than those without and have an increased probability of leaving behind more offspring (survival of the fittest). These offspring will inherit many of the properties of the parents, including the advantageous traits. Because individuals with beneficial traits leave more offspring, the beneficial traits are passed on more frequently and come to dominate the species (natural selection). Over time, this process will change characteristics of the species to such an extent that after very many generations the distant offspring may be classified as a different species altogether.

Having developed his theory of evolution, Darwin ([1871] 1896) took up the topic of "higher mental powers" in chapter 3 of *The Descent of Man and Selection in Relation to Sex*, "Comparison of the Mental Powers of Man and the Lower Animals." His goal in the chapter was to show that there are no fundamental or qualitative differences in the mental faculties of men and higher mammals. He felt that if this could not be done, then the theory of evolution would need to be dismissed (p. 65):

> If no organic being excepting man had possessed any mental power, or if his powers had been of a wholly different nature from those of the lower animals, then we should never have been able to convince ourselves that our high faculties had

been gradually developed. But it can be shewn that there is no fundamental difference of this kind.

Darwin used two strategies to show no fundamental difference of kind and fuse the Christian divide between man and beast: (1) he argued that nonhuman animals are much more sophisticated and intelligent than we recognize, and, equally important, (2) that the "lower men" or "savages" (i.e., non-Europeans) are much less intelligent and more animal-like than we assume (p. 65):

> We must also admit that there is a much wider interval in mental power between one of the lowest fishes, as a lamprey or lancelet, and one of the higher apes, than between an ape and man; yet this interval is filled up by numberless gradations.
>
> Nor is the difference slight in moral disposition between a barbarian, such as the man described by the old navigator Byron, who dashed his child on the rocks for dropping a basket of sea-urchins, and a Howard or Clarkson; and in intellect, between a savage who uses hardly any abstract terms, and a Newton or Shakespeare. Differences of this kind between the highest men of the highest races and the lowest savages, are connected by the finest gradations. Therefore it is possible that they might pass and be developed into each other.

To satisfy the first part of the strategy, Darwin begins by noting that there are no or few differences between man and lower animals when it comes to basic instincts having to do with self-preservation, sexual attraction, parental care, and so on. He then turns to basic emotions and concludes that "the fact that the lower animals are excited by the same emotions as ourselves is so well-established, that it will not be necessary to worry the reader by many details" (p. 69). His evidence consists of stories about dogs, monkeys, and elephants exhibiting revenge, love, remorse, kindness, and happiness in interaction with conspecifics or man: "Even insects play together, as has been described by that excellent observer, P. Huber, who saw ants chasing and pretending to bite each other, like so many puppies" (p. 69).

According to Darwin, even the more complex emotions, such as jealousy, wonder, curiosity, and humor, are shared among human and nonhuman animals. For example, a dog playing stick-retrieval with his master will seize the stick and run with it, wait until the master approaches to take it, "seize it and rush away in triumph, repeating the same manoeuvre, and evidently enjoying the practical joke" (p. 71).

Darwin acknowledges reason as the crowning achievement of human minds. He admits some difficulty in differentiating reason from instinct or "mere association" but nonetheless makes his case for the ubiquitousness of reason with examples such as the following (p. 78):

Col. Hutchinson relates that two partridges were shot at once, one being killed, the other wounded; the latter ran away, and was caught by the retriever, who on her return came across the dead bird; "she stopped, evidently greatly puzzled, and after one or two trials, finding she could not take it up without permitting the escape of the winged bird, she considered a moment, then deliberately murdered it by giving it a severe crunch, and afterwards brought away both together. This was the only known instance of her ever having willfully injured any game." Here we have reason though not quite perfect, for the retriever might have brought the wounded bird first and then returned for the dead one.

In a similar vein, Darwin considers the abilities of abstraction, formation of general concepts, and self-consciousness. With respect to abstraction and the formation of general concepts, he sees rudimentary forms in the behavior of his dog. When Darwin says "hi, hi, where is it?," his terrier excitedly rushes to the nearby bushes and thickets, scenting for game, and finding nothing, looks up in a tree, perhaps looking for a squirrel. For Darwin, "these actions clearly shew that she had in her mind a general idea or concept that some animal is to be discovered and hunted?" (p. 83).

The issue of self-consciousness is more challenging to ascertain. Darwin concedes that insofar as it involves reflecting on the nature of life and death, it may be beyond animals, but if it is restricted to reflecting on one's past experiences, then "how can we feel sure that an old dog with an excellent memory and some power of imagination, as shewn by his dreams, never reflects on his past pleasures or pains in the chase?" (p. 83).

When it comes to human language, Darwin recognizes some of its unique and interesting properties but nonetheless sees rudiments of it in bird songs, monkey calls, and even in the communication of ants. For "sense of beauty," he notes that "yet man and many of the lower animals are alike pleased by the same colors, graceful shading and forms, and the same sounds" (p. 93).

His final topic is belief in God or religion. Darwin recognizes the feeling of religious devotion as a "highly complex one, consisting of love, complete submission to an exalted and mysterious superior, a strong sense of dependence, fear, reverence, gratitude, hope for the future, and perhaps other elements," but once again he sees primitive remnants of these qualities in the love and submission his dog feels for him.

To lower the bar from the human side, Darwin explicitly postulates that "savages" or members of "lower races" (non-Europeans) may be equally limited in their ability to reason, contemplate God, and even engage in self-conscious reflection (p. 83): "On the other hand, as Buchner has remarked, how little can the hard worked wife of a degraded Australian savage, who uses very few abstract words and cannot count above four, exert her self consciousness, or reflect on the nature of her own existence."

In the Darwinian account, many aspects of the Christian hierarchy of the Great Chain of Being remain, but it is possible to pass from one rung of the ladder to the next through gradual, natural, evolutionary processes. In fact, it is necessary for the theory of evolution.

Interestingly, Darwin recognizes the difference between "instincts" and the "higher intellectual faculties" such as reason. He does not argue that instincts can explain reason and other higher intellectual faculties (as do some modern evolutionary psychologists we will encounter), so there must be differences. He doesn't say what the differences are but human intelligence or reason is not a spiritual gift from God. Reason is widely distributed among the animal kingdom (to variable degrees) and is explained with earthly mechanisms. The articulation of the mechanisms underlying both reasoning and nonreasoning behaviors has been taken up by many others, as we will encounter in due course.

Darwin's reconciliation of the mental faculties of man and lower animals is not his finest work. With the benefit of 150 years of research on animal and human intelligence, it is difficult to take his "data" seriously, irrespective of where one stands on the question of human intelligence. It is important to remember that Darwin has a vested interest in a particular theory and believes that it requires gradual linear transformations between species. Therefore, unsurprisingly, there is no attribute of man that he cannot find some semblance of in monkeys, dogs, and even ants. Vested interests are a natural part of reasoning (chapter 13).

Today we understand that Darwin was wrong on at least three counts.[1] First, he was wrong to trivialize the differences between human and nonhuman mental faculties. Second, he was wrong to view "lower men" as intermittent links between Europeans and apes. Third, while it is true that the evolutionary story cannot have any unbridgeable gaps, this does not imply linear, incremental changes at the level of phenotype (phyletic gradualism). Let's consider the first and third mistakes, leaving the second for historians and sociologists.

Rationality and Cognitive Science: Maybe a Little Bit Special

The behaviors of human and nonhuman animals have come under considerable scrutiny since Darwin. Within the modern cognitive science community, the consensus is that while animals are surprisingly clever, there are sharp discontinuities in their cognitive abilities and ours. Many of these discontinuities seem to converge on abilities to deal with abstract, higher-order nonperceptual relations such as spatial relations, sameness relations, logical and conceptual coherence relations (including analogical and causal relations), and deception (Penn, Holyoak, & Povinelli, 2008; Premack, 2007; Shettleworth, 2010). These clusters of unique abilities are all thought to draw on mental representations that have propositional content with a subject/predicate structure (Fodor & Pylyshyn, 1988). Such mental representations allow humans to make a distinction between the world and a representation of the world. They also allow us to hold certain mental representations or thoughts in logical and conceptual (including causal) relation to other mental representations or thoughts, as in the following examples. Chapter 6 considers this machinery in some detail.

A simple logical relation is illustrated by the transitivity relation, which allows us to engage in hierarchical ordering. For example, if I know George is taller than Mary and Mary is taller than Michelle, then without being told I know that George is taller than Michelle. It follows as a matter of logical necessity from the first two premises. Despite some claims to the contrary, I will argue in chapters 5 and 6 that such logical inferences are unique to humans.

An example of a conceptual coherence relation is analogical inference. At some point in high school, most of us were introduced to the structure of the atom with the following analogy: "A nucleus is to an atom as the sun is to the solar system; and the electrons are to an atom as the planets are to the solar system." Analogies require finding and mapping relevant and salient structural (usually nonperceptual) similarities in objects and relations from a source domain (better understood, the solar system in this example) to a target domain (less well understood, the atom in this example). Humans do this effortlessly, indeed incessantly; nonhuman animals do not (Penn et al., 2008).

The ability to deal with causal relations is another differentiating criterion. The causal knowledge of nonhuman animals is tightly coupled to specific perceptual and task parameters. David Premack (2007) distinguishes between causal association, causal illusion, and causal reasoning. Causal association is when actions or stimuli that are followed by a desired result become associated with the result, such that, given the former, the individual will seek out the latter. Most species will be capable of making this connection only after due reinforcement. Fewer species (e.g., primates) will have access to causal illusion, which requires knowing that any goal-directed action followed by a desired result leads to the illusion that the former caused the latter. However, causal reasoning requires more than association or illusion. For example, causal illusion may allow an animal to recognize that a large stone will crush a walnut more effectively than a small stone will. However, observing a large stone lying on top of the crushed walnut, and understanding that the stone crushed the walnut, requires causal inference, which again seems to be specific to humans (Premack, 2007).

Another important distinguishing feature between human and nonhuman animals seems to be the ability to engage in deception. This requires making a distinction between the world and a representation of the world. David Premack differentiates between deception involving false positives and deception involving false negatives. A false positive is a signal that indicates a food or predator when in fact none exists. The famous fable "The Boy Who Cried Wolf" is an example of this type of deception. A false negative is a failure to give a signal when food or a predator is present. An example would be when a monkey, unobserved by conspecifics, may fail to signal the presence of food. Among nonhuman animals, false negatives are much more common than false positives. A robust nonhuman example of a false positive might be when a plover bird leads intruders away from its nest by feigning a broken wing and then, when the intruder is beyond the nest, takes flight. However, because the bird can only employ this strategy in a specific instance, this type of behavior is best explained as an adaptation or instinct rather than deception. A more interesting example may be when capuchin monkeys, who sound alarm calls for predators and food, sometimes give false alarms. That is, they will signal food or a predator when there is none (Wheeler, 2009). But there is no evidence that this is intentional. The capuchins may have done this once by accident and learned to repeat the calls through some form of positive reinforcement (see chapter 5 for a discussion of positive reinforcement).

Genuine deception involves intention. Error recognition and correction are prerequisites for intentional actions. For example, vervet monkeys have unique cries for predators such as leopards and snakes. These cries are different from one another and are uniquely recognizable by other vervet monkeys. Suppose a monkey that receives the call for "leopard" mistakenly takes action to protect itself from a snake. If the signaling monkey notices this and does nothing, the original cry is just an instinctive adaptation. If, however, the signaler takes measures to correct the behavior of the recipient, then this would indicate that the original call was intentional, with the goal of protecting the recipient. There are no unambiguous examples of such behavior in the animal literature.

The point of this section is to highlight what was obvious to Alexander Pope and his contemporaries: that we may not be fallen angels but there are real qualitative differences between human and nonhuman minds. Modern cognitive science accepts this characterization or, more precisely, it accepts that reasoning is unique to humans and proposes a mechanism less mysterious than God to explain it. This is a significant advance. Ironically, the proposed mechanism (psychological states with propositional content with a subject/predicate structure realized in computational systems) is characterized independent of biology and still seems to float on top of the four Fs, as did the Holy Ghost. This has allowed cognitive science to forget about the four Fs; we are only reasoning minds. I'm arguing that this is simply untrue and hinders development of accurate models of human behavior.

Rationality: Special but Tethered

We need to reclaim the insight that human behavior is a function of both reason and animal passions, and the former is largely unique to us. One step in this direction is to adopt an updated version of the theory of evolution that contains a richer repertoire of mechanisms than just natural selection, has a worked-out concept of inheritance in terms of DNA and gene mutation, and is not necessarily committed to phylogenetic gradualism. Applying such a theory to brain development may help get us where we need to be.

A major update to the theory of evolution was the incorporation of knowledge of Mendelian inheritance in the form of genes and DNA, known as the modern synthesis, or neo-Darwinism (Dobzhansky, 1950; Mayr, 2000). Traits are determined by genes carried by individual organisms, and inheritance occurs through gene selection mechanisms. Genetic material is isolated from the organism and its environment. Inheritance of acquired characteristics is impossible. Evolution is defined as "changes in allele frequency within populations." The following four factors contribute to change in allele frequencies: (1) random changes in allele frequencies in a population; (2) gene flow resulting from immigration and emigration of individuals from a population; (3) mutation pressures from repeated occurrence of the same mutations; and (4) natural selection favoring the best adapted organisms.[2]

These updates to the theory of evolution allow for recognition that gene mutation and differential expression may affect developmental plans and processes, and slight changes at this level can lead to abrupt changes at the phenotype level. For example, three genes unique to humans appeared in the human genome about three million to four million years ago and are thought to be responsible for the dramatic increase in our brain size over this time period. Genetic defects through deletion of these genes lead to microcephaly (abnormally small brain size), while duplications result in macrocephaly (abnormally large brain size) (Fiddes et al., 2018). Small *quantitative* changes in genes controlling developmental programs can result in large *qualitative* changes in terms of phenotype and behavior.

Such insights have allowed biologists to accommodate step functions in evolutionary processes (Gould & Eldredge, 1977). In one such account, referred to as punctuated equilibria, once species appear, they are generally stable, showing little variation in the fossil record. When evolutionary change does occur, it is abrupt and rapid (in geological terms), leading to branching and speciation. These developments allow for large, "sudden" differences across species.

More specifically, we will see in chapter 10 that the modern version of the theory of evolution allows us to tell a story of qualitatively different behaviors, underwritten by qualitatively different neural systems. A critical part of the story is the branching of the phylogenetic tree, where brains in newer parts of the tree retain ancestral structures, with modifications and additions. That is, certain brain structures (and corresponding behaviors) are conserved and propagated while others are added, modified, and expanded in new directions. This naturally results in the tethering of various behaviors and systems, including reasoning systems, to evolutionarily older systems.

*   *   *

Alexander Pope left man straddling "on this isthmus of a middle state," precariously teetering between heaven and earth, God and beast, thought and passion. The value in this formulation was the recognition of both components of human nature, but it did require God to underwrite reason. The Darwinian revolution dramatically reformulated the intellectual landscape and diffused the dilemma created by the Western-Christian worldview by forcefully arguing for continuity of all life-forms. Darwin seems to accept a difference between reasoning and nonreasoning systems, but his commitment to phyletic gradualism forced him to see reason everywhere, blurring important distinctions. The cognitive revolution has restored some special status to the rational mind but at the expense of ignoring the important role of nonreasoning systems. This is unfortunate because it results in cognitive theories unable to explain the full range of human behaviors. It is also inconsistent with the modern theory of evolution and our current knowledge of neuroanatomy and neurophysiology.

The goal of this book is to bridge the chasm between reasoning and non-reasoning systems by *tethering* reasoning systems to evolutionarily older biological systems and demonstrating how these systems modulate rational choice (and vice versa). What we have been variously referring to as the four Fs, animal passions, or nonreasoning systems are in modern parlance the autonomic, instinctive, and associative systems. These are characterized in part II (chapters 3–5). The reasoning mind is developed in chapter 6, while tethering, issues of communication between the different systems, and the underlying control structure are addressed in part IV.

II Kinds of Minds

Instinct and reason are the attributes of two different entities.
—Blaise Pascal

Consider the behaviors in the following scenario. A mother instructs her son, "Michael, quickly finish digesting your breakfast and get off to school on time. Pay attention in class and don't get distracted by that pretty girl that you are always with. Think of the long-term consequences if you let your grades slide. Also, when you go to your driving lesson, learn to parallel park." The various behaviors involved in these instructions include digesting breakfast, leaving for school on time, ignoring the pretty girl, considering consequences of poor grades, and learning to parallel park a car. Each is interestingly different.

While digesting breakfast is a natural consequence of ingesting it, it is a very odd request to make of someone. The digestive process is not under conscious control. Unlike leaving for school on time, one cannot choose to digest or not to digest one's breakfast. The request not to get distracted by the pretty girl is different from the request to digest breakfast. It certainly sounds like a conscious choice, but it is not clear that it is fully under one's control to the same extent as leaving for school on time. Learning to parallel park is different still, as is the inference of the future consequences of poor grades. Only the latter clearly calls on the reasoning mind. I want to suggest that each behavior is sufficiently different that it may be appropriate to associate each with a different "kind of mind."[1]

Behaviors are organismic responses to environmental change or perturbation, where the environment can be either internal or external. Internal environments are largely "known" and predictable. External environments are largely unknown and usually unpredictable. I propose to identify and classify "different types of behaviors" based on the answers to the following five questions:

(1) What is the function of the behavior?
(2) How tight is the causal coupling between stimulus and response?
(3) What is the origin of the behavior?
(4) What are the underlying mechanisms of the behavior?
(5) What brain structures realize these mechanisms?

Classifying along these dimensions allows us to identify at least four distinct types of behaviors—autonomic, instinctive, associative, and rational—and correlate them with four different kinds of minds. I will not only classify behaviors (and minds) along these five dimensions but will also organize the resulting categories into a hierarchy beginning with those systems that appear earlier and are most widely available on the evolutionary tree, followed by those of more recent origin, and more narrowly available. While I may refer to the systems as "lower" and "higher" level systems, this hierarchy is understood in the context of a phylogenetic tree (see chapter 10, box 10.1), not the ladder envisioned by the metaphor of the Great Chain of Being.

Even though autonomic, instinctive, learning, and reasoning systems will be classified as distinct behaviors and mechanisms—even associated with different kinds of minds—an overarching theme of this volume is that they are not isolated systems. Evolutionarily more recent systems are always tethered to earlier systems such that they are constrained by them and can in turn modulate them to *some* extent. I will begin with the simplest, earliest evolved behaviors, and their underlying mechanisms, and continue to more complex, more narrowly available behaviors until we reach reason in chapter 6.

These four systems, or kinds of minds, will be familiar to all readers. They have been extensively studied for the past hundred or more years. The objective of this part of the volume is to provide an accessible but substantive description of each system. Some readers may find this laborious, but having a common understanding of these terms is critical for circumventing misunderstandings as we develop the model of tethered rationality. I will also point out examples of tethering or interaction between the different kinds of minds along the way, but a more substantive discussion of this issue is relegated to part IV of the volume.

3   Reflexes, Homeostasis, and the Autonomic Mind

Rational behavior requires theory. Reactive behavior requires only reflex.
—W. Edwards Deming

What human beings consciously wish is often quite at variance with the results
their reflex patterns automatically create for them.
—Timothy Leary

Given an organism with a nervous system, the simplest behaviors involve
reflexes. If you are sitting in your doctor's office with your feet dangling from
the examination table, and the doctor taps on the patellar tendon below your
kneecap, your lower limb will jerk forward. We have a good understanding of
the underlying mechanism. An afferent (sensory) neuron from the quadricep
muscles carries the signal (from the knee tap) to the spinal cord and passes
it directly to an efferent (motor) neuron, causing the stretched muscles to
contract (figure 3.1). This is a simple monosynaptic reflex arc, where only a
single synapse connects the stimulus and response nerve pathways.

A slightly more complex example of a reflex is the corneal blink. If I snap
my fingers in front of your eyes, you will blink. The underlying mechanism is the same as for the patellar reflex, except that this is a polysynaptic
reflex. That is, there are intervening neurons between the sensory input
and motor output neurons, called interneurons. An afferent (sensory) neuron carries a signal to interneurons in the brain stem, and the interneurons
pass the signal to efferent (motor) neurons, causing the blink (to protect the
eye from external threat).

Whether the doctor taps on your kneecap or snaps their fingers in front
of your eyes, the stimulus is causally sufficient for the response (leg swing
or eye blink). In both cases, I can offer you a large reward for not swinging
your leg when tapped on the knee or not blinking your eye when a finger

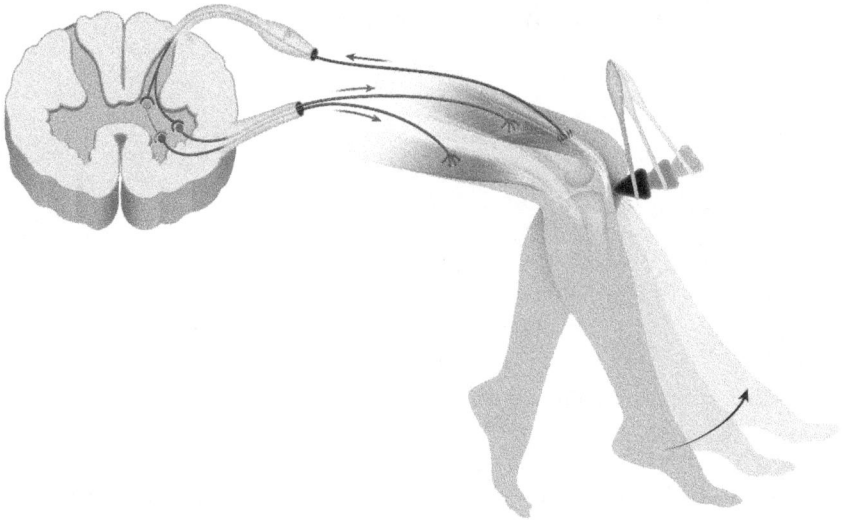

Figure 3.1
A monosynaptic reflex arc. The sensory input (tapping) generates the motor response directly through a monosynaptic reflex arc. There are no intervening neurons between the sensory nerves communicating the tapping signal from the quadricep muscles back to the spinal cord, and the motor neurons from the spinal cord innervating quadricep muscles. The figure also shows the presence of an inhibitory interneuron that serves to relax the hamstring muscle. Drawing by Aldona Griskeviciene.

is snapped in its vicinity. Despite your best efforts, your leg will swing out and you will blink when the respective stimuli are presented. You do not have a choice; the behavior is involuntary and automatic. Reflexes cannot be overruled by reason because the sensory neurons either connect directly or via interneurons to motor neurons without the information going to the brain, specifically the cerebrum, for interpretation. Where these behaviors exist in other species, they are underwritten by similar mechanisms.

However, these systems can also connect with higher-level systems in the cerebral cortex. For example, when a person steps on a sharp object, a sensory neuron carries the signal to an interneuron in the spinal cord. The interneuron communicates with motor neurons, which pull the foot away from the sharp object (polysynaptic reflex). However, if this was all there was to it, the action would probably result in imbalance and falling. So the interneuron also connects to motor neurons controlling the muscles of the other leg to allow for any adjustment needed to maintain balance. The interneuron may also connect with other neurons in the cerebellum and the cerebrum, which will allow for conscious awareness of the event and subsequent voluntary action.

Reflexes have also been co-opted for more complex functions involved in monitoring and regulating internal bodily functions controlled by the autonomic nervous system (ANS). The ANS innervates the smooth muscles (internal organs), cardiac muscles, and glands. It monitors blood pressure, salinity, core body temperature, blood pH, and blood glucose levels and controls heart rate, digestive activity, breathing, salivation, and sexual arousal, among other things. Two competing subsystems of the ANS are the sympathetic and the parasympathetic systems. The sympathetic nervous system is arousing. It prepares the body for the fight-or-flight response when under threat. It will increase the heart rate, dilate blood vessels and air passages, and stop energy-intensive processes such as digestion and urination. The parasympathetic system is calming. It is sometimes referred to as the "feed and breed" or "rest and digest" system. It maintains the body at rest, decreases blood pressure and heart rate, and induces digestion and bodily secretions. A third subsystem is the enteric nervous system, which controls the gastrointestinal tract.

In the case of respiration (figure 3.2), chemoreceptors monitor CO_2 levels (and to a lesser extent O_2 levels) of blood in the aorta and pH levels in the cerebrospinal fluid surrounding the brain and send the information to the breathing control center in the brain stem. Efferent nerve impulses from the brain stem, sent via the intercostal and phrenic nerves, result in contraction of inspiratory muscles and expulsion of CO_2-laden air. This stimulates the stretch receptors in the lungs. These receptors are part of a reflex arc. The stimulation inhibits inspiration to prevent damage to the lungs and chest cavity from excessive stretching and allows for expiration. After each inhaling breath, the inspiration center is reflexively inhibited, automatically allowing exhalation, so breathing is maintained in part by series or chains of automatic reflex actions, not reasoned decision-making.

Blood glucose levels are maintained largely through biochemical reactions (figure 3.3). Eating a meal increases blood glucose levels beyond the normal homeostatic "set" point (approximately 90 mg / 100 ml). This causes the pancreas to release insulin into the bloodstream and signals the liver to take up any excess glucose and store it as glycogen, causing the blood glucose levels to drop. If glucose levels drop below the set point (e.g., due to not having eaten for a while), the pancreas will secrete glucagon into the bloodstream, which will signal the liver to start converting stored glycogen into glucose and releasing it into the bloodstream to restore levels to normal.

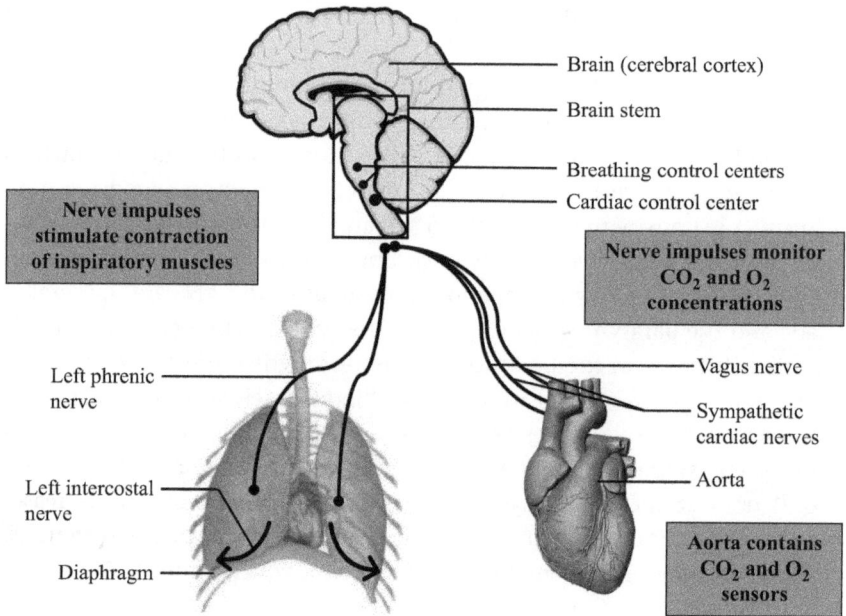

Figure 3.2
Control of respiration by the autonomic system, in part by a series of reflex actions with control centers in brain stem structures.

In terms of the five dimensions being used to classify behaviors, the function of the autonomic system is to monitor the internal environment of organisms and maintain it within a certain optimal range. The acceptable range and respective countermeasures are preset. An organism cannot *decide* via reason to release insulin into the bloodstream only when glucose levels exceed 50 mg / 100 ml or 150 mg / 100 ml or choose not to increase heart rate while undertaking physical exertion. The autonomic system responds automatically. Furthermore, an organism cannot *learn* to breathe or release insulin into the bloodstream. That knowledge is encoded into the genome and hardwired into the ANS. Systems simply come online as needed.

Most importantly, there is a very tight causal coupling between stimulus and response. Normally, certain stimulus changes are causally sufficient (and sometimes necessary) for certain behavioral changes. More accurately, given the constraints of normal biology, and appropriate *ceteris paribus* clauses, certain events and changes will be causally sufficient (and sometimes necessary) for other events or changes. In the case of the patellar reflex (knee-jerk), the tapping of the tendon activates a sensory neuron

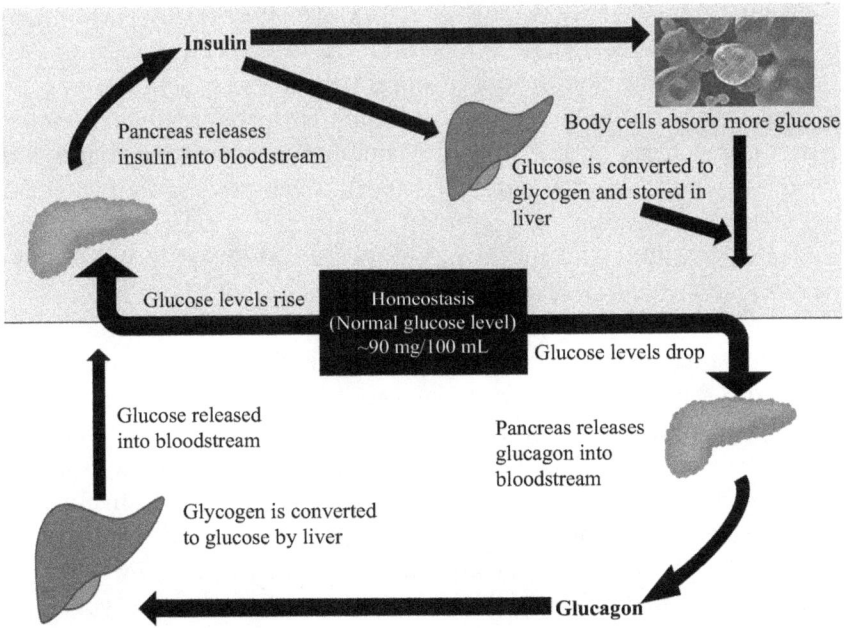

Figure 3.3
Homeostatic control of blood glucose levels by the autonomic system.

which in turn directly activates a motor neuron, resulting in a stretching of the quadriceps muscles and subsequent knee-jerk. How the sensory signal is generated can of course vary, but its generation is necessary and sufficient for the motor response. (One can consciously move one's leg through different neural pathways, but the resulting motion will be different.) Similarly, if glucose levels drop below a certain point, other things being equal, the pancreas will secrete glucagon into the bloodstream. If glucose levels do not drop below a set point, it will not do so, other things being equal.

Where exceptions occur, they can be explained either in terms of multiple pathways to a behavioral response or in terms of complex causal relations involving multiple factors and individual differences in these factors. The corneal blink provides an example of the first exception. Snapping my fingers near your eyes is sufficient but not necessary for you to blink. You can also blink as a result of dry eyes and conscious effort. An example of the second exception is provided by viral infections. The presence of the appropriate virus is necessary for someone to display the symptoms of COVID-19. However, not everyone exposed to the virus comes down with the symptoms of COVID-19. This seems to violate the sufficiency condition and looks like an exception to the tight causal leash. Not so. It means we do not fully understand the causal story. Symptoms of COVID-19 require the presence of the virus, but may also require the presence of properties X, Y, and Z, which can exhibit differences among individuals. If the virus plus the necessary properties are present, that will be necessary and sufficient for displaying the symptoms of COVID-19.

The mechanisms underlying the autonomic system are a combination of reflex arcs, chemical reactions, and homeostasis, which we understand at the level of neuronal signaling and underlying biochemistry. In terms of brain systems, the brain stem and hypothalamus structures, widely available across the evolutionary spectrum, are involved (see chapter 10).

This characterization of the autonomic system as autonomous of instinctive, associative, and reasoning systems is incomplete. For example, the level of satiety or hunger will also affect the engagement of instinctive, learned, and (in humans) reasoned food-foraging behaviors (chapters 11 and 12). Some autonomic systems can be modulated through volitional choice, to a limited extent. In the case of breathing, for instance, voluntary signals from the cerebral cortex to the medulla and pons can modulate the automatic breathing process, *within a certain range*, when it comes to such actions as singing or swimming. The extent of this control varies among individuals. Most of us certainly cannot will ourselves to stop breathing altogether or will ourselves to feel warm in an ice bath, but some people, such as the "Iceman," William Hof, make credible claims of being able to do the latter (Carney, 2017). There is also some controversial evidence that meditation techniques allow certain individuals to "reach down" with their voluntary systems into the autonomic systems and control body temperature, heart rate, breathing, pain, and even immune system responses (Benson et al., 1982; Heathers et al., 2018; Kox et al., 2014).

Autonomic systems can also feed information upward to modulate what should be volitional processes, including reasoning processes. Here is a personal anecdotal example. When I was a graduate student, I would come home late in the evening and several times a week my wife and I would then walk to the supermarket to purchase groceries. In the midst of this routine, my wife would sometimes have occasion to ask me, "Why are you upset?" I would reply, sincerely, that I was not upset. Despite my denials, she noticed that I was snapping at her. This is conscious, volitional behavior. If I have a reason to be upset with her, it could even be considered rational behavior. For example, if I did not really want to go shopping and I believed that she was unnecessarily pressuring me to do so, that would be a reason for my snapping behavior. But I had no such reasons, so my behavior cannot be explained using the machinery of rationality. After several repetitions of these episodes, she figured out the problem and the solution. I was cranky because I was hungry. Indeed, it is reported that cravings for carbohydrates result in feelings of anxiety, fatigue, and tension. Satisfying the craving increases energy levels, resulting in feelings of happiness. No beliefs or reasons were involved in my snapping behavior. It was just low blood sugar level (figure 3.3). My wife began carrying chocolate bars with her, and every time I became cranky, she would give me one, and all would be well. This is an example of lower-level systems modulating higher-level systems and reminds us that the cognitive system is tethered to these simpler systems. My wife's actions also serve as an example of how higher-level cognitive systems can modulate lower-level systems.

Continuing with the low-blood-sugar example, a study of experienced judges making parole decisions reported that favorable rulings directly after a food break were 65% and dropped to nearly 0 prior to a food break (Danziger, Levav, & Avnaim-Pesso, 2011).

Many of the processes controlled by autonomic systems also have affective components associated with them; that is, they are associated with *feelings*. One obvious example is sexual arousal, but it is also pleasant to breathe, to quench thirst and hunger, and to maintain a normal body temperature. Disruptions in autonomic systems lead to unpleasant feelings. For example, low blood sugar can lead to hunger pangs and irritability, not being able to breathe leads to feelings of suffocation, and heart failure can lead to severe chest pains. I will signal the role of affective arousal or feelings in each of the four types of behaviors that we will discuss and then address it more comprehensively in chapter 11.

*   *   *

The autonomic system constitutes a coherent and interesting, albeit simple, kind of mind, widely available on the phylogenetic tree and quite adequate for monitoring and regulating predictable internal environments, but its adequacy for monitoring and responding to unpredictable *external* environments seems less obvious. It is certainly a nonstarter for explaining rationality, but it does have the potential to modulate rational processes, as illustrated by the anecdotal example of my own low blood sugar level and the data from the study of parole judges. In the next several chapters, we will consider the kinds of minds more suitable for navigating changing external environments.

4 The Instinctive Mind

Citing concerns over historically high seasonal traffic and the resulting potential flight delays, a Canada goose was thinking of migrating home 2 to 3 weeks early in order to avoid the crowds, avian sources confirmed Friday.

—*The Onion* (March 31, 2019)

Autonomic systems are great for regulating internal bodily functions. They can signal hunger or sexual arousal, but to satisfy these and other needs, the organism must act in the external world. It must take steps to procure food or a mate. These behaviors call on instinctive, associative, and (where available) reasoning systems. This chapter considers the instinctive system.

The presentation is divided into three parts. The first part offers an intuitive characterization of instinctive behavior, followed by a more detailed account, including discussions of mechanistic models. The second part of the chapter considers several types of human behaviors—overeating, cooperation and cheating, in-group and out-group biases, sexual arousal and mating, and gender-specific behaviors—and asks whether they can be explained as instincts, as they are in nonhuman animals, or need to be explained as social (i.e., belief-based or reasoned) constructs. This is an empirical scientific question but also a political powder keg. I will address both issues in the third part of the chapter, but my primary focus will be the science. I will take the example of stereotypical gender behaviors and use the scientific data to tease apart biological and social contributions. The value of this exercise is to show how we can determine whether a given behavior is instinctive or a social construct. Later in the volume, the same strategy will be applied to the other behaviors.

My favorite example of instincts is perhaps the behavior of the capricorn beetle reported in Fletcher (1957). A female capricorn beetle lays its eggs in the ridges of the bark of an oak tree. A larva emerging from a hatched egg enters the trunk of the tree when it is the diameter of a piece of straw. It will live in the tree trunk for three years, growing and transforming into a fully developed capricorn beetle. As a larva, it eats its way through the wood, with the undigested wood passing through its body into the tunnel behind it. It is not in contact with any other creature during its entombment. To prepare for its exit from the tree, the larva eats its way toward the perimeter of the trunk, to the bark, often leaving a thin film of bark unbroken. It then retreats back into the tunnel and hollows out a large chamber to inhabit. Interestingly, the chamber is spacious enough to accommodate the fully formed beetle, with some room for the action of its legs. The larva then constructs a door at the opening of the chamber with a chalky white substance disgorged from its stomach. Upon completion of the door, the larva rasps material off the sides of the chamber to ensconce itself in the wood fibers. It then sheds its skin and becomes a pupa and positions itself so that it is facing the entrance to the chamber. (Without this advance positioning, there would not be enough room in the passage for the full-grown beetle to turn around to exit the chamber.) When it is ready to leave, it breaks through the chalky white substance covering the doorway, pushes aside any refuse material, and exits the tree trunk.

How do we explain this behavior? On the one hand, it seems replete with foresight of the later stages of its own development. Planning seems to be involved (e.g., making the chamber large enough for its adult size and turning itself around to face the exit while still small), but appearances can be deceiving. We are, after all, talking about a beetle larva, a creature with very limited neural resources. The exhibited behavior is also very specific and rigid. The larva cannot exhibit this "foresight" or "planning" in any other context. The behavior must be inborn or innate to the larva (i.e., encoded in its genome), as it has not been in contact with any other creature to have learned it. The behavior is more complex than a simple reflex arc but presumably still involuntary, implying a tight causal coupling between stimulus and response.

This latter point is further highlighted in a story about squirrels related by psychologist and philosopher William James. James makes some observations about squirrels in the wild burying nuts in the ground for the winter. This behavior also seems to involve foresight and planning (anticipating the dearth of food in the winter and stockpiling it in the autumn), but as James continues the story in the context of a pet squirrel, it turns out to be otherwise (James, 1890, p. 400):

> Now, as regards the young squirrel [which he has tamed], which, of course, never had been present at the burial of a nut, I observed that, after having eaten a number of hickory-nuts to appease its appetite, it would take one between its teeth, then sit upright and listen in all directions. Finding all right, it would scratch upon the smooth blanket on which I was playing with it as if to make a hole, then hammer with the nut between its teeth upon the blanket, and finally perform all the motions required to fill up a hole—*in the air*, after which it would jump away, leaving the nut, of course, uncovered.

The behavior is elicited by, indeed causally connected to, certain superficial features of the environment—the presence of a nut, an appeased appetite, and turn of the seasons—and the actions are executed even when they are unnecessary and ineffective. The behavior seems to share more properties with reflex arcs and autonomic processes than with reason and foresight. As James goes on to say, "The cat runs after the mouse, runs or shows fight before the dog, avoids falling from walls and trees, shuns fire and water . . . [n]ot because he has any notion either of life or of death, or of self, or self-preservation . . . [b]ut simply because he cannot help it" (p. 34). But unlike most reflexive and autonomic behavior, this behavior is directed at the external environment.

Interestingly, the term *instincts* has largely disappeared from the psychology literature.[1] It is difficult to find an article on instincts in American Psychological Association journals after the 1960s. In my more than 20 years of teaching cognitive psychology and cognitive neuroscience, I have never encountered the term in any textbook that I have used. However, from the latter half of the nineteenth century to the mid-twentieth century, instincts were important topics in animal and human psychology. They were discussed extensively by Charles Darwin ([1859] 1995), Herbert Spencer (1882), William James (1890), Lloyd Morgan (1903), and William McDougall (1923), among others, before falling into disrepute (Kuo, 1921) with the behaviorist takeover of psychology in the United States discussed in chapter 5. Their importance in explaining nonhuman (and human) animal behavior was reaffirmed in the 1930s to 1960s by the European ethologists, led by Konrad Lorenz (1952, 1958), Nikolaas Tinbergen (1951, 1953), and Karl von Frisch (1962), who shared the 1973 Nobel Prize in Physiology for their remarkable work.

I want to resurrect the term *instincts* and use it in a nonmetaphorical, technical manner. For this purpose, nuance and details matter. In particular, I want to ensure that the reader understands what instincts are, how they differ from reflexive, autonomic, associative, and reasoning processes, the types of mechanistic models needed to account for them, and the types of behaviors they can and cannot explain. The following discussion relies largely on the work of William McDougall and Konrad Lorenz. While there were important differences in the two accounts, having to do with the former's commitment to psychologism (i.e., appeal to purposeful behavior) and the latter's ardent commitment to mechanism, there was also considerable overlap in their characterizations of instincts (Kalikow, 1975, 1976; Richards, 1974).

What Are Instincts?

There is considerable agreement that instincts are species-specific, adaptive behavioral dispositions, encoded into the genome by evolution, that benefit the survival and reproductive success of the organism. That is, they are inherited, not learned, as the capricorn larva example illustrates. Instincts are common to all members of a species, at least those of the same sex. For example, it is the male three-spine stickleback fish (not the female) that builds the nest, and once the female has laid eggs in it, aggressively defends them from other male sticklebacks (identified by their red breast) or indeed any red-colored object. Instincts cannot be eradicated from the behavioral repertoire of the species in which they are innate elements or acquired by the individuals of other species. Accordingly, the aggressive nest-protecting behavior of the male stickleback cannot be modified or eliminated through training, nor can it be acquired by salmon or trout through learning.

Some instincts have limited developmental windows, others are seasonal, and still others persist throughout life. For example, the suckle response in mammals is limited to newborns and is extinguished when no longer necessary. Another example of a developmental window for instincts is the predator reaction in greylag goose goslings illustrated in an experiment where Lorenz rigged a rope across two trees and moved a fake predator along the rope. The goslings reacted to the shadow of the predator but not until eight weeks after hatching. Before that, they responded only to their parents' warning call. The reaction matured at a certain time, unaffected by learning through repeated occurrences (Richards, 1974). Rutting behavior in animals such as white-tailed deer is seasonal. Fight-or-flight responses, once developed, persist throughout life.

Instincts are triggered by specific (external) environmental cues and entail specific motor responses. The feeding behavior of the herring gull chick is triggered by the red dot on the parent gull's beak. A red dot painted on a yellow stick will solicit the same behavior (ten Cate, 2009; Tinbergen,

1951). There is a reasonably tight causal connection between the stimulus and the response (with some noted exceptions) and some limited scope for modulation via learning and reasoning (where available).

The phenomenon of habituation provides an apt example of learning modulating instincts. Prairie dogs sound an alarm at the presence of a predator. They will typically give an alarm call at the detection of human footsteps, but if no actual danger befalls them after repeated exposure to human footsteps, they become habituated to the sound of footsteps and do not give the alarm. However, the alarm call continues to be sounded in the presence of nonhuman footsteps, suggesting that habituation is stimulus specific and there is no generalization to other stimuli.

There are also examples of more explicit learning, where environmental feedback is needed to fine-tune the stereotypical instinctive behavior. For instance, while male zebra finches are genetically predisposed to produce a song, they must undergo a period of listening to and practicing the song of their fathers, which will then determine the particulars of their own song. More controversially, the same point can be made about human language. It has been argued that humans are born endowed with a language-acquisition device that requires exposure to human speech at a certain stage of development (the window of opportunity) to set parameters specific to local languages (Chomsky, 1972).[2] These examples illustrate interactions between instinctive and learned behaviors that we will return to several times.

Described in this manner, instinctive behavior is not only automatic and deterministic but also unconscious and robotlike. Wallace Craig (1917) was perhaps the first to question this robotic conception of instincts. Based on his studies of the blond ring dove, he argued that instinctive behavior was not like unfelt reflexes but involved "an element of appetite, or aversion, or both" (p. 91). By "appetite" he meant a continuous state of agitation in the absence of the stimulus. The receiving of the stimulus is "consummatory" or satisfying for the animal—that is, it relieves the agitation and returns the animal to rest. An "aversion" is a state of agitation resulting from the presence of a certain stimulus and ceases when the stimulus is withdrawn, so we have here an insertion of the notions of affect or feelings (with positive or negative valence) between the stimulus and the behavioral response. William McDougall (1923) also adopted this insight and explicitly inserted an affective link to mediate between the stimulus and the response, noting that "every instance of instinctive behavior involves a knowing of some thing or object, a feeling in regard to it, and a striving towards or away from that object."[3] Whether this applies to the larva of the capricorn beetle is a moot point, given its limited neural endowment; but once we move up the phylogenetic tree to birds and mammals, it becomes a much more plausible conjecture that will play a pivotal role in the tethered rationality model. I return to it several times and deal with it more thoroughly in chapter 11.

Mechanistic Models of Instincts

Perhaps the earliest proposed account of instincts was the chain-reflex theory. It was advocated by Spencer (1882), Sherrington (1952), and Pavlov (1928) and extensively developed and articulated by Konrad Lorenz (1937). The basic claim was that instinct can be explained as a complex bundle or chain of reflexes connecting the environment to the animal's sensory and motor systems. Particular environmental stimuli, in particular situations and particular developmental stages, call forth particular actions, as when the snapping of my fingers in front of your eyes inevitably results in the simple reflex of an eye blink. Rather than a simple reflex, instincts need to be accounted for by *chains* of reflexes found in autonomic systems, such as those responsible for breathing. The chain-reflex idea was not a metaphor but rather was meant to be taken literally.

Lorenz viewed the organism as a mechanism prepared to display specific stereotyped behavior in response to specific environmental stimuli. His initial chain-reflex model introduced the concepts of the "releaser" and the "innate releasing mechanism." The specific environmental stimulus is the releaser (such as a sound or color). The releasers are external to the animal and are few and specific. The innate release mechanism, like a tightly wound spring, determines the potential or readiness of the animal to respond to the releaser. For example, the swollen abdomen and the posture of the female stickleback is the releaser that unlocks the innate release mechanism of the male's mating behavior.

Lorenz argued that there were no overarching instincts, such as the "parental instinct" or the "reproductive instinct." Where seemingly complex behaviors were involved, they could be broken down into more specific simple behaviors, each associated with a releaser and an innate release mechanism. In this way, the reproductive instinct in the stickleback fish might be comprised of fighting, building, mating, and caring for offspring. Mating, for example, could further be differentiated into a zigzag dance, leading a female to the nest, showing the entrance, quivering, and fertilizing the eggs. Each of these simpler behaviors would then be connected at the neural level by perception-action reflex arcs. The nerve signal generated by the presence of the stimulus would travel down fixed nerve pathways to automatically trigger the motor response.

The chain-reflex theory encountered a number of empirical difficulties. Two such problems were "intention movements" and "vacuum activities." Intention movements are incomplete performances of a behavior or a chain of behaviors. For example, a cat may bare its teeth and raise its paw as if to attack without actually attacking (i.e., completing the behavior). This is inconsistent with the workings of reflexes. Reflexes are all-or-none; you cannot half blink your eye in response to a stimulus. Vacuum activities are behavior patterns that are initiated in the absence of the usual releaser. One example would be the attempt of the pet squirrel described by William James to bury a nut in the blanket. Another example is provided by captive raccoons. In the wild, raccoons will hold their food underwater, and make washing movements, prior to eating it. Captive, caged raccoons will engage in the same behavior with their food in the absence of water. Again, this is problematic for the chain-reflex theory because the reflex cannot be triggered without the stimulus.

Two important influences resulted in the abandonment of the literal chain-reflex model for a more metaphorical energy-based model. The first was Wallace Craig's critical insight regarding the role of feelings in instinctive behavior that we encountered earlier. In accepting Craig's insight, Lorenz and Tinbergen (quoted in Kalikow 1976, p. 18) noted that "subjective experience is not a chance side effect or 'epiphenomenon' of physiological processes! Without the 'sensual pleasure' which presumably represents the experiential aspect of every instinctive behavior pattern, performance of the pattern would only take place when the organism entered the elicitatory stimulus situation purely by chance." The incorporation of feelings into the theoretical account introduced the idea that the animal *wants* to engage in the behavior for the satisfaction, or positive affect, that it releases and allowed Lorenz to begin differentiating instinctive behavior from reflexes.

The second insight came from Erich von Holst's observations that some activities thought to be caused by chain reflexes, such as the crawling movement of earthworms, are actually the result of internally produced stimuli. Holst demonstrated that if all nerves in an earthworm responsible for its creeping reflex are severed, so that no stimulus can result in a reflex action, the earthworm's ganglia still send out the signals for the creeping motion, suggesting that the creeping movements are innate, fixed motor patterns. That is, rather than being a function of dormant reflexes triggered by a series of external stimuli, these behavior patterns are generated endogenously (Kalikow, 1975).

These insights ultimately led to the development of the metaphorical but mechanistic energy model of instinctive behavior, presented in figure 4.1.

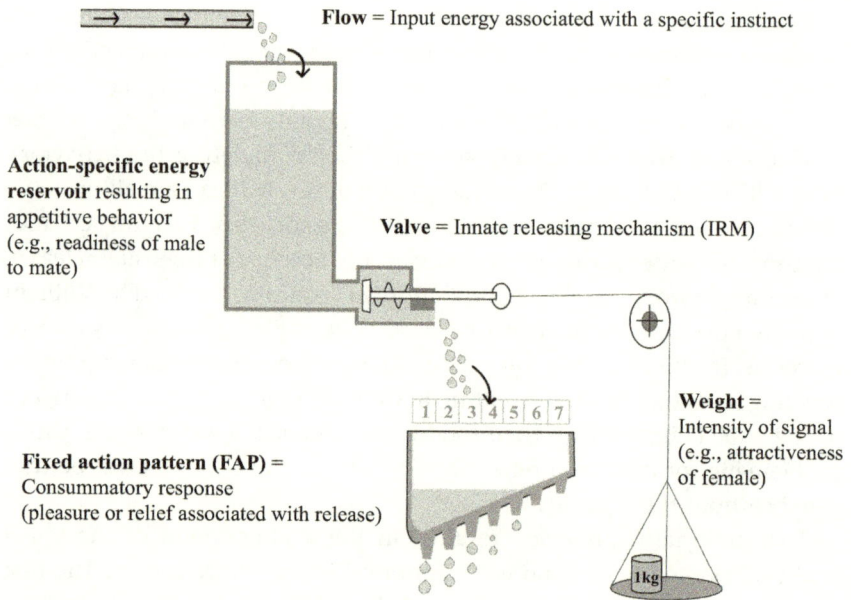

Figure 4.1
The Lorenz hydraulic or energy model of instincts. Adapted from Lorenz (1950).

The central nervous system of an organism provides it with certain reservoirs of action-specific energy. Each instinct has its own reservoir. Energy is directed into the reservoir but is blocked or inhibited by a valve corresponding to the innate release mechanism, resulting in a buildup. This energy drives appetitive approach behavior in the animal (e.g., arousal and readiness of the male to mate). The valve or innate release mechanism is attached to an opening mechanism controlled by weights. The mass of the weights corresponds to the intensity of the stimulus signal in the environment (e.g., the attractiveness of a female). The opening of the valve is a function of the pressure in the energy reservoir and the mass of the weights. When the valve opens, the energy "drips" into a hierarchically organized, fixed action pattern "template," which determines the pattern of the resulting behavior. Whether the full pattern or a partial pattern of behavior is displayed is a function of the volume of pent-up energy in the reservoir and the intensity of the stimulus (i.e., how much the valve opens). There is an affect-laden drive correlated with the volume of pent-up energy, resulting in the need to discharge, and a relief affect associated with discharging the behavior (the consummatory response). This reduces the pressure of the pent-up energy and brings the animal back to equilibrium, explaining why an animal will actively seek environments in which a behavior or fixed action pattern can be discharged.

Despite being a metaphor, this is an important model for our purposes. It provides a tight causal connection between stimulus and response, similar to reflexes but with increased degrees of freedom. The behavior is a function of the volume of action-specific energy and the intensity of the stimulus in the environment. Each factor can have variable values. This provides flexibility but also ensures that the instinctive behavior can usually be carried out under appropriate circumstances. If an action is not initiated, the intensity of the stimulus may be insufficient and/or the energy store of the specific reservoir inadequate. An action that is initiated but not completed (i.e., an intention action) can be explained as having insufficient energy in its specific reservoir to complete the full hierarchy of responses. Vacuum behaviors can be explained as having an excessive volume of energy in the corresponding reservoir that forcefully leaks through the valve even in the absence of the stimulus. But given normal amounts of specific energy reservoirs and stimulus intensity, and all other things being equal, the stimuli are usually causally necessary and sufficient to release the stereotypical behavior.

I believe this model captures important insights about behavior that have been ignored by subsequent computational models. The work of Jaak Panksepp and other neuroscientists, which will be introduced in chapter 11, shows how Lorenz's metaphorical model can be conceptualized as a neuroscientific model by incorporating arousal and reward systems and mapping specific instincts onto specific brain stem, diencephalon, and subcortical neural circuits associated with specific neural chemistry.

Are Instincts Enough?

How far can instincts take us in explaining human behavior? That is, how much of human behavior meets the constraints of instinctive behavior and can be explained by such a mechanism?

It should be uncontroversial that some aspects of human behavior are controlled by instincts, but there is much disagreement on which ones, and how many, meet the strict criteria. Darwin ([1859] 1995) thought there was an inverse relationship between instincts and intelligence. The higher an animal was on the evolutionary scale, the fewer instincts it would have and the more intelligent behavior it would exhibit. William James (1890) thought there was no inverse ratio between instincts and intelligence. Indeed, he argued that humans have more instincts than any other species. William McDougall (1923) listed 18 human instincts in his social psychology text, including the parental instinct; the sex instinct; the instinct of pugnacity; the gregarious instinct; the instincts through which religious conceptions affect social life; the instincts of acquisition and construction; and the instincts of laughter, imitation, play, and habit.

Some human behaviors, such as the suckle response in newborn babies, are undisputedly considered instinctive, involving mechanisms similar to those in other mammals, but as we go through even a brief selective list of human behaviors, things quickly become less clear-cut.

What about eating behavior? Many animals, particularly carnivores, overeat. Lions will eat up to one quarter of their body weight after a large kill. During a salmon run, grizzly bears eat until they can eat no more. Given that they live in a feast-or-famine environment, it is adaptive to maximize caloric intake when food is available and store the excess as fat deposits for times of scarcity. It may be many days or weeks before the lion makes another big kill and another year until the next salmon run. These behaviors are undoubtedly instinctive, but what about when humans overeat? Given that our ancestors also evolved in a feast-or-famine environment during the Pleistocene period, can my propensity to overeat also be accounted for by similar adaptive, instinctive mechanisms, or is it to be explained as a social or cultural manifestation (e.g., driven by the insidious advertising of the fast food industry)?

Some animals seem to live in socially organized cooperative groups based on reciprocity. For example, vampire bats feed on blood. If a bat returns to the communal roost without having fed, it may be in danger of starving unless another bat regurgitates blood to it. Bats that have successfully fed will regurgitate blood to those that have not (Wilkinson, 1984). It has been argued that such altruism is fitness enhancing only if the recipient bat reciprocates at some point in the future. But it is to the advantage of the recipient bat not to reciprocate (i.e., to cheat). Therefore, social animals must evolve mechanisms for detecting and punishing cheaters as an adaptive strategy to maintain the fitness of the group. Can these same mechanisms account for our railing against "welfare cheats" and my American friend's rejection of universal healthcare, or is there more to the human story?

Many species, from bees to baboons, live in organized groups and cooperate more favorably with members of the group than with outsiders. Some of these species, such as wolves and chimpanzees, are also territorial. They mark and defend their home range against conspecifics. This behavior secures food, mates, and child-rearing resources and is therefore presumably adaptive. Humans likewise form coalitions and are territorial animals. We insist on exclusive possession for ourselves, our families, and our communities (i.e., the "in-group") and will fight to exclude outsiders ("out-group"). Are the same instinctive mechanisms at work in humans as in wolves and chimpanzees when people characterize immigrants as "rapists and drug dealers" and chant "Build that wall! Build that wall!" during Trump rallies, or is the story more complex?

All animals have specific behaviors associated with sexual arousal and mating. For example, during the rut, the testosterone levels in red deer bucks increase a thousandfold (Lincoln, 1971). Secondary sexual characteristics become more prominent. The bucks become less cautious than usual (making them more susceptible to hunting and motor vehicle accidents). They mark their territory and fight other bucks, sometimes to the death, to display their dominance. Their sole focus is to find, chase, and impregnate as many estrus does as possible. Different species partake in different stereotyped behaviors but to the same end. Is this also the case for human males? Can the Edwards example from chapter 1—or more generally, the numerous instances of powerful men sexually pursuing, harassing, and even assaulting women—be explained in terms of similar instinctual mechanisms or is something more in play?

In all sexually reproducing species, there are some stereotypical behaviors associated with each sex regarding courtship, territorial aggression, mating, and parental care. Human societies also assign sex-specific roles to members based on stereotypes such as "women are more emotional, caring, and nurturing" while "men are more competitive, aggressive, stronger, and less emotional." In the case of nonhuman animals, we explain these behaviors as instinctually determined. Do the same explanations apply to humans, or is there some other explanation, such as social construction?

Darwin, McDougall, and Lorenz would certainly not hesitate in agreeing that the instinctive mechanisms that provide such convincing explanations of animal behaviors in these examples provide equally compelling explanations of human behavior. Modern evolutionary psychologists, whom we will encounter in chapter 9, would wholeheartedly concur, arguing that it is only our bruised vanity that prevents us from accepting the obvious. However, most adherents of the cognitive and social sciences intellectual framework (the standard model of rationality), which has come to dominate our thinking about human social behavior, would vociferously disagree, emphasizing socialization and reason instead.

To anticipate my own answers to these questions, I will argue that each of these five human behaviors—overeating (chapter 12), reciprocal cooperation and cheating (chapter 9), in-group/out-group bias and territoriality (chapter 13), sexual arousal and mating (chapter 11), and gender-specific behaviors (discussed below)—have instinctive or biologically innate components. However, that does not mean that either the behaviors or the underlying mechanisms are similar across species. In fact, we will see that in some cases, human instinctive behaviors are much more elaborate than those found in other species and that in other cases they are unique to us. But critically, in all human cases, instinctual systems interact with and are modulated by reasoning systems. No other species can claim as much.

In the balance of this chapter, I would like to address two questions. The first has to do with politics, the second with science. One would think that the question of whether these behaviors are explained by innate mechanisms or as social constructs is an empirical scientific question, but such questions have become highly charged social and political powder kegs and lie at the heart of current "political correctness" debates.

The twentieth century was marked largely by a loud and bitter pushback against the view that human social behaviors are biologically or instinctively determined, or even just modulated by these systems (Ruse, 1985). This is illustrated by the reaction to the 1975 publication of *Sociobiology: The New Synthesis* by eminent Harvard entomologist Edward O. Wilson. The volume outlined the biological basis of social behaviors such as aggression, sex, parental care, territoriality, and caste roles, among others. The first 26 of its 27 chapters, dealing with nonhuman animals from ants to elephants, were universally hailed as an intellectual tour de force. However, in the last chapter, "Man: From Sociobiology to Sociology," Wilson extended his analysis to humans and all hell broke loose. He was attacked by a broad coalition of students and academics, including friends and colleagues, in a particularly vicious, vitriolic manner. He quickly became one of the most vilified scientists of his time (Ruse, 1985). Why was this?

The second set of questions concerns the science. How do we determine whether a behavior is instinctive or a reasoned social construct? What type of evidence can be brought to bear on this question? I will use the example of gender-specific behaviors to explore and answer questions relating to both the politics and the science.

Extended Example: Are Gender-Specific Behaviors Instinctive or Social Constructs?

Society has long believed that gender identity is biologically determined by external sexual characteristics (and, more recently, chromosomes) and that sex and gender are one and the same. Implicit in this assumption is the additional assumption that there are intrinsic behavioral differences between males and females. If societal norms are relevant in determining gender differences, it is only because boys and girls exhibit innate behavioral differences, which trigger differential treatment by parents.

Recently, a number of groups in Western societies have begun arguing that gender identity is socially constructed, independent of biological sexual characteristics (Lorber, 1995). For example, the American Psychological Association (2014) notes that "gender refers to the socially constructed roles, behaviors, activities, and attributes that a given society considers appropriate for boys and men or girls and women." Any emerging behavioral differences between boys and girls are explained by differential treatment by parents who hold societally enforced gender stereotypes; for example, giving boys trucks and soldiers and girls dolls to play with. This differential treatment results in gender stereotypes such as women are emotional, caring, nurturing, dependent, and physically weak, while men are more competitive, aggressive, physically strong, and less emotional than women. The only correct way to assign gender is based on one's "internal sense of being male, female, or something else" (American Psychological Association, 2014) and presumably change it as necessary. First, we will consider the politics and then the science of this fierce debate.

Politics of Gender

Many issues regarding social behaviors become politicized because they have social policy implications. Gender is one such issue. On the one side, there is outrage because of the fear that to accept a practice or norm as embedded in, or as an outgrowth of, our nature is to sanction it, even though it may be inconsistent with current social norms. If what are perceived as social flaws and inequities of our society are determined by human nature, attempts to change them will be difficult or futile. On the other side, there is fear that the failure to acknowledge any constraints on the world order "gives rise to relativism, in which everything that exists is of equal value and at the same time undifferentiated, without any real order or purpose" (Congregation for Catholic Education for Educational Institutions, 2019).

In February 2019, the Vatican released the document "Male and Female He Created Them: Towards a Path of Dialogue on the Issue of Gender and Education," which voiced these concerns (Congregation for Catholic Education for Educational Institutions, 2019):

> Gender theory (especially in its most radical forms) speaks of a gradual process of denaturalization, that is a move away from nature and towards an absolute option for the decision of the feelings of the human. In this understanding of things, the view of both sexuality identity and the family become subject to the

same "liquidity" and "fluidity" that characterize other aspects of post-modern culture, often founded on nothing more than a confused concept of freedom in the realm of feelings and wants, or momentary desires provoked by emotional impulses and the will of the individual, as opposed to anything based on the truths of existence. . . .

This ideology inspires educational programmes and legislative trends that promote ideas of personal identity and affective intimacy that make a radical break with the actual *biological difference* between male and female. Human identity is consigned to the individual's choice, which can also change in time. These ideas are the expression of a widespread way of thinking and acting in today's culture that confuses "genuine freedom" with the idea that each individual can *act* arbitrarily as if there were no truths, values and principles to provide guidance, and everything were possible and permissible.

While ostensibly (and ironically) appealing to biology and reason, the Catholic Church has a vested interest in upholding the world order as articulated in its religious texts. It did not take long for the document to be condemned by LGBTQ groups as harmful and encouraging hatred and bigotry (DeBernardo, 2019): "The document associates sexual and gender minorities with libertine sexuality, a gross misrepresentation of the lives of LGBT people which perpetuates and encourages hatred, bigotry, and violence against them." These views too ostensibly appeal to science, but also have a vested interest in particular conclusions.

Who is correct? More generally, how do we differentiate a social construct from an instinct or "biological construct?" Looking *disinterestedly* at the science is always a good start. What does the science say?

Basic Science of Sex and Gender: Genes and Much More
Most species reproduce sexually by fusing together genetic materials from a male and female individual. The female is defined as the individual who contributes the physically larger gamete (ovum), and the male is the one who makes the smaller contribution (sperm). In many (but not all) species, sex is determined by chromosomes. In mammals, the XX and XY chromosomes specify female and male, respectively. All mammalian fetuses begin as females, though. This is the default mode of embryonic development. Fetuses with a Y chromosome undergo a process of masculinization as outlined in figure 4.2. It begins with the Sry (sex-determining region Y) gene initiating the formation of testes (during the sixth week in humans). Two hormones, antimüllerian and testosterone, are secreted by the testes within a certain critical window during gestation (6 to 12 weeks in humans). The antimüllerian hormone suppresses development of the female reproductive tract and genitalia. The testosterone is converted to

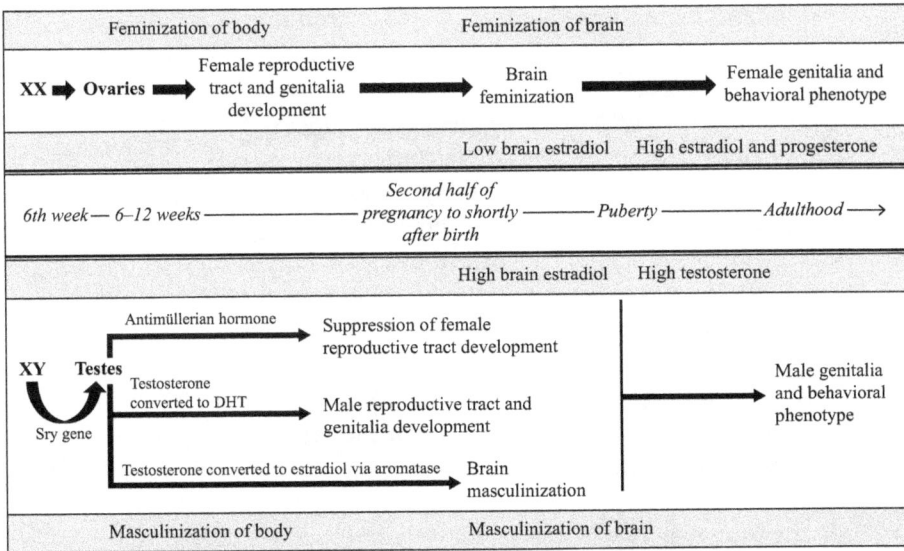

Figure 4.2
Differentiation of sex and gender in mammals.

DHT (dihydrotestosterone) by the enzyme 5-alpha-reductase and results in masculinization of the fetal body, that is, development of the male reproductive tract and genitalia (Breedlove, 1994; de Vries et al., 2014; Morris, Jordan, & Breedlove, 2004; O'Shaughnessy & Fowler, 2011; Swaab, 2007). Disruption in androgen action during the critical window will result in impaired development of the testes and penis (Matsushita et al., 2018; Place & Glickman, 2004; Welsh, Suzuki, & Yamada, 2014). There is another hormonal surge during puberty that completes sexual differentiation (de Vries et al., 2014; Lenz, Nugent, & McCarthy, 2012; MacLeod et al., 2010). The basics of this story were largely worked out in the 1940s and 1950s (Josso, 2008), but the sexual differentiation of the body is only half the story.

We accept that all nonhuman, sexually reproducing species exhibit innate gender-specific behaviors associated with courtship, territorial aggression, mating, and parental care. Insofar as all behaviors are determined by brain systems, these behavioral differences must also be underwritten by neural sexual dimorphism; that is, by structural and neurochemical differences in male and female brains. The process of masculinizing the mammalian brain involves the conversion of testosterone to estradiol by the enzyme aromatase. The critical window for this process is the second half of pregnancy in humans. These hormonal processes serve to permanently sculpt developing neural systems by, among other things, either inhibiting or facilitating neural apoptosis (cell death) and modulating the formation and elimination of synaptic connections, in certain key subcortical brain regions such as the sexually dimorphic nucleus of the preoptic area (SDN-POA) in the anterior hypothalamus. One easily observable consequence of this process is that the SDN-POA area of the brain is several times larger in male rats than in females, and lesions to the POA eliminate copulatory behaviors in male rats. These brain differences are driven by perinatal secretion of testosterone during the critical window. Introduction of testosterone outside this window has no effect on the size of the SDN-POA in rats (Lenz et al., 2012; MacLusky, Naftolin, & Goldman-Rakic, 1986; Matsuda et al., 2011; Morris et al., 2004; Nugent et al., 2015; Sato et al., 2004; Wu & Shah, 2011; Zuloaga et al., 2008). Thus, the normal unfolding of the gestation process results in a fetus with a male brain and a male body or a female brain and a female body (figure 4.2).

In the vast majority of cases, these basic processes unfold normally.[4] However, notice the role of the Y chromosome; it determines the release of androgens at two different time points for two different purposes: masculinizing the body and masculinizing the brain. The independence of these processes means that there are two ways in which the modulation of the timing and/or quantity of hormone release can result in transsexuality (Swaab, 2007). Similarly, even in the "default" female developmental processes, hormonal imbalances during critical periods can disrupt normal sex and gender development (Nordenström et al., 2002). Furthermore, the process can also break down at the chromosomal level, as in the case of Klinefelter syndrome (Smyth & Bremner, 1998).

The development of sex and gender has been extensively studied and experimentally manipulated in animal models. Androgen modulation prenatally, and even neonatally, can affect sex-specific behaviors such that females exhibit greater same-sex aggression and males exhibit greater female-specific behaviors, such as lordosis (downward curvature of spine) (Clemens, Gladue, & Coniglio, 1978; Clemens & Gladue, 1978; Edwards & Burge, 1971; Gladue & Clemens, 1980; Huffman & Hendricks, 1981; Palanza et al., 1999; Rines & vom Saal, 1984; Schechter, Howard, & Gandelman, 1981; Tobet & Baum, 1987; vom Saal, 1979; Ward & Renz, 1972). In the first such experiment, the female offspring of guinea pigs that were administered testosterone during pregnancy and again as adults displayed reduced lordosis behavior and increased male-like copulatory mounting behavior. Control guinea pigs that received testosterone only as adults did not display this reversal in sexual behavior (Phoenix et al., 1959). In certain birds, song vocalizations are associated with males. There are corresponding neural differences underlying this vocal dimorphism. Female zebra finches will develop a male-like song system (along with corresponding neural substrate) after treatment with estradiol as nestlings (Pohl-Apel, 1985; Simpson & Vicario, 1991).

Most of this research has been done with nonhuman animal models. Among these animal models, there is considerable agreement on the overall processes of sexual differentiation of bodies and brains, though details continue to be revised and fine-tuned (Cahill, 2006; McCarthy, 2016). Two basic problems arise in applying the animal models to humans. (1) The manipulations and experimental techniques used to generate the results in rats, guinea pigs, and ferrets cannot, for obvious ethical reasons, be applied to humans. Therefore, we cannot directly investigate whether the same systems are at work in humans. (2) In these animal models, the notion of "male" and "female" behaviors is tightly constrained to courtship, mating, territorial aggression, and parental care. Human behavior is much more nuanced and extensive. In fact, insofar as it is a product of the rational mind, I am arguing that it is qualitatively different from nonhuman behaviors. So perhaps gender-specific behaviors in humans are built by the rational mind based on societal and cultural expectations rather than via sculpting of brain systems by hormonal processes. The way to approach this issue is, in the first instance, to restrict ourselves to the same basic behaviors that have been studied in the animal models and ask whether there are similar typical "male" and "female" behaviors in humans and, if there are, whether they are socially and culturally learned or have a biological basis.

There are data that speak to these issues. As an example, I will briefly consider the data on children's play and toy preferences. In every culture, there are differences in the types of play and toys that interest girls and boys. This is largely uncontroversial. What is controversial is whether these differences are determined by learning—to ensure that boys and girls grow up to accept their socially and culturally mandated gender roles—or have some gender-specific biological basis (Connor & Serbin, 1977; Goldberg & Lewis, 1969; Hines et al., 2016; Taylor, Rhodes, & Gelman, 2009). Disentangling the two is not trivial, but neither is it impossible (Eliot, 2011).

One strategy is to test infants prior to extensive socialization. In eye-tracking studies of young infants from three to eight months old, girls showed a visual preference for dolls, while boys showed a visual preference for trucks, suggesting differential inborn sensitivity to low-level perceptual features associated with the different objects prior to awareness of gender categories (Alexander, 2003; Alexander, Wilcox, & Woods, 2009). Given such a finding, one can also test nonhuman animals for similar dimorphic perceptual preferences. These sex-dimorphic preferences for object features have been detected in vervet and rhesus monkeys, using objects similar to those in the studies of children, suggesting it is an evolutionary adaptation that arose earlier than the hominid line (Alexander & Hines, 2002; Hassett, Siebert, & Wallen, 2008).

Another strategy is to see whether these preferences change as a result of different levels of androgens during early fetal development. In a longitudinal study involving 342 male and 337 female children, levels of testosterone in mothers were measured during pregnancy and were later correlated with the gender role behavior of the children as they developed. Higher levels of prenatal testosterone in the mother were positively correlated with young (3.5 years) girls being more interested in toys, games, and activities typically associated with boys. This relationship held even when social factors such as the presence of male or female siblings, parental commitment to traditional sex roles, and the presence of a male partner in the household were taken into consideration. This relationship also held in a group of young adults whose exposure to prenatal androgen levels was estimated using the "digit ratio marker" technique. Young women with more masculinized digit marker differences showed more preference for male-typical toys and activities (Alexander, 2006). There was no relationship between a mother's prenatal testosterone levels and behavior in boys (Hines et al., 2002).

Girls with congenital adrenal hyperplasia (CAH) are exposed to excessive levels of androgens in early fetal development. This provides another opportunity to test the hypothesis of hormonal versus social environment contributions to gender determination. In one study, CAH girls three to eight years old showed a preference for boys' toys and a reduced preference for girls' toys compared to their unexposed female relatives of similar age and raised in similar environments (Berenbaum & Hines, 1992). Another sample of young CAH girls displayed a greater preference for playing with boys compared to their unexposed female relatives of a similar age (Hines & Kaufman, 1994). These preferences are even modulated by the *level* of fetal androgen exposure (Nordenström et al., 2002). By contrast, boys exposed to increased levels of fetal androgens do not display any such differences.

These data, as far as they go, are consistent with the animal data. To interpret these data as indicating that gender differences are socially constructed in the absence of biological, hormonal factors is to misconstrue the science. To interpret these data as indicating that chromosomal differences constitute gender differences is also to misconstrue the science. Both sides of the political divide are (intentionally?) getting it wrong.

The Church is wrong because the science is saying that gender ambivalence is very real. Those claiming that gender is a social construct are wrong because the studies show that the sex-dimorphic preferences under consideration can be detected very early in human infants prior to any social gender category formation, are present in other species, and vary as a function of fetal androgen exposure. These data are speaking to the importance of biological factors, but they are *not* precluding the role of environmental factors. In fact, many mediating environmental factors are actually biological, including prenatal exposure to hormones, medications, environmental chemicals, and stress on the mother during pregnancy (Coolidge, Thede, & Young, 2002; Dessens et al., 1999; Zucker et al., 1996). Environmentally mediated events in the unfolding of the two separate processes triggered by the presence of the Y chromosome can result in gender ambivalence.

Perhaps what is really at issue is the extent to which postnatal social and cultural environmental factors—based on beliefs and reasons—will interact with the biological factors. We understand much less about this. The most charitable counterinterpretation of the data is that even if prenatal biological and environmental factors do not mandate specific gender identities, sexual dimorphism surely predisposes humans to be more receptive to certain socially presented gender-specific cues than to others (McCarthy, 2016).

It is also important to acknowledge the huge gap between these biological predispositions (which should be understood as statistical distributions) and societally constructed gender norms and expectations discouraging and even prohibiting women from being doctors, judges, engineers, or from voting, running for political office, and other traditionally male pursuits. These prohibitions are not mandated by biological predispositions. They are largely social constructs that typically ensure men's dominant role in society. They have been and will continue to be questioned and corrected in response to social, educational, economic, and technological factors. This is the defining characteristic of social constructs.

Such an account is consistent with the two themes of this book: (1) that human behavior is qualitatively different from the behavior of other animals, meaning we are endowed with the ability to reason; and (2) that we have been generated by the same evolutionary processes as all other animals so we are not exempt from the laws of biology. This is another way of saying that the reasoning mind is tethered to simpler associative, instinctual, and autonomic processes.

The purpose of this extended example is fourfold. First, I use it to indicate that whether a certain behavior is a social construct or has a biological basis is an empirical not political issue. Answers to the following five questions can help us distinguish between the social and biological: (1) Is the trait universally present in human societies or is it culture specific? (2) Is it available on other branches of the phylogenetic tree? (3) Does it emerge early in human infants, prior to any opportunity for extensive socialization? (4) Is it underwritten by implicit, automatic, low-level mechanisms? (5) Is it possible to trace specific subcortical neural circuitry and neurochemistry devoted to it (and find homologous behavior and circuitry in other species)? Affirmative answers to all or most of these questions are indicative of instinctively determined systems. Affirmative answers to these questions do not preclude a modulating role for reason and socialization in humans.

Second, to say that something has a biological basis is not to say that it is simply a matter of genetics. Genetics may be the starting point, but the prenatal and postnatal environments in which the neural development unfolds has an enormous impact on the resulting brain organization. The social environment will also matter, though only postnatally. It will also impact behavior by sculpting neural systems. In this sense, being a social construct or having a biological basis are not mutually exclusive categories. Both result in changes to brains, which in turn changes behavior. This issue is further considered in chapter 14.

Third, I'm not interested in dictating social policy or advocating for one position or the other. My own personal view is to live and let live. But if you are in the business of advocating for specific policies and changing human behavior to conform to those policies, it is in your interest to get the science right. You will have a better chance of modifying behavior if you have an accurate model of what is actually driving it.

Fourth, nothing I've said here will actually matter to the two sides of this debate. Just as the science of climate change failed to sway skeptical minds in chapter 1, the basic science of sex and gender identity presented here will fail to change many minds on either side of this particular battlefield. This is not rational. Why not revise false beliefs in the face of counterevidence? Why this should be the case is a fascinating question that will be addressed in chapter 13, once the machinery of tethered rationality is in place. Ironically, the answer itself will involve constraints on the system of rationality by lower-level biological constructs.

The extent and nature of biological constraints on rational choice lie at the core of the argument being developed in this volume. The politicization of these issues has not only generated fear and rage but also prevented us from developing more realistic models of rationality. Wanting the world to be a certain way doesn't make it so. Human behavior is not on a tight biological leash like that of the larva of the capricorn beetle, but neither does our system of rational choice float unfettered above the biology. We need a commonsense model of tethered rationality.

*   *   *

The examination of the characteristic features of the instinctive mind illustrates both its strengths and its weaknesses in accounting for human behavior. Instincts are an inexpensive solution to guiding behavior that is essential, is needed prior to any opportunity for learning, has high cost associated with error, and does not need to change across generations. In such circumstances, instincts are the preferred solution. But their very strengths—automaticity, innateness (availability from birth), the stimulus being (usually) causally necessary and sufficient for triggering a response, and largely realizable in hardwired brain stem, diencephalon, and subcortical systems (see chapter 10)—also constitute their limitations. Instincts do not allow for learning. They do not allow for novel or flexible responses to stimuli. They do not allow for reason. We are not born knowing that fruit bats spend afternoons hanging from tree branches or that the Earth is undergoing a general warming trend. We learn these things through observation and reason. Given the limitations of the instinctive mind, we must look for additional mechanisms to explain learning and rationality.[5]

5   The Associative Mind: More than Instincts, Less than Reason

Give me a dozen healthy infants, well-formed, and my own specified world to bring them up in and I'll guarantee to take any one at random and train him to become any type of specialist I might select—doctor, lawyer, artist, merchant-chief, and, yes, even beggarman and thief, regardless of his talents, penchants, tendencies, abilities, vocations, and race of his ancestors.

—John B. Watson

Human nature is like water. It takes the shape of its container.

—Wallace Stevens

While I believe that Lorenz was right about instincts being the foundation of human behavior, a complete edifice is much more than a foundation. The foundation, while primary, only allows for limited predictions about the supervening structure. Instincts are a neuronally inexpensive solution to guiding behavior that is essential, does not need to change across generations, and may be needed prior to any opportunity for learning. In such circumstances, instincts are the preferred mechanism. But without additional resources, instincts cannot accommodate within-generation environmental fluctuations. The most widespread evolutionary solution for this is the incorporation of a mechanism that learns from its interaction with the environment. Most of us can drive bicycles and/or cars. We were not born being able to do so. Many of us have seen the famous white Lipizzaner stallions go through their paces and dolphins perform in a marine show. They were not born being able to do so. These are all examples of learned behaviors.

The classical mechanism for learning is association, which in the first instance simply means "connection between two or more things." The origin of associations as the key to understanding learning (and thinking) can be traced from Plato to Locke, Berkeley, Hume, John Stuart Mill, and

William James, among many others (Mandelbaum, 2020). The general theme across the writings of these philosophers was that associations can be formed between events, actions, perceptions, impressions, and ideas. The process of learning and thinking is one of attending to the association relationships among these entities, whether it is to combine them, break them down into finer particulars, or see how one can lead to another.

Association is the glue that holds all mental stuff together. The specific principles of association vary among authors but generally include the principles of similarity, contrast, and contiguity, with contiguity being common across all authors.[1] The central concept of contiguity is defined as co-occurrence in time and/or space between events, actions, perceptions, impressions, or ideas. That is, minds are built in such a way that if a certain event, action, perception, impression, or idea occurs in the spatial and/or temporal vicinity of another event, action, perception, impression, or idea, the one will be associated with the other.

Imprinting is a simple, very specific form of associative learning through contiguity, famously studied by Lorenz (1970) in greylag geese. Newly hatched chicks of many bird species will accept the first moving stimulus they encounter within a "critical period" of a few hours after hatching as their "mother," whether it is the actual mother, a human substitute, or even a pair of boots worn by the human substitute! Filial imprinting is so robust that it was used by Angelo d'Arrigo, an Italian aviator, to teach captive-born eagles and Siberian cranes traditional migratory routes of their conspecifics as they followed him for thousands of miles in his ultralight aircraft across the Mediterranean, the Sahara, and even over Mount Everest (Daniszewski, 2002). The phenomena of imprinting and timing are genetically determined and specific to some species, but the stimuli on which a newborn will imprint are learned and the mechanism of learning is association through contiguity relations.

Classical Conditioning

Classical conditioning also provides many instances of learning through contiguity relations. A personal anecdote provides a relevant example. Approximately 15 years ago, the night before traveling out of town to begin a new brain-imaging study, I ate a chocolate protein bar and came down with nausea, followed by vomiting, fever, and chills. In retrospect, I'm quite certain that the illness was not actually caused by the consumption of the protein bar. I was probably already infected with a flu virus and would have come down with these symptoms whether I ate the protein bar or not.

However, to this day, I have to avoid these particular protein bars because the smell still triggers a vomiting reflex.

Ivan Pavlov stumbled on the phenomenon of classical conditioning while investigating the physiology of digestion. Classical conditioning is a system of association between an autonomic, uncontrolled behavior, such as salivation upon the presentation of food, and some other arbitrary event. For example, a dog, like many animals, will naturally and automatically salivate when presented with food. If the presentation of food is paired with some other stimulus, such as the sound of a bell, after a number of presentations of the bell sound followed by the food, the dog will begin to salivate when the sound is presented, even in the absence of food.[2] The dog has unconsciously (and involuntarily) learned an association between the sound of the bell and the presentation of the food. The association can be strengthened by repeated pairing of the unconditioned stimulus (food) and the conditioned neutral stimulus (bell sound) and weakened (extinguished) by the presentation of the conditioned stimulus (bell sound) alone. An animal may also generalize the association and respond to stimuli that are similar to the conditioned stimulus (e.g., the sound of a phone ringing) and even discriminate between the sound of the bell and, for instance, a clapping sound (Clark, 2004).

Behaviorism and Operant Conditioning

While imprinting seems to be species specific, classical conditioning occurs across species and involves similar mechanisms. Both are examples of involuntary forms of associative learning but do not address the more interesting questions of modulating *voluntary* behavior in response to environmental cues. The study of shaping voluntary behavior through contiguity associations was taken up in earnest by behavioral psychologists using a framework known as operant or instrumental conditioning (Skinner, 1953). This framework dominated American psychology for the first half of the twentieth century.

We all intuitively understand operant conditioning. Imagine you are trying to get a good night's sleep and as soon as you begin to doze off, your baby daughter, whom you have just safely tucked into her crib, cries and requires your attention. Your parental care instincts kick in and you get up to hold and soothe the child, tuck her back into her crib, and return to your bed to resume your sleep. A little while later, the child cries again and the procedure is repeated and repeated, with the duration between successive cycles of you comforting the child and tucking her back into her crib and her crying again getting shorter and shorter until she finally ends up in your bed. What is going on? Every time you get up to comfort the child, you reward her and *positively* reinforce the connection between the crying and being comforted. The child learns to associate crying with receiving comfort. But the child has also altered your behavior through *negative* reinforcement, by ceasing to cry every time you get up to hold her, thus reinforcing the connection between holding the child and the cessation of the annoying crying.

The basic principle of operant learning is that if an action is followed by a satisfying (pleasant) change in the environment (e.g., presentation of a reward or end of punishment), the chances of it being repeated in similar circumstances will *increase*, whereas if it is followed by an unsatisfactory (unpleasant) change (e.g., removal of reward or start of punishment), the chances of it being repeated will *decrease*.[3] The reinforcers can be either primary, satisfying basic needs (e.g., food, water, sex), or secondary, having acquired value through association (e.g., money). Punishment can involve inflicting pain or withholding a reward (e.g., food) to inhibit behavior. However, it was reinforcement of behavior through reward that garnered much of the attention of the operant conditioning research program.

Both learning by classical conditioning and operant conditioning involve acquisition, extinction, spontaneous recovery, generalization, and discrimination. The main distinction is that while in classical conditioning the organism learns to associate automatic, uncontrolled behaviors (e.g., nausea, salivation) with seemingly arbitrary stimuli, operant conditioning works on voluntary behaviors (e.g., jumping, singing, problem solving) and the organism learns to modulate these behaviors in response to rewarding and punishing environmental consequences. Let's consider some examples of varying degrees of complexity.

As fencing in Africa becomes more ubiquitous, it is resulting in the fragmentation of wildlife habitats. To maintain some connectivity between habitats, strategic gaps are often provided to allow the passage of wildlife. It has been observed that when several kilometers of this fencing were temporarily removed at a field site in Kenya, potentially allowing for much more flexible movement across habitats, the wildlife continued to follow their established paths as if the fencing was still in place. This may result from negative reinforcement—the absence of predation on previous crossings along established routes (Dupuis-Desormeaux et al., 2018).

For a more complex example, consider the "obstruction problem" from the animal learning literature. This is a problem in which a lure is placed outside a cage but within reach of the animal. The animal can easily access it in this condition. In a subsequent condition, an obstruction is placed in front of the lure, preventing the animal from accessing it. The task requires the removal of the obstacle to access the lure. Interestingly, many animals, such as chimpanzees and gorillas, have difficulties with this problem. Other animals, such as orangutans, crab-eating macaques, rhesus monkeys, carrion crows, and jackdaws seem to naturally solve it (Nakajima & Sato, 1993). It has been suggested that a key factor in certain animals' failure to solve this problem may be their personal history and experiences. If this is the case, it should be possible to use operant conditioning techniques to reshape their overall experiential repertoire and train them to solve this problem.

To test this hypothesis, positive reinforcement (food) was used to train pigeons to peck at a key placed outside the cage but within reach of their beak. The key was then obstructed with a block. None of the pigeons were able to push the block aside to continue accessing the key (and hence the reward). A similar block was then placed inside the cages of half the pigeons. Every time a pigeon moved the block with its beak, and only with its beak, the behavior was reinforced. Eventually, the reinforcement led the pigeon to move the block around the cage with its beak. The obstruction experiment was then repeated. The pigeons that had been trained to both peck the key and move the block around their cage were able to move aside the obstruction with their beak and access the key. The pigeons trained only to peck at the key but not move the block were unable to solve the problem (Nakajima & Sato, 1993).

For a more natural human problem-solving example, remember back to when you learned to drive a car. Seated in the driver's seat, the actions available to you included pressing the accelerator, pressing the brakes, shifting gears, steering left, and steering right, among others. Each action has a consequence, desirable or undesirable. Press the accelerator pedal, and the car starts moving. Press it too hard and long and the car moves dangerously quickly. Turn the steering wheel and the car will turn. Fail to turn the steering wheel and the car will not turn. You sculpted your actions in response to their positive and negative consequences. This is clearly a very common form of learning for humans and is explained reasonably well by operant conditioning.

Now let us consider simple language-learning examples, beginning with songbirds. In many songbird species, such as the male zebra finches discussed in chapter 4, song acquisition and development have both instinctive and learning components. While in the male zebra finch learning involves imitating a tutor, in some other species, such as brown-headed cowbirds, song development seems to be guided by operant conditioning mechanisms (Sturdy & Nicoladis, 2017). It is reported that when female cowbirds hear a preferred song, they produce a "wing stroke" movement of their wings, and such songs lead to more precopulatory displays by females. Thus, the female is reinforcing the production of certain songs by the male (West & King, 1988). The "wing stroke" is the secondary reinforcer, while sex is the primary reinforcer.

Operant conditioning techniques may also (arguably) play a role in human language learning.[4] In one study, mothers socially interacted with their eight-month-old infants (e.g., smiling, moving toward them) in two conditions. In one condition, the social cues were produced directly after infants' vocalizations. In the other condition, the mothers offered the same social cues, but they were unconnected to the infants' vocalizations. The mothers in the linked condition were reinforcing the vocal output of their infants, while the mothers in the unlinked condition were not. As predicted by operant conditioning, infants in the contingent reinforcement (i.e., linked) condition produced more and higher-quality vocalizations than infants in the unlinked condition (Goldstein, King, & West, 2003). Operant conditioning techniques are also being utilized in certain therapies; for example, teaching certain language skills to autistic children (Hewett, 1965; Howlin, 1981).

Operant conditioning can even be used to train animals to exhibit behaviors that look like human reasoning. Suppose I tell you the following: the red block is heavier than the blue block; the blue block is heavier than the green block; the green block is heavier than the yellow block; and the yellow block is heavier than the purple block. I then ask you, which block is heavier, the blue block or the yellow block? You will have no difficulty in telling me that the blue block is heavier than the yellow block. This judgment is an example of reasoning. It involves the coherency (or "making sense") relation that we encountered in chapter 1.

Can a chimpanzee select the blue block over the yellow block in the preceding example? Can a pigeon? What about a goldfish? It turns out that all can be trained to do so using the operant conditioning paradigm (Delius & Siemann, 1998; Gillan, 1981; Grosenick, Clement, & Fernald, 2007; McGonigle & Chalmers, 1977). The animals are presented pairs of stimuli such as *red block / blue block* and rewarded every time they select red over blue. Then they are presented the pair *blue block / green block* and rewarded every time they select blue over green and so forth for each pair (*red block + blue block −; blue block + green block −; green block + yellow block −; yellow block + purple block −*, where the plus sign indicates a reward if that block is chosen and the minus sign indicates no reward if that block is chosen). After repetitive training involving 1,500 to 15,000 trials, depending on the species, many animals are able to select the *blue block* over the *yellow block*, even though they did not explicitly learn that specific pairing. On the surface, it certainly looks like transitive inference, but looks can be deceiving. We will revisit this issue more critically in chapter 6.

Based on such data, the behaviorists confidently proclaimed that the same, singular mechanism of association (via contiguity or co-occurrence) paired with positive and negative consequences was adequate to explain not only the behaviors of rats, pigeons, and dolphins but also our own abilities, ranging from learning to ride a bicycle, to learning language, even to reasoning, solving novel problems, long-range planning, design, and scientific discovery (Skinner, 1953). The proponents of this research program believed that they had discovered a simple but universally applicable learning mechanism that applied across all species. Gregory Kimble (1956, p. 195) noted that "just about any activity of which the organism is capable can be conditioned and . . . these responses can be conditioned to any stimulus that the organism can perceive." That is, the principles of operation of all minds are the same across species. B. F. Skinner (1984, p. 609) noted that "the pigeon is more than a model. . . . It has supplied terms and principles of great practical value and, I believe, of equal value in interpreting human behavior observed under less favorable circumstances outside the laboratory." The only factors that differentiated the pigeon from the man were their respective environmental histories (and presumably the size of their neural endowment). This is all nicely consistent with the Darwinian worldview.

In terms of the five dimensions we are using to differentiate kinds of minds, the function of the associative mind is to allow individuals to monitor and respond to local within-generation environmental changes by modulating behaviors in response to positive and negative consequences. Where the consequences are positive, the behavior is reinforced. Where the consequences are negative, it is discouraged. As these associations are a matter of individual environmental history, they will vary across time and across individuals. This means that, unlike with instincts, there can be no fixed, species-wide causal coupling between a given stimulus and response.

Different causal relations can be identified for the training and execution phases of learning. During the training of specific individuals, the positive and negative feedback will be causally necessary and sufficient for learning (within biological constraints). For example, training a dog to walk at heel will involve rewarding it when it does so at command and withdrawing reward (or introducing punishment) when it fails to do so.

The reinforcement is necessary and, if successful, sufficient for shaping the behavior. During the execution phase, the stimulus will usually (but not always) be causally sufficient. For example, once your dog has learned to "walk at heel" at a specific command, given the command, it will usually do so but not always (it may, for example, be distracted by a squirrel). Furthermore, the command may not be necessary, as the dog may engage in the behavior for some other reason.

Such behaviors are clearly learned, and in terms of brain systems, subcortical (hippocampus) and cortical structures are engaged. Interestingly, the behaviorists never asked questions regarding possible underlying neural mechanisms of associative learning. But by the 1940s enough was known about the biological and computational properties of neurons to give a viable account. Some of the basics are outlined in the appendix to this chapter.

Minds without Mental States

Not only did behaviorism explain all behavior across the phylogenetic tree with a single, widely available mechanism—association through co-occurrence—it did so without any commitment to mental states. The ban on postulating psychological states was driven by the argument that mental states are inherently subjective and *in principle* not observable or measurable. The behaviorists reasoned that if mental states could not be overtly detected and measured, they could not play a role in the science of behavior and certainly not an efficacious role. Appealing to mental entities such as beliefs and desires was the same as appealing to a homunculus. It may be okay in informal conversation but has no place in scientific discourse. This is an epistemological objection to the use of mental state terms.

There was also an ontological reason for banning reference to mental states. The behaviorists were convinced that even if there are mental states, they are waxlike; that is, undifferentiated and extremely malleable. We can shape them in any arbitrary fashion by controlling the organism's history of environmental stimuli (e.g., make an animal thirsty by depriving it of water, feeding it salt, or bleeding it). Therefore, it is the animal's history of interaction with the environment that is primary (Skinner, 1984). The mental state, even if it exists, is secondary.

The behaviorists banned not only reference to psychological states but also any appeal to neurological states. One reason for this ban may have been that in the 1920s and 1930s techniques for measuring neurological states were in their infancy. But there was also another reason. The behaviorists were concerned not simply with predicting behavior but also controlling it. They argued that even if we could measure internal neurological states, we still would not be able to manipulate them (for either ethical or technological reasons) in order to control the behavior of the organism. It was best to focus on overt behavior, which they thought could be arbitrarily shaped by operant conditioning techniques. In later chapters I will suggest that we make a similar mistake today when we continue to insist that belief revision is sufficient to modify any behavior.

Returning to the main point, explanations of behavior involving mental states were not to be taken seriously. They were deemed to be either mistaken, unsubstantiated, or simply a shorthand for scientific behavioral descriptions.[5] I will argue in chapter 6 that mental states are the building blocks of the rational mind. If they are not "real," then there is no rational explanation to be offered for our behavior; indeed there is no rationality. On this radical behaviorist account, our actions are controlled not by reason but rather by our respective unique histories of environmental stimuli and reinforcement (and perhaps some instincts). From a commonsense perspective, it does stretch one's credulity.

To illustrate the significance of this point, I will appeal to another personal anecdote involving single-trial operant learning. I consider myself largely impervious to most commercial advertising. If you ask me why, I will tell you of my experience as a 12-year-old. Like many 12-year-olds in the early 1970s, I had a paper route. It allowed me to earn $3 to $4 per week. With access to such funds, I bought myself a racing car set. I had been watching the advertisements on television for months while saving up my money and dreaming about the car set. One could lay out the track in innumerable patterns, the cars would go superfast, would careen around corners, slide under bridges, and even do gravity-defying loops in midair. After I had saved up $26, I was in a position to purchase this amazing car racing set that did everything except make breakfast. When I brought it home and set it up in the basement, it didn't work quite as in the TV advertisement. The track was difficult to lay out. The cars would not stay on it. They did not move all that fast. The lights didn't work. There were no loops or even bridges. My sense of disappointment was so profound and lasting that I have never again been taken in by deceptive advertising.

This will strike most readers as a straightforward, reasonable explanation of my immunity to advertising and one that appeals to operant conditioning. However, a behaviorist like Skinner cannot abide this explanation, because it is replete with the use of mental state terms (desired, watched, believed, dreamed about, longed for, disappointed in), and these mental state terms are causally implicated in my behavior. The behaviorists would say that my so-called mental states are just surrogates for the contingencies of reinforcement (i.e., the interrelationship between the detected environmental stimulus, my behavioral response, and the consequences of the response) that have shaped my behavior. There's nothing corresponding to mental states in my head and there is no causal story to tell involving them. Even my memory of these events is not something inside my head. My behavior has just been altered as a function of contingencies of reinforcement.

In banning reference to mental states in psychological explanations, Skinner (1984, p. 608) believed he was doing for psychology what Darwin did for biology: "Operant ('instrumental') conditioning is a kind of selection by consequences, and like natural selection it replaces a creator by turning to a prior history." Behaviorism looks for "antecedent events in the history of the individual to account for the origin of behavior, as Darwin looked for antecedent events to account for the origin of species."

While psychological explanations appealing to mental states seem intuitively obvious, it is certainly worth considering the possibility that the behaviorists were right. Maybe mental states are too mysterious to be discharged by the natural sciences. There is no gain in explaining one mystery (human behavior) with an equivalent or greater mystery (mental states). For example, what causes thunder and lightning? The ancient Greeks explained it by saying, "Zeus is angry and darting his thunderbolts." This explanation appeals to Zeus and his anger (a mental state). But to complete the explanation we have to explain Zeus and his mental states. Here is an explanation of lightning and thunder that does not appeal to a mysterious being and his mental states: lightning occurs when excess positive and negative charges build up in clouds in water droplets and ice particles. The rapid movement of negatively charged particles toward positively charged ones (and occasionally vice versa) results in electrostatic discharges, which we see as flashes of visible light. The energy discharge heats up the surrounding air, resulting in its rapid expansion at supersonic speeds, which is heard as thunder.

The former explanation introduces a greater mystery (Zeus and his anger) than the phenomenon to be explained, while the latter explains the phenomenon in less mysterious terms. The second explanation follows from our understanding of positively and negatively charged particles, electrostatic buildup and discharge, air pressure, temperature, and other physical phenomena. Progress in science over the past 500 years has resulted in part from the insight that we can, indeed must, explain the world—whether it be the falling of an apple or the origin of species—without appealing to a mysterious creator and his mental states.

But will this dictum also apply to human behavior, where the mental states are our own and we seem to have firsthand experience of them and their causal efficacy?[6] This is an open question, and at least three considerations can be brought to bear on it: (1) cost-benefit analysis, (2) empirical evidence, and (3) possibility of mechanistic explanations of mental states.

With respect to the cost-benefit analysis, while our intuitions about the world have often proven wrong, we do not want to give them up cheaply. In the case of physics, we have reaped enormous knowledge and technological advances (whether it be the Large Hadron Collider or the cell phone in your pocket) for giving up our folk intuitions of physics. In the case of psychology, this is hardly the case. The behaviorists asked us to abandon our most basic intuitions but offered little of lasting value in return.

In terms of the empirical evidence, the question is whether we can understand or explain our behavior and the behavior of our conspecifics *without* appealing to mental states. Can we understand why the Prince of Denmark brooded incessantly over the death of his father with such tragic consequences? As an exercise, try to construct an explanation of Hamlet's behavior without using mental state vocabulary. You will not succeed.

These two considerations give us reasons for not discarding our intuitions about mental states prematurely. But what if, on top of these reasons, we can provide a mechanistic explanation for mental state terms, rendering them totally nonmysterious and part of the natural order of things? This would be a game changer and justify their use in scientific explanations. Appealing to mental states (and discharging them as computational states) is at the heart of the cognitive mind, which we take up in chapter 6.

For behaviorists, banning all references to mental states also meant banning references to affective states or "feelings." This strikes me as extremely problematic for the theory. If learning is a function of reinforcement, then it follows that any animal capable of altering behavior in response to a reinforcement schedule must be able to *discriminate* between positive and negative reinforcement. We already introduced the affect issue with respect to autonomic and instinctive minds in previous chapters. While it is possible to imagine that a spider might weave its intricate web by executing an algorithm in a robotlike manner, it is much harder to imagine how an organism is to modify its behavior in response to positive and negative reinforcement if it cannot actually differentiate between them. The behaviorists skirted the problem by operationalizing reward as "any appetitive stimulus given to an organism that serves to reinforce the occurrence of behavior," thus simply begging the question. Ignoring the issue does not make it go away. In chapter 11, I will argue that the evolutionary solution to this problem is neural mechanisms that allow the animal to *feel* the difference between positive and negative reinforcement.

Associationism without Behaviorism

On the behaviorists' account, the only accepted association relation was contiguity (or co-occurrence) between stimulus and response (shaped by reinforcement)—not between ideas, impressions, and perceptions—because as mental concepts, the latter did not really exist. If we step back and cast our net more widely, however, most historical philosophical accounts allow for associations between actions, perceptions, impressions, and ideas. They also appeal to association relations beyond contiguity, particularly similarity and causation. I will consider one such account offered by William James, who overlapped with Darwin and predated the behaviorists.

James (1878) distinguished intelligence from instincts and believed that both humans and animals were capable of thinking, but with an important difference. All thinking involved the formation of associations. Associations formed because "actions, sensations, and states of feeling" that occur simultaneously or in close temporal or spatial approximation tend to coalesce together. When one of these is then subsequently considered, the others are also brought to mind in some form. However, James distinguished between two types of associations: association by contiguity and association by similarity. Nonhuman animals are only capable of association by contiguity; humans are capable of both types of associations.

In the case of association by contiguity, all the elements are operated on together. For example, in figure 5.1a, if A and B have been previously experienced together, then the subsequent presence of A will bring to mind all of B. For instance, thinking about a recent drink with my neighbor brings to mind the conversation about his newly planted garden and his tale of the wild rabbits impatiently waiting for the harvest!

By contrast, association by similarity is a more complex notion, though it does also involve contiguity. Similarity associations are not formed simply through co-occurrence of ideas, perceptions, or sensations. They must be broken down into shared common elements. For example, in figure 5.1b, A and B are associated by virtue of sharing the property m. Both A and B possess the property m (e.g., "bachelors" and "husbands" possess the property of being "male") and share it through the relation of identity. We can say that A and B are "similar" because they are in part identical (m). The other parts of A and B are associated by virtue of contiguity to m. For example, reading William James on similarity brings to mind David Hume. This

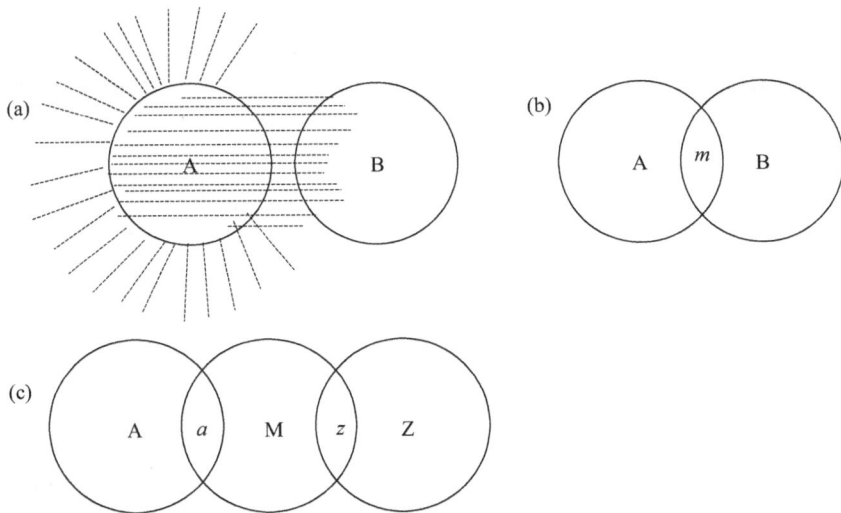

Figure 5.1
(a) All animals can make contiguity-based associations. (b) Only humans can make similarity-based associations. (c) Similarity-based associations can lead to reasoning. Based on drawings by William James (1878).

is not because they appear together in textbooks or have co-occurred in my experience. Rather, the two are connected by the problem of induction that they both needed to confront. Thus, all associations require contiguity, but similarity is more complex in requiring both contiguity and identity.

Reasoning involves finding an intermediate representation M linking together two or more events or ideas, A and Z, as in figure 5.1c. M would be the reason for inferring Z from A. (Notice that it could equally be a reason for inferring A from Z, which may not follow.) The superiority of reason lies in the fact that we can infer Z from A even though the two have never co-occurred in our experience. For example, if I know that C$1.30 equals US$1.00, and £0.75 equals US$1.00, then I can infer that £0.75 is equivalent to C$1.30. The property M (value of the US dollar) unites the values of Canadian and British currencies and allows us to determine equivalence. Such inferences can be expanded into chains of considerable complexity.

A moment's reflection will reveal that in advancing this system of reasoning, James is going to quickly stumble on the problem of induction (discussed more fully in chapter 8). Any two ideas, events, or objects will share innumerable properties. Most of these properties will not warrant any interesting or relevant inference. In the monetary example, if instead of picking out the values of the Canadian and British currencies with respect to US currency, I pick out the fact that they are both printed on paper or in green ink, the inference of interest does not follow. For the system to work or be useful, the "correct" shared property must be selected. How this is done remains a mystery. James (1878, p. 246) puts it this way:

> We may say that the particular part which may be substituted for the whole, and considered its equivalent in an act of reasoning, only *depends on our purpose, interest, or point of view* at the time. No rules can be given for choosing it except that it *must lead to the result*, and to follow this rule is an *affair of genius*.

For James, the instinctive mind and the associative mind accounted for all behaviors across the phylogenetic tree. What separated the reasoning of the "brute" from the reasoning of man was that the latter, but not the former, was capable of associating ideas via similarity of internal components or properties. Like Darwin, William James also shared the worldview of his contemporaries, so the gap between man and "brute" was bridged by noting that "the lowest men" (non-Europeans) occupy intermediate positions between "the highest men" (Europeans) and "brutes."

Are Associations Enough?

Is the associative mind enough to explain the full range of human behaviors? Based on the preceding optimistic assessments of B. F. Skinner and Gregory Kimble, one might think that the answer was an unequivocal "yes." But we have encountered such enthusiasm and certitude before with the founders of ethology, in regard to instincts. In both cases, we have extremely intelligent scientists looking at the phenomenon of human and nonhuman behavior, offering an explanation, but feeling compelled to make it the *only* explanation for all behaviors rather than identifying which behaviors might be better explained by which mechanisms and how various systems might interact to explain the range of behaviors we exhibit. At least Lorenz eventually accepted a (modest) role for learning and incorporated it as part of the model. Skinner (1984, p. 609) maintained that instincts were largely unimportant, saying, "Genetic examples are not very important in the human species; indeed, they are particularly hard to modify through operant reinforcement. Fortunately, the species possesses a large pool of uncommitted behavior available for quick shaping."

An examination of the data and arguments yields a more modest answer to our question about the scope of behaviorist explanations. First, the larva of the capricorn beetle reminds us that not all behaviors are learned, so association on its own cannot be sufficient. Associations are an important evolutionary development on top of the instinctive brain and indeed integrated with it at certain points, such as in the cases of imprinting, development of certain bird songs, and perhaps human language. But are associative mechanisms, even in conjunction with autonomic and instinctive systems, sufficient to explain all behaviors, including rational behavior? If we restrict ourselves to Skinner-type associationism, the answer is a simple, unequivocal "no." If we construe associationism more broadly to encompass relations beyond co-occurrence and have these relations associate mental states, as William James and his philosophical predecessors did, then the answer is a more interesting "no."

Limits of Associationism without Mental States

There are two major shortcomings in explaining rationality via Skinner-type associationism. First, there is no acknowledgment of the building blocks of rationality: mental states and their contents. Unless we can totally reconstruct rational behavior without appealing to beliefs and desires, this account is a nonstarter for explaining rationality.

Second, the universality of the associative learning mechanism rests on the claim that any activity an animal is capable of can be conditioned to any perceivable stimulus.[7] This turns out to be false. A number of consequential experiments have shown that associative learning only applies within a narrow range of an animal's biological dispositions.

In one important illustrative experiment, Garcia and Koelling (1966) exposed rats to a particular taste, sight, or sound, followed by surreptitious exposure to radiation or drugs, which led to nausea and vomiting. According to classical conditioning, the rats should have learned to associate taste, sight, and sound stimuli with nausea and vomiting and avoided all three stimuli. This is not what happened. The rats developed an aversion only to the taste stimulus, not to the sight or sound stimuli. Furthermore, they developed an aversion to the taste stimulus even if they became sick several hours after receiving the stimulus (Garcia, Ervin, & Koelling, 1966).

These results contradict basic classical conditioning assumptions that any perceivable stimuli (taste, sight, sound) can act as a conditioned stimulus and that the unconditioned response (nausea and vomiting) must immediately follow. The behavior of the rats is actually consistent with my own behavior involving nausea and vomiting and the taste and smell of the chocolate protein bar related earlier. I developed an aversion to the taste and smell of the protein bar but not to the sight of the wrapper or the drawer that I kept it in. Also, I happened to be wearing blue trousers that day but did not develop an aversion to trousers or the color blue.

Nausea and vomiting are the body's responses to potential poisoning. To be effective (adaptive), they need to be associated with mechanisms relevant to the selection of food, not with any arbitrary stimuli. Humans and mice are omnivores guided by the taste and smell of foods, so it is reasonable that humans and mice develop taste and odor aversions. In quail, food selection is based on color. They would presumably develop a color aversion in the preceding experiment. Vampire bats have only one food source: blood. It is never poisonous. Therefore, they have not evolved taste aversion learning. That is, they cannot be taught to equate food intake with stomach illness. It is not evolutionarily useful for them (Prescott, 2012).

These principles that apply to classical conditioning seem also to be true of operant conditioning. Food can be used as a reinforcer to teach a hamster to dig or rear up because these actions are part of the repertoire of the animal's normal food searching behaviors. But you cannot use food as a reinforcer to shape face washing and other behaviors not normally associated with food or hunger (Shettleworth, 1973). As Robert Heinlein (1973) noted in a different context, "Never try to teach a pig to sing; it wastes your time and annoys the pig." The lesson here is the obvious one: innate biological constraints predispose animals to learn associations that piggyback on their adaptive behaviors, a point that Lorenz and Tinbergen emphasized in the 1930s.[8] This is actually a positive feature of the system and only becomes a shortcoming if one thinks that all behaviors must be shaped by co-occurrence associations.

Limits of Associationism with Mental States

One might think that if we adopt perceptions, thoughts, and ideas, as William James and many philosophers did, we can go much further in explaining rationality. We *can* go further, but not far enough. The problem is that contiguity or co-occurrence relations are often a coincidental feature of our environment and inadequate mental glue to build up the complex, abstract relations involved in reasoning and thinking. Consider a simple example involving memory. Human memory is often modeled as nodes and arcs in a graph structure. The nodes represent percepts or concepts, while the arcs indicate spatial or temporal contiguity relations.[9] This organization can be demonstrated by simple association and priming experiments. If I give you the word completion task

gambler-c

bone-m

you will probably complete it as

gambler-card

bone-meat

This illustrates the associative nature of memory. "Card" co-occurs with "gambler" and "meat" co-occurs with "bone" in our experience. That is, "activating" the "gambler" node automatically activates adjacent nodes, such as "cards," "deck," and "roulette," through spreading activation, allowing for their retrieval. The activation of the "bone" node will similarly activate "meat" and "marrow," among other nodes. But if instead you are given the word completion task

dog-c

bone-m

you will probably complete the first as "dog-cat" for the same associative reasons as earlier. However, in this particular case, you will be much faster in completing the second word pair than in the first case. That is because the activation of "dog" in the second case also activates (primes) the node for "bone," and that allows you to make the association between "bone" and "meat" more quickly in the second case.

But if the associations are confined to co-occurrence, then one cannot distinguish the different relations that the associations are comprised of. For example, in figure 5.2, "dog" and "cat" are associated because dogs *chase* cats; "dog" and "meat" are associated because dogs *eat* meat; "dog" and "bone" are associated because dogs *chew* bones; and "bone" and "meat" are associated because bone is *part of* meat. Models of memory, while associative, are highly processed and must differentiate between such relations, taking us well beyond co-occurrence relations.

Associations can also be spurious. For example, a reduction in the consumption of margarine is associated with a reduction in divorce rates, at least in the state of Maine. An increase in US spending on science, space, and technology is associated with an increase in suicides by hanging, strangulation, and suffocation (Vigen, n.d.). In both cases, the correlation rate is over 99%. What are we to make of these correlations? Nothing. We dismiss them because they are crazy, spurious associations. They make no sense because we can fathom no underlying cause or reason for them to be related, so "making sense" requires something more than mere association. It requires relations that participate in semantic, conceptual, logical, and/ or causal coherence (see chapter 6). In most cases such relations cannot be captured by co-occurrence. As we saw in the memory example, simple

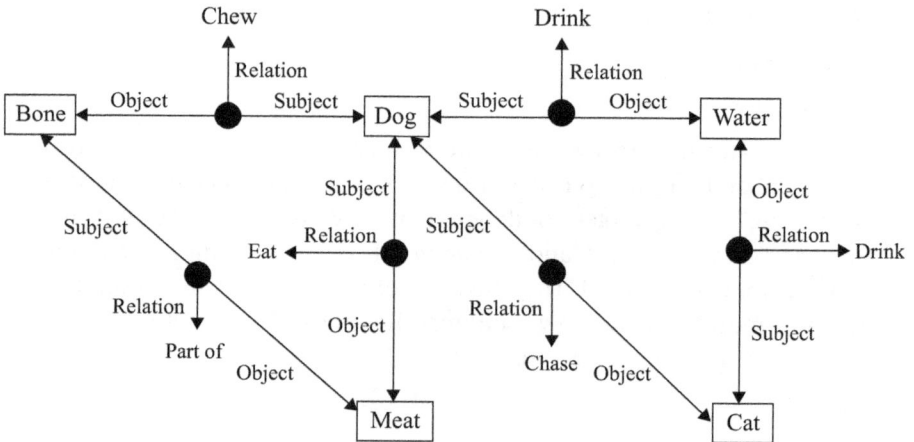

Figure 5.2
Fragment of a semantic network representation of "dog."

co-occurrence of "dog" and "cat" and "dog" and "bone" cannot differentiate between the relation *chase* (dogs, cats) and the relation *chew* (dogs, bones). Most relations are more than co-occurrence.

What if we replace the behaviorists' single contiguity relation with the richer repertoire used by William James and many philosophers? This is easier said than done. The one thing the behaviorists were right about is that contiguity in space and/or time is the only nonmysterious mechanism of association. We saw earlier that William James, who introduced the notion of similarity to differentiate certain associations made by humans from those made by other animals, had to admit that for the system to work, the similarity relation had to be between the "correct" properties, the determination of which was a "matter of genius," thus begging the crucial question. That is, the "similarity" relations are the "correct" relations and the "correct" relations are either undefined or refer back to "similarity" relations. Others who have appealed to conceptual and/or causal relations have encountered similar difficulties.

For example, David Hume ([1748] 1910) famously defined a cause to be "*an object followed by another, and where all the objects, similar to the first, are followed by objects similar to the second. Or, in other words, where, if the first object had not been, the second never had existed*" (p. 348). Hume starts with contiguity ("an object followed by another"), but it is not sufficient for the task. He then brings in the notion of similarity and finally ends up using a counterfactual.[10] If these relations are essential to human thought and cannot be reduced to contiguity, one needs to look for additional constructs and mechanisms.

*   *   *

This assessment is not intended as a dismissal of the associative mind. The associative mind is as critical a part of the overall picture as the autonomic and instinctive minds, but it is not an alternative to them. Its function is to fine-tune behavior to local contingencies of the external environment that could not have been anticipated and programmed into the genome. This can range from the development of taste aversion to specific foods, to animals adhering to certain known pathways even when more direct routes are available, to my learning to drive a car, or concluding that all advertising is deceptive. Creatures that can modulate their behaviors to accommodate local environmental fluctuations will be better off (i.e., in terms of enhanced fitness) than those that cannot.

Not only is associative learning a well-defined system both conceptually and in terms of the underlying mechanisms, it is also, of course, tethered to autonomic, instinctive, and reasoning systems. The protein bar incident related earlier is an example of tethering between classical conditioning and autonomic systems: the smell of a specific type of protein bar triggered a vomiting reflex many years after the conditioning. Using biofeedback techniques to regulate heart rate or blood pressure is an example of operant conditioning reaching down to modulate autonomic systems.

Examples of interactions between associative learning and instincts were provided by the imprinting examples, learning of songs in zebra finches and brown-headed cowbirds, and perhaps human language learning. There are also relevant experimental data involving nest building: even though nest building among birds is instinctive, there is some limited scope for modification of the instinctive behavior through learning. In a study of nest building in zebra finches, it is reported that when first-time nest builders observed familiar male birds building nests with a material of a certain color for which they had shown no preference, the first-time nest builders nonetheless incorporated material of that color into their own nests. They did not do this when observing unfamiliar males selecting material of the same color (Guillette, Scott, & Healy, 2016). But despite some learned variation in selection of materials, a zebra finch nest will always look like a zebra finch nest. A zebra finch cannot be taught to build a nest like a robin or an oriole. Learning can also be modulated by instinctive behaviors. A trained dog will walk at its master's heels. However, if a cat or squirrel appears, its instinctive urge will be to give chase, and it may or may not be able to restrain itself.

An example of associative learning interacting with reasoning systems (at least systems involving mental states) is the anecdotal story of my disappointment in the racing car set and my subsequent belief that all advertising is deceptive. This illustrates the associative mind modulating the reasoning mind. It also works in the other direction. Suppose I see an advertisement for a particular product that I'm interested in purchasing. This particular brand promises X, Y, and Z. Immediately, my learned hype detectors are engaged and lead me to conclude that the product will not do X, Y, and Z. I then read a review of this product in *Consumer Reports*, which I believe to be an independent testing service, and it confirms that this particular product can indeed do X, Y, and Z. This knowledge overrules the learned aversion, and I go out and purchase the product.

In the associative mind, we have a mechanistic account that can explain behaviors beyond those explained by the instinctive mind. It does so by postulating a single mechanism of association based on co-occurrence, which supposedly accommodates much (if not all) of learning and thinking in nonhuman animals, and purportedly in man. However, as we have seen, the claim about explaining thinking in man does not hold up. While part of the story, contiguity or co-occurrence relations can be spurious, underdetermine many relations necessary for human thought, and cannot account for the coherency relation central to rationality. Coherence is a relationship between mental states with propositional contents. Additional machinery is required to account for it. We now turn to the cognitive mind to provide this machinery.

Appendix: Mechanistic Accounts of Associations

Neural networks provide a natural mechanistic account of the associative mind. I briefly describe them here both at the biological and computational level.

Neurons as Biological Systems

Neural networks are built from collections of neurons. A neuron is a cell. Like all cells, it has a body containing a nucleus and the machinery necessary for the production of proteins and other cellular functions, all suspended in an intracellular fluid called the cytoplasm and enclosed in a membrane composed of a lipid bilayer. Unlike other cells, neuronal cells also have an extension called an axon, which terminates in processes called synapses (figure 5.3a). The cell body contains processes called dendrites. The inside of the cell is negatively charged compared to the outside. This difference is called the resting membrane potential. It is a function of the properties of the neuronal membrane and the distribution of ions across it. What is special about neuronal cells is that they receive information at the dendrites in the form of chemical, physical, or electrical signals from sensory and somatosensory receptors and neurotransmitters. These incoming signals disrupt the resting membrane potential of the cell. If the input signals reach a certain threshold, the neuron will "spike" or "fire," resulting in an action potential that will travel down the axon propagated by differential passive and active distributions of sodium and potassium ions across the cell membrane. The arrival of the action potential at an axon terminal causes a voltage-gated calcium ion channel to open. This triggers the release of a neurotransmitter into the synaptic cleft (the space between synapse and dendrite). The neurotransmitter diffuses across the cleft and binds to receptor molecules in the postsynaptic membrane, completing the process of transmission. There is a lock-and-key relationship between receptors and

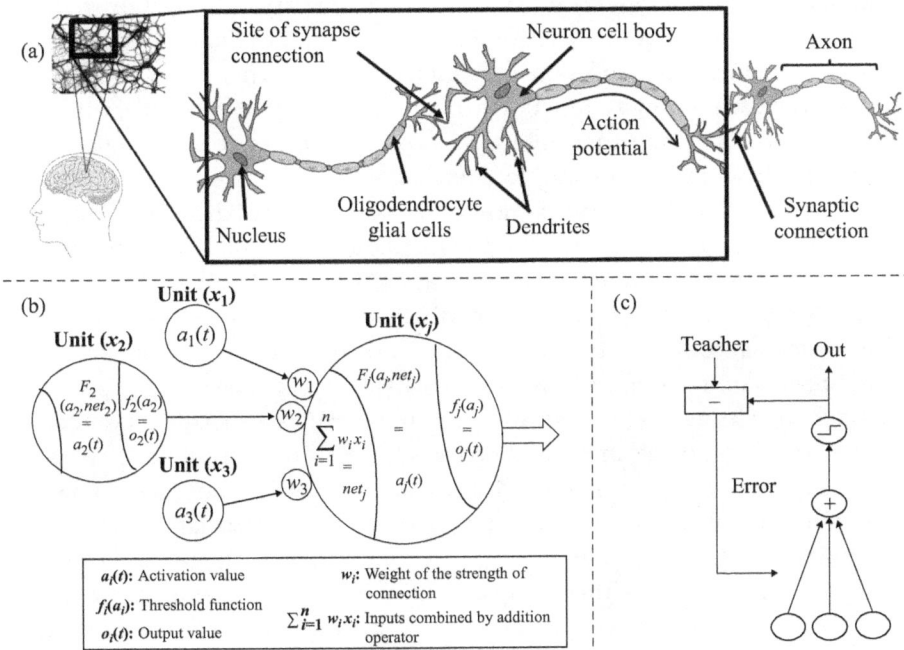

Figure 5.3
(a) The biology of neural networks. (b) Axiomatization of neural networks: converting a biological problem into a computational problem. (c) Training of neural networks. Modeled after Rumelhart and McClelland (1986).

neurotransmitters. The successful transmission of the neurotransmitter and its binding with the receptor cell results in changes in the resting potential of the receptor cell, possibly generating an action potential that will cascade down to other cells.

The other type of cells found in the nervous system are glial cells. They provide various support functions for neurons. Oligodendrocyte glial cells wrap around axons, much like the plastic wrapping on copper electrical wiring, and provide a myelin (fatty) coating to insulate axons and increase electrical transmission and other critical support services.

Neurons as Computational Devices

We can study this biological system as a mathematical or computational system by converting each cell into a black box functionally characterized by numerical input and output values and then focus on the properties that emerge from the interaction between a collection of such units (figure 5.3b). For this purpose, cells can be enumerated as a set of processing units $x_1, x_2, x_3 \ldots$, each designated as x_i and having a varying activation value denoted by $a_i(t)$; that is, an activation value of unit i at time t. Units interact by sending activation signals to their neighbors. This activation value $(a_i(t))$ is passed through a function f_i to produce an output value $o_i(t)$, which, if it exceeds a certain threshold $f_i(a_i)$, is communicated to other units via unidirectional links. The unit does not fire if the threshold is not reached. Each connection between one unit and another is mediated by a weight (or strength), designated as W_{ij}, represented by a real number. This number determines the effect the first unit (i) has on the second unit (j). This effect can be either excitatory (positive) or inhibitory (negative). All inputs to the unit are then combined by some operator (usually addition). The combined inputs to a unit, along with its current activation value, determine (via the function F) its new activation value. The weights (i.e., association strengths of interconnections) can be modified as a function of learning.

These networks can be trained using operant conditioning techniques. When initially set up, the weight distributions (i.e., association strength between units) will be random. This will result in a random response to any input. The response is then compared to the correct or desired response, and the difference between the generated response and the correct response is fed back into the network as an error. This error term is used to update the weights by strengthening connections that have contributed to decreasing error and weakening those that have contributed to increasing error. The trial is repeated until the correct answer is generated. The training results in three types of weight modifications: (1) development of new connections (changing a zero value to a positive or negative value); (2) loss of an existing connection (changing a positive or negative connection weight value to zero); and (3) increasing or decreasing the strength of an existing connection.

There are many learning algorithms for these networks, but for illustration purposes a simple candidate is the Hebbian learning rule, where a change in the strength of a connection is a function of pre- and postsynaptic neural activities (i.e., neurons that fire simultaneously or successively become associated). If x_j is the output of the presynaptic neuron, x_i the output of the postsynaptic neuron, W_{ij} the strength of the connection between them, and lambda (γ) some learning rate constant, then the learning rule could be written as

$$\Delta W_{ij}(t) = \gamma * x_j * x_i.$$

In chapter 10 we will briefly consider how these simple neural networks introduced by McCulloch and Pitts (1943) and explored by Rosenblatt (1958) in the form of the Perceptron scaled up from doing simple linear classification to recognizing faces from YouTube videos and driving cars on city streets (Le, 2013).

6   The Reasoning Mind: Propositional Attitudes and Coherence

Man is made by his belief. As he believes, so he is.
—Vedic proverb

Reason is the power or capacity whereby we see or detect logical relationships among propositions.
—Alvin Plantinga

The associative mind, built on top of and integrated into the instinctive and autonomic minds, seems to explain much behavior, but it cannot fully explain the part of human behavior we attribute to rationality. In the 1950s and 1960s, a young cohort of psychologists, computer scientists, and philosophers arrived at an even more drastic conclusion and, with the iconoclastic zeal and confidence of the youthful behaviorists 50 years earlier, began to rebuild the human mind from ground zero, throwing out the proverbial baby with the bathwater.

This chapter tells the story of the reasoning mind as construed within the cognitive mind. The story tugs us in the opposite direction from Darwin, the ethologists, and the behaviorists. According to the cognitive framework, there is something qualitatively different, or special, about us. Chimpanzees are not convening conferences to deal with global warming or trying to determine whether tree rings or ice core samples provide more accurate proxies for prehistoric global temperatures. The dinosaurs did not predict the approaching asteroid that resulted in the Chicxulub impact that destroyed them, much less take any steps to avert this outcome. The cognitive science consensus is that what allows us to do these things is that we have mental states with propositional contents with the properties of productivity, systematicity, compositionality, and inferential coherence, as I will explain shortly. The great achievement of the cognitive revolution was to provide a mechanistic (computational) explanation for mental states, thus legitimizing the widely accepted intuitive constructs of beliefs and desires as being necessary for rationality. One of its greatest shortcomings, however, is that the computational mechanism has been developed as a self-sufficient, ghostly apparition floating above the biology, unfettered by constraints of the associative, instinctive, and autonomic minds. My contention is that this renders the reasoning mind incomplete for the task at hand.

We begin this chapter by examining the flexibility and generality of human behavior. I endorse the widely accepted view that it results from the representational or symbolic capabilities of the human mind, which play a critical role in determining who and what we are as a species. I then look more closely at the representational capability of our minds (introducing the technical terms *intentionality* and *propositional attitudes* along the way) and identify a very specific type of representation—propositions or sentence-like structures—as being key to rational behavior. Propositions have a number of important properties, not the least of which is that they can be related to other propositions by coherence relations, making them essential for inference. Twentieth-century developments in formal logic and computation have shown us how these types of representations can be captured in computational systems, placating concerns about postulating entities that cannot be explained mechanistically.

We conclude the chapter by asking whether the cognitive mind is sufficient to explain human behavior. I answer this question in the negative. Ironically, my concern is that it is inadequate because (on its own) it may be too "powerful" and that we need to appeal to interactions with simpler systems (the autonomic, instinctive, and associative minds) to explain many real-world human behaviors, particularly those involving food, sex, and politics.

Animal Symbolicum

The counterintuitive picture of man as a purely instinctive or associative animal without a rich mental life was, of course, never universally accepted, especially within the European philosophical traditions. German philosopher Ernst Cassirer asked himself the same questions that were asked by Alexander Pope in chapter 2: What is the nature of man? What is our place in the universe? What allows us to live in a world so different from that of other animals? Cassirer focused on the qualitative difference in the *generality* and *flexibility* of human behavior in comparison to the behaviors of nonhuman animals. While the latter could be accommodated by autonomic, instinctive, and associative minds, human behavior could not. It called for a qualitatively different kind of mind (Cassirer, 1944, p. 24):

> Every organism, even the lowest, is not only in a vague sense adapted to but entirely fitted into its environment. According to its anatomical structure it possesses a certain *Merknetz* and a certain *Wirknetz*—a receptor system and an effector system. Without the cooperation and equilibrium of these two systems the organism could not survive. . . . They are links in one and the same chain and constitute a functional circle.

Cassirer acknowledged that the human world cannot be exempt from the biological laws that govern all life, but he also argued for a symbolic link to mediate between the sensors and effectors based on the insight that man's behavior is not just a response to the environment per se, but involves a complex interaction between the environment, his goals, and the contents of his internal knowledge states (pp. 24–25):

> Between the receptor system and the effector system, which are to be found in all animal species, we find in man a third link which we may describe as the symbolic system. . . . No longer in a merely physical universe, man lives in a symbolic universe. Language, myth, art, and religion are parts of this universe. They are the varied threads which weave the symbolic net, the tangled web of human experience. All human progress in thought and experience refines upon and strengthens this net.

Cassirer redefined man from the "animal rationale" to the "animal symbolicum." Indeed, the modern cognitive science claim is that symbolic processing is necessary for rationality. It is the insertion of symbols between the receptor and effector systems that breaks the tight causal chains of reflex arcs, instincts, and even associations, and allows for a "gap" between stimulus and response, meaning that the antecedent condition (stimulus) is never sufficient for the consequent condition (response). This is a widely accepted necessary condition of rationality. Cassirer, however, focused on external symbol systems. Cognitive science, by contrast, postulates *internal* symbol systems. The distinction is important, but not necessarily for present purposes, where we are concerned with the structure of these symbol systems, be they internal or external.[1] We now examine the representational capacity of human mental states in greater detail.

Propositional Attitudes and the Structure of Representational Mental States

German philosopher and psychologist Franz Brentano, who preceded Cassirer, also focused on representational capacity as the defining feature of human mental states. He referred to it as *intentionality* (Brentano, [1874] 2012, p. 68):

Every mental phenomenon includes something as object within itself, although they do not all do so in the same way. In presentation something is presented, in judgement something is affirmed or denied, in love loved, in hate hated, in desire desired and so on. This intentional in-existence is characteristic exclusively of mental phenomena. No physical phenomenon exhibits anything like it. We can, therefore, define mental phenomena by saying that they are those phenomena which contain an object intentionally within themselves.

Intentionality is a technical term in philosophy that looks and sounds like the English words *intentional* or *intending*. Among the uninitiated, this can result in considerable confusion. While my having an intention means I intend to do something, *intentionality* refers to the representational capacity, or "aboutness," of certain mental states.[2] Mental states can refer to things beyond themselves, whereas tables, rocks, and trees cannot.[3] For example, I can have the belief that it is currently raining outside. My belief in this case refers to the state of affairs just beyond my office window. I can also have the beliefs that Julius Caesar crossed the Rubicon in 39 BC and that someday a woman will walk on the moon. In these cases, I am reaching beyond the vicinity of my office window, both temporally and spatially, referring to distant places in the past and the future.

As representational states, intentional states presuppose a distinction between the world and a representation of the world (a necessary condition for the concepts of error and deception) and have a "direction of fit" (Searle, 1983). Beliefs are said to have a *mind to world* direction of fit, meaning that if a belief is false, it must change, not the facts in the world. For example, my belief about "Julius Caesar crossing the Rubicon in 39 BC" is actually false. To make it concur with the facts in the world, I cannot change the facts; I must change my belief (to "Julius Caesar crossed the Rubicon in 49 BC"). By contrast, mental states such as hopes and desires are not true or false but can be either satisfied or unsatisfied. If my desire to photograph the Milky Way galaxy from Banff National Park remains unsatisfied, to bring it to fulfillment I must change the state of affairs in the world (by getting up and doing it) rather than changing my desire. This is referred to as *world to mind* direction of fit.

It is also worth noting that not all mental states are intentional states but all intentional states are mental states. If I have a pain in my left knee, there are qualia or feelings associated with it, and while it feels like it is in my left knee, it does not refer to my left knee. It does not refer to anything—it is just a pain—whereas if I come to you proclaiming that I'm *afraid* or in *love* and cannot tell you what it is that I'm afraid of or who or what I love, then

I'm misusing those terms, and if I persist, some psychological or psychiatric intervention may be required.

Intentional mental states can be analyzed as two distinct components, the psychological state and the content, as

Psychological_State (content)

or, more concretely,

Belief (that I will complete this book manuscript in two months)
Desire (that I will complete this book manuscript in two months)

Psychological states are the familiar, indispensable constructs such as belief, desire, hope, fear, love, hate, grief, envy, and jealousy. They are innate and finite in number but can be associated with indefinitely many contents. So, I can *believe* that it is raining outside, that my student will defend her thesis next week, that Canada is a wonderful place to live, and so on. Similarly, I can also *fear, hope,* or *wish* that it is raining outside, that my student will defend her thesis next week, that Canada is a wonderful place to live, and so on.

The psychological states of desire and belief play a special role in cognitive science because they constitute the goals and world knowledge of organisms, respectively. They are typically not considered to have affective components, or such components are ignored (Searle, 1992), which makes them ideal constructs for a theory of information processing. Other intentional states, such as hope, fear, and love, are emotional states and have affective arousal and valence components associated with them; thus they are largely ignored within information processing theory. We return to the issue of affect associated with intentional states in chapter 11.[4]

Intentional states are often referred to as *propositional attitudes. Attitude* refers to the psychological state, while *proposition* denotes a form of content in a discursive language with a subject-predicate structure, taking the form *that* such and such is the case. What distinguishes a reasoning creature from a nonreasoning creature is the possession of propositional attitudes. The philosopher Donald Davidson stated it as follows (2004, p. 136): "Dumb beasts see and hear and smell all sorts of things, but they do not perceive *that* anything is the case. Some nonhuman animals can learn a great deal, but they do not learn *that* something is true."

To be clear, the claim is not that we are thinking creatures because we have psychological states such as beliefs, desires, and fears. The claim is that we are thinking creatures because our psychological states have propositional contents.

All intentional states, whether in humans or nonhuman animals, are directed at something. For example, imagine a cat tracking a mouse. Its gaze follows the mouse as it scampers across the floor, perhaps even when it is out of sight, behind the couch, eventually emerging on the other side. What form this mental state takes remains unclear. However, its content is not in the form of a proposition (Bermudez, 2002). What is unique about human intentional states is the proposition or sentence-like structure of their contents. Propositions have several interesting properties that make them indispensable for rational behavior.[5]

First, propositions have a subject-predicate structure. For example, in the sentence "George is tall," the subject is "George" and the predicate is "is-tall," and the propositional form might be written as follows: *is-tall* (George). "Is-tall" is a simple one-term relation being attributed to the subject "George." The sentence that "George is taller than Michael" incorporates a more complex two-term relation that relates two objects and might be written as follows: *taller-than* (George, Michael). The subject-predicate structure is thought to mirror the object-property structure of the world and is considered both necessary and sufficient to capture it.[6] Here Bertrand Russell (quoted in Langer, 1942, p. 82) articulates why:

> It may well be that there are facts which do not lend themselves to this very simple schema; if so they cannot be expressed in language. Our confidence in language is due to the fact that it . . . shares the structure of the physical world, and therefore can express that structure. But if there be a world which is not physical, or not in space-time, it may have a structure which we can never hope to express or know. . . . Perhaps that is why we know so much physics and so little of anything else.

This is a much more formal and narrower conception of symbolic content than entertained by Cassirer but one that does a considerable amount of work and crucially allows for mechanization of the system (discussed in the appendix to this chapter).

Second, the subject-predicate structure of propositions also subserves the semantic relation between propositional attitudes and the world, at least insofar as this relation is restricted to truth. It allows us to predicate properties of objects and to state propositions that can be true or false. For example, the proposition "Thomas Jefferson owned slaves" predicates the property of "owning slaves" to Thomas Jefferson and can be true or false, but the subject "Thomas Jefferson" on its own is neither true nor false; neither is the predicate "owned slaves."

Third, propositions are individuated in part by their relationship to the world and in part by their relationship to each other. The nature of their relationship to the world is a contested matter, with many philosophers preferring a causal relationship. However, there is agreement that thoughts are related to other thoughts (or propositions to other propositions) by virtue of semantic, logical, and conceptual coherence relations (not simply by co-occurrence, as in the associative mind).

Semantic relations hold by virtue of the meanings of words. For example, a widow is a woman whose husband has died. A bachelor is an unmarried man. We do not need to look to the world to confirm these propositions.

Logical relations hold by virtue of the "closed-form" terms in a language, such as *and, or, if then, all, some, not,* and prepositional phrases such as *greater than* and *inside of.* Each is associated with certain fixed patterns of inference. Figure 6.1 provides one representation of how transitive relations may be internally represented. It depicts a certain state of affairs in the world regarding Bob, Tom, and Peter, specifically that Peter is taller than Tom and Tom is taller than Bob. This state of affairs is mapped onto the schema *above* with the order top-middle-bottom, which constitutes an inference pattern (figure 6.1a). The inference is associated with the schema, not with the symbols "Bob," "Tom," and "Peter," or even the specific relation "taller than." The symbols "Bob," "Tom," and "Peter" are simply placeholders. They can be replaced with any subjects. Even the relation "taller than" can be replaced with any other transitive relation, such as "shorter than," "heavier than," or "more expensive than" and the inference pattern will still hold as long as the structural correspondence from the relation to *above* is consistent and the ordering of the elements is preserved. This is the case in the mappings in figures 6.1a and 6.1c but not in the mappings in figures 6.1b and 6.1d. In mapping figure 6.1b, structural consistency is not maintained between *taller* and *shorter*. In mapping figure 6.1d, structural consistency is maintained but the mapping of *shorter* onto *above* without reordering the elements will not preserve the truth of the premises. Other inference schemata are available for other types of relations. As with semantic relations, logical relations do not involve any knowledge of the world, only the inference schemata associated with the closed-form terms of the language.

Conceptual relations involve evaluation of propositions in light of our understanding of the world, including co-occurrence experiences and causal knowledge (also represented as propositions). For example, I may conclude that all dogs have tails because all dogs that I have encountered have had tails (co-occurrence), or I may conclude that the seasons are caused by tilting of the Earth on its axis, from having a (causal) understanding of the Earth's orbit around the sun. I cannot draw these inferences based solely on the structure and meanings of propositions.

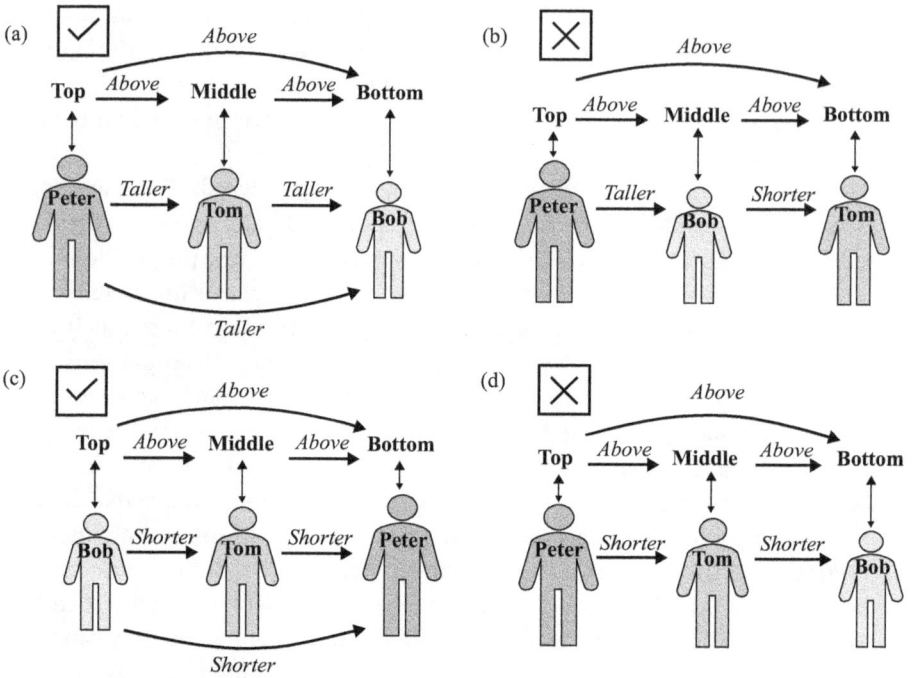

Figure 6.1
A transitive reasoning schema modeled after Halford, Wilson, & Phillips (2010).

For philosophers such as Donald Davidson, a propositional language rich enough to have at least the structure provided by the logic of quantification—meaning a subject/predicate distinction, variables, quantifiers, and recursion—is necessary for a human mind. Let us make sure we understand what this means.

We have already discussed the subject-predicate distinction. Variables are placeholders and have appeared in several of our earlier examples. In the representation "the dog chased the cat" from figure 5.2, "dog" is the agent, "cat" is the object, and "chase" is the relation. The terms *agent, object,* and *relation* are examples of variables. One can plug many different specific agents, objects, and relations into the same structure, to similar effect. In figure 6.1, the *above* ordering schema has a number of variable slots into which various specific relations and objects can be mapped. Variables allow for the reuse of a limited number of structures through the replacement of specific contents.

Quantifiers are linguistic/logical terms such as *all, some,* or *none.* They are necessary to specify the scope of predicates. Consider the examples "apples are nutritious," "all apples are nutritious," "some apples are nutritious," and "no apples are nutritious." In the first case, we do not know which apples are nutritious. In the second case, we know that the predicate nutritious applies to all apples. In the third case, it applies to at least one apple. In the fourth case, it does not apply to any apples.

All natural languages are composed of a finite number of phonemes and words. The phonemes are on the order of a few dozen, and the words may be on the order of one hundred thousand. Both are finite. But despite the finite number of building blocks, all natural languages have the ability to generate an infinite number of sentences, of indefinite length. This has astounded and puzzled philosophers for centuries.[7] As Wilhelm von Humboldt ([1836] 1999) noted, *language makes "infinite use of finite means."* One of the significant advances of the twentieth century was to explain this property of language through the mechanism of recursion.

Recursion is illustrated in box 6.1 with a concrete example of generating an infinite number of sentences in the language of the simple propositional calculus, given a finite vocabulary and a handful of rules. Similar recursively specified rules are postulated to underlie our capacity to generate an indefinite number of thoughts and natural language sentences.

Many cognitive scientists have arrived at similar conclusions: propositional contents of our intentional states have a subject-predicate structure, use variables and quantifiers, and are recursively generated. Jerry Fodor (1975) famously dubbed our internal system of mental representations the "language of thought" and argued (with Zenon Pylyshyn) that it must have the properties of productivity, systematicity, compositionality, and inferential coherence, all of which follow from the preceding characterization (Fodor & Pylyshyn, 1988).

Productivity refers to the unbounded generative capacity of human thought and language. It is made possible by the recursive application of a finite set of rules to a finite set of symbols (box 6.1). Such a system allows for compositionality and systematicity. *Compositionality* requires that a primitive symbol make approximately the same semantic contribution to the meaning of every complex expression in which it appears. For example, in the sentences "George ran to school" and "George ran home," the word *ran* means roughly the same thing in both sentences.[8] This allows for the meaning of a complex expression to be a function of the atomic symbols and composition rules of the language. *Systematicity* is connected to compositionality. It is the property of language (and thought) that ensures that our ability to produce or understand certain sentences is intrinsically related to our ability to produce or understand certain other sentences. For instance,

Box 6.1
Illustration of recursion

Vocabulary:

Brackets: (,)

Connectives: →, ¬, ∧, ∨, ↔

Variables: P, Q, R, . . .

Recursive rules:

(a) Any variable is a well-formed formula (wff).

(b) Any wff preceded by ¬ is a wff.

(c) Any wff followed by → followed by any wff, the whole enclosed in brackets, is a wff.

(d) Any wff followed by ∧ followed by any wff, the whole enclosed in brackets, is a wff.

(e) Any wff followed by ∨ followed by any wff, the whole enclosed in brackets, is a wff.

(f) Any wff followed by ↔ followed by any wff, the whole enclosed in brackets, is a wff.

(g) If something is not a wff by virtue of clauses (a)–(f), then it is not a wff.

This given finite vocabulary and finite set of recursive rules will generate all, and only all, the well-formed formulas (wffs) (i.e., grammatical sentences) in the language of the simple propositional calculus. Rules (a)–(f) determine that certain symbol sequences *are* wffs, but they don't say anything about whether *other* symbol sequences, or indeed arbitrary other objects, are wffs. That is the function of rule (g): nothing else is a wff.

The first rule defines the simplest, "atomic" wffs, not built from other wffs (i.e., given as part of the vocabulary). Every other rule refers back to a simpler notion of wff and specifies a more complex construction. This is the notion of recursion: the concept of a wff is partially defined via clauses already containing it and thus *recurring* to that notion. This can go on indefinitely and allows for the generation of an infinite number of "sentences" of arbitrary length, which are all well-formed formulas in the language of the propositional calculus. Circularity is avoided because the definition ultimately cashes out in terms of primitives, given as part of the vocabulary.

---

if we understand the sentence or thought "John loves Mary," then we must also understand the sentence or thought "Mary loves John." Finally, *inferential coherence* requires that all instances of a given logical form be processed by the same inferential machinery. Thus, if we are prepared to infer "John went to the store" from the sentence "John and Mary and Peter went to the store," then we must also infer it from "John and Mary went to the store." The inference simply follows from a schema built into the structure of the language (figure 6.1).

The cognitive science claim is that it is because we have propositional attitudes, or a "language of thought," with the preceding properties that we live in a qualitatively different world than other animals. This system of representation is reflected in the structure of all natural languages, which possess each of these properties to some extent.[9] If other animals are to qualify as thinking or reasoning creatures like us, they will likewise need to evolve such an apparatus.

Propositional attitudes are not an incidental feature of our mental lives. They are causally efficacious in our behavior, but not in the tight sense in which particular stimuli set off a patellar reflex, or trigger the mating behavior in red deer bucks, or cue a trained dolphin to jump through a hoop at a marine show. They are interestingly different. Consider why Hamlet kills his uncle Claudius. It could be because he believes that Claudius killed his father, usurped the throne, and married his mother, Gertrude. These beliefs and the action do not simply co-occur. The former probably resulted in desire for revenge and eventually the action. However, these beliefs and desires are neither necessary nor sufficient for the action. They are not *necessary*, since Hamlet may want to kill his uncle for many reasons, for instance, because his uncle (accidentally) poisoned Gertrude; or perhaps his uncle beat him when he was a child; or perhaps he simply wants his uncle out of the way so he can be king himself. They are not *sufficient*, because he may choose to act in some other way, such as making the evidence public and charging Claudius with murder in the courts. Many (but not all) reasons can justify an action. Any given reason can justify many (but not all) actions. This is the "gap" between stimulus and response that Cassirer postulated as the sine qua non of rationality.

Propositions and Coherence

Many cognitive scientists argue that with this conceptual apparatus in hand, we can account for most of the differences between humans and other animals, identified in chapter 2. Given our topic, I will confine myself to how this apparatus accommodates the reasoning mind.[10] In particular, we want to see how logical, semantic, and conceptual relations form the basis of coherency judgments and thus of the reasoning mind. Consider the following example:

> Eve is 42 years old. She is a serious and orderly woman. She loves a glass of good wine and playing chess. She tries to watch the news on foreign TV stations every day. She was happily married for 20 years before abruptly losing her husband to cancer.

Based on this description of Eve, you will concur that Eve is not 20 years old, is over 40 years old, is a widow, keeps abreast of world affairs, is more likely to be a librarian than to be a truck driver, and may volunteer at the cancer society. You have no personal knowledge of Eve, and none of this information is explicitly stated in the description. So how do you know? Being 20 years old is logically excluded (with some arithmetic) by the fact that she is 42 years old. Being over 40 years old is logically consistent with being 42 years old. The logical machinery that propositions come with works nicely here. Your acceptance of the fact that she is a widow is a semantic inference based on word meaning. In fact, these semantic and logical inferences can be largely drawn in a vacuum, without any world knowledge.

You may also concur that she keeps abreast of world affairs because it is implied by the statement that "she tries to watch the news on foreign TV stations every day." We would call this an inductive or conceptual inference, and it may be arrived at in one of several ways. You may have a belief that the news contains information about current events. This could be retrieved from semantic memory. If Eve is watching foreign TV stations, she is exposed to information about current events in foreign countries. This information could also be contained directly in semantic memory and retrieved, or it could be logically inferred from the knowledge that all countries prioritize the reporting of their own current news events; if she's watching foreign news stations, she is exposed to information about current events in many countries. From this, one can generalize (inductive inference) that Eve is up-to-date on current world events.

What about the inference that Eve is more likely to be a librarian rather than a truck driver? Neither is explicitly mentioned in the description. One could have explicit beliefs with the propositional contents that many librarians have properties of being quiet, orderly, and enjoying cerebral pursuits and that many truck drivers have properties of being boisterous, enjoying less cerebral pursuits, and being beer drinkers, and from these beliefs, and the given description of Eve, conclude that her described properties are more consistent with our beliefs about librarians than with our beliefs about truck drivers. Therefore, she is more likely to be a librarian.

The cancer society inference illustrates that an important part of world knowledge that comes into play in conceptual inferences is causal knowledge. We know that Eve has lost her husband to cancer. Her age and years of marriage allow us to infer that it was recent. If so, she may still be mourning the loss. Based on personal experience and/or world knowledge, we might develop a causal model whereby bereavement sometimes leads people to seek assistance and eventually offer assistance to others in similar situations. Given that her husband died from cancer, it would be appropriate for Eve to reach out to the cancer society, both to seek assistance and eventually offer assistance. Hence it would not be unreasonable to infer that Eve likely volunteers at the cancer society.

To emphasize the ubiquitousness of such causal attributions in conceptual inferences, I offer another anecdotal story. Canada Day is usually celebrated with an air show in Ottawa. A few days prior to the show, the pilots practice over the town where I live. I was watching the practice show on a cloudy, showery afternoon with some friends. When an F-18 emerged thundering from the clouds, shaking the house, the drizzle turned into a full-fledged rain. My friend Lynda wondered out loud whether the vibration of the powerful jet engines was shaking the rain out of the clouds, just like sitting in a vibrating car shook her bladder, resulting in the need to relieve herself! Not only does this inference draw on a causal model, it does so through analogy. This is a critical part of conceptual inference.

Causal relations are among the most contested in the philosophical literature. My own intuitive view is that causal relationships reflect how our minds structure and comprehend the world. Causal relations may or may not exist in the actual world, but they are certainly imposed on the world by the human mind. We cannot understand the world except in causal terms. For example, it is an interesting fact that mathematical formulas used by physicists to describe the physical world contain no causal relations.[11] However, when physicists describe the same phenomena verbally, causal relations are invariably part of the description. In a famous series of experiments, psychologist Albert Michotte ([1946] 2017) asked participants to observe geometric figures moving on a screen and describe what they saw. When certain constraints of speed, position, and timing were satisfied, subjects invariably used an intentional vocabulary implicating causal relations, such as "the square is chasing the rectangle. The triangle pushed the circle. The pentagon is running from the circle." They could not help it. Linguists, such as Len Talmy (1983), have argued that basic causal relations (along with spatial relations) are built into the "language of thought" and reflected in the structure of natural languages. Neuropsychological data also suggest a close relationship between language and simple causal relations, along with simple logical and conceptual relations (Goel, 2019; Roser, Fugelsang, Dunbar, Corballis, & Gazzaniga, 2005).

Returning to Eve, we can know all these things about her by virtue of coherence relations, ranging from formal logical relations, to semantic relations, to conceptual (including causal) relations,[12] that is, the machinery of reason. Thoughts and propositions—with the structural properties outlined here—are the only types of constructs that can be related through logical, semantic, and conceptual relations. These relations are not available to the autonomic mind, the instinctive mind, or the associative mind, all of which lack access to propositional attitudes.

Can My Goldfish Reason?

When discussing the associative mind in chapter 5, we encountered some data suggesting that many animals, including apes, monkeys, pigeons, rats, tree shrews, and even goldfish—all presumably without propositional attitudes—with access only to instinctive, associative, and autonomic minds, appear to engage in rudimentary forms of inference, particularly transitive inference. If this claim is correct, it is inconsistent with all I have said in this chapter. A more careful examination is warranted.

The reader will recall that in these experiments animals are presented with pairs of stimuli, for example *red block* and *blue block*, and rewarded only when they select the stimuli the experimenter wants them to select. Consider an experiment presenting the pairs of blocks *red block + blue block −; blue block + green block −; green block + yellow block −; yellow block + purple block−,* where the plus sign indicates a reward if that block is chosen and the minus sign indicates no reward if that block is chosen. Many animals can eventually be trained to select the *blue block* over the *yellow block*, even though they have not been explicitly trained on this pair. We know the mechanism at work here is associative learning, because the animals have undergone thousands of trials of reinforcement training. Some researchers claim that this is equivalent to our inferring that Mary is taller than George, given that Mary is taller than Michael and Michael is taller than George. Is an associative mind sufficient for rudimentary inference? Is there nothing more to transitivity than a particular pattern of association? I argue that whatever the animals are learning and doing, it is not reasoning.

An examination of the animal data on transitive inference quickly highlights a number of dissimilarities between human reasoning and nonhuman animal behavior. The first dissimilarity is that human transitive inference actually requires relations that are transitive. For example, the relation "taller than" is transitive but the relations "father of" and "lover of" are not. We differentiate between them and will only draw the transitive inference in the former case. In the preceding animal learning example, color preferences are not transitive. The animal's preference has become transitive through differential reinforcement without there actually being a transitive relation among the task items. So what has the animal understood about transitivity as a logical property (e.g., taller than, shorter than)? It is not clear. At best, we can say that the animal has learned something about the (transitive?) structure of the rewarding regime but nothing about the relations among the stimuli. What the animal is learning is that it is more likely to be rewarded by choosing *blue block* over *yellow block*.

The second dissimilarity is that co-occurrence is a simple symmetrical relation; transitive hierarchical ordering is not. More generally, logical relations have specific, nonarbitrary structure. For example, suppose I have noticed that every time it rains, the grass turns green. So, if green grass and rain co-occur in my experience, they become associated. However, the co-occurrence cannot differentiate between "if it rains, then the grass will turn green" and "if the grass turns green, then it will rain." This issue also came up in discussion of William James's associative reasoning example in figure 5.1c, where the association warrants the inference from A to Z, but equally from Z to A (which may not follow). Co-occurrence can no more encode logical relations than it can encode the causal relation that Hume struggled with (see chapter 5).

Here the reader may object and point out that neural networks working on associative principles (see chapter 5 appendix) can be used to build logic gates. If this is the case, it trivially follows that they can encode logical relations. Animal brains work on associative principles, just like human brains, and therefore may be just as capable of reasoning. I will grant that associative neural networks can be used to build logic gates. I will also grant that animal neural networks work on the same associative principles as human neural networks. But from these two true premises it does not follow that associative relations are equivalent to logical relations. Consider the following analogy: Sand, gravel, and cement are used to make concrete bridge girders, but a girder is much more than sand, gravel, and cement. Girders have to be designed along certain principles. Sand, gravel, and cement have to be mixed in a very specific ratio, in a very specific manner, shaped in very specific forms, and placed in very specific situations to function as girders. In a similar fashion, neural associations may be the building blocks of logical relations, but this is not to say that they *are* logical relations. To act as logical relations, they have to be constrained, organized, and wired in certain ways. In human brains, associative neural networks *are* so constrained, organized, and wired; in nonhuman brains, they are not. In chapter 10, we will examine some of the differences between human and nonhuman brains that may account for this.

The third dissimilarity between logical relations and animal transitive behavior learned through associations is that logical relations have necessary entailments. That is, if Mary is taller than George and George is taller than Michael, then it is a matter of necessity that Mary is taller than Michael or indeed (given some semantic knowledge) that George is shorter than Mary. These are not simply coincidental associations.

The fourth dissimilarity is that logical relations can be systematically generalized. If we understand the transitivity of *taller than*, we will also understand the transitivity of *shorter than*, *more expensive than*, *heavier than*, *sweeter than*, indeed all transitive relations. We do not have to relearn it in each case. The animal training does not result in such systematic generalization. It will take the nonhuman animal thousands of trials to learn the transitivity of each relation anew (Allen, 2006).

A fifth and final dissimilarity, rarely discussed (or even reported) in the literature, is in the number of trials required to train humans and nonhumans to respond transitively. For pigeons or baboons, it can range anywhere from 1,500 to 15,000 trials, and even after this training, the behavior may only be demonstrated by 60% of animals. But a more interesting question is, how many trials does it take us? People always hesitate when asked this question. Surely the obvious answer is "one." What is the trap? The trap is that this answer is both right and wrong. If you and I are presented with the experimental material in the same manner as it is presented to a pigeon (e.g., in the context of a videogame), it will take us on the order of 800 trials to behave in a transitive manner, with an accuracy rate similar to that of the pigeon, and we may be unaware that we are responding transitively (Delius & Siemann, 1998).

Eight hundred trials can be considered better than the 1,500 to 2,500 trials required by pigeons, so perhaps we are smarter than pigeons. We may even be using neural machinery similar to the pigeon's in these cases. But we can also learn to do simple transitive inference in one trial—even without a single learning trial—with a 100% success rate. This suggests that, while we can learn to do "transitive inference" like a pigeon or baboon, we can also draw upon some totally different machinery that allows us to do transitive inference in a single trial. This is the mechanism of language, and perhaps more specifically, propositional attitudes. Transitive relations (more generally, simple logical relations) and inferences are encoded into the fabric and structure of propositional attitudes and language and come for free with this structure.

The only exceptions to this critique of the data on nonhuman transitive inference are the naturally occurring "inferences" involving dominance relations that have been documented in many animals, including birds and fish (Bond, Kamil, & Balda, 2003; Grosenick et al., 2007; Paz, Mino, Bond, Kamil, & Balda, 2004). In one experiment, pinyon jays were manipulated roughly as follows. There are three birds, A bird, B bird, and C bird. A bird and B bird know each other, and B bird accepts A bird as dominant. C bird is introduced and allowed to interact with A bird as B bird watches from a separate enclosure. C bird and A bird interact, and C bird emerges as dominant in the interaction. B bird has observed this but has not directly interacted with C bird. When B bird is finally allowed to interact with C bird, B bird assumes a submissive position to C bird. The interpretation of the study was that B bird inferred that C bird was dominant over him because A bird was dominant over him and C bird was dominant over A bird. This is a transitive inference. But this behavior constitutes a very specific adaptation and fits the characterization of an instinct unless it can be shown to generalize beyond dominance hierarchies. If it can be shown to generalize to many other situations, then the argument I am making will need to be reconsidered. (An alternative explanation of these data is that B bird was simply following the rule "behave like the bird that is dominant to me," A bird, which does not involve a transitive inference.)

Let me clarify how I am construing logical relations. Basic logical relations are intuitive and built into the cognitive machinery. Our ability to do complex inferences is not. The latter is highly correlated with education and IQ. Consider the following examples. Suppose I say to you, "Either Socrates is mortal or it is not the case that Socrates is mortal." This statement must be true (law of excluded middle). But suppose you refuse to accept its truth and ask me to prove it. What do I do? How can I possibly prove it to you?[13] I can't. As in the bathtub overflow example from chapter 1, either you see it or you don't. But all of us with normal cognitive capacity will see it. Another way of making the same point is to suppose that you were raised in a home where you were consistently taught that "Felix is a cat and Felix is not a cat." As a child, you may learn to repeat this, but as your cognitive faculty develops, you will recognize the inconsistency and reject it, without external correction. So what is intuitive are basic notions such as law of excluded middle and contradiction (as earlier),[14] the law of identity (everything is identical to itself), things identical to the same thing are identical to each other ($A = C = B \rightarrow A = B$), and so on. These are all simple, intuitive, self-evident notions that no one can prove to us but that we universally accept. They constitute the building blocks of reasoning. I believe our associative neural networks are so structured and constrained as to be naturally sensitive to these and other primitive logical relations. Some of these structured networks may be the same as those involved in natural language processing. This is why basic coherence relations are intuitive and obvious. Our formal systems of reasoning are built on top of these innate, intuitive structures, with considerable effort.[15]

From the Cognitive Mind to the Computational Mind

Scientific explanations must be mechanistic—that is, consistent with the laws of physics, chemistry, and biology. Both the behaviorists and ethologists understood this and did not engage in talk of mental states, much less propositional attitudes. We have now explored these mechanisms and, finding them wanting for the explanation of rationality, return full circle to our intuitions about the reality of mental states. If, however, we are going to use propositional attitudes or the "language of thought" to explain rationality, we need to understand how they can be instantiated in a physical device in a manner consistent with the natural sciences.

The cognitive mind as described by Cassirer, with his focus on the generality and flexibility of human thought—by which he meant to highlight the *gap* between stimulus and response in human behavior—and his appeal to symbolic processing as a possible solution, was picked up by a number of computer scientists, such as John McCarthy of Stanford University, Marvin Minsky of MIT, and Herbert Simon and Allen Newell of Carnegie Mellon University, working in the nascent field of computer science and artificial intelligence. All made important contributions toward developing mechanistic, nonmysterious accounts of symbolic processing.

Allen Newell and Herbert Simon were concerned with explaining rational, conscious, human problem-solving behavior and set out to show that computers could also display generality and flexibility—understood as the ability to solve many different problems and respond in many different ways to the same problem—in their problem-solving behavior. (Notice that this is an overlapping but different notion of flexibility and generality than Cassirer's.) In 1959, they developed a computer program called the General Problem Solver, which used a simple, means-ends analysis strategy to solve problems ranging from proving theorems in logic and Euclidean geometry to solving the Tower of Hanoi task (Newell, Shaw, & Simon, 1959). It was an early demonstration of separating the program or problem-solving rules from the knowledge or data structure of the task domain. By substituting different data structures corresponding to the different tasks, the same general-purpose program was able to solve many different problems. Adjustments to the computer program allowed for the generation of different solutions to the same problem. This exploration of flexibility and generality in the problem-solving capabilities of computers allowed Newell and Simon to imagine how we might account for the flexibility and generality of human behavior by thinking of our reasoning machinery as a computer program or information processing system.

Newell and Simon proposed and developed the idea of the human reasoner as an information processing system or, as they called it, a physical symbol system—that is, a standard (Turing machine type) computer with an input, an output, long-term memory, (limited) working memory, and a central processing unit with a handful of operators built into the hardware (Newell, 1980). They reasoned that if the human cognitive system is a symbol processing or information processing system (the two terms are interchangeable for our purposes) and we also consider computers to be information processing systems, maybe the human cognitive system works on the same principles as a computer.

What is an information processing system? Is your (old-style) television an information processing system? What about your (old-style) radio? What about your (old-style) telephone? In each of these devices, an electromagnetic signal is received, undergoes a transformation, and is displayed as an audio and/or video signal. Is this sufficient to qualify these devices as information processing systems? Not for our purposes. The key feature of information processing in this cognitive account is that the transformation that the signal undergoes must be a function of its content. In the case of the television (and the radio), it does not matter to the transformation function whether I'm watching (or listening to) the eleven o'clock news or the hockey game. In the case of my old wall telephone, its operation is indifferent to whether I'm using it to argue with my daughter or explain formal systems to a student. The content of the information does not affect the transformation of the signal.

With your computer, on the other hand, the state changes of the machine are sensitive to the content of input signals. For example, if you are using Microsoft Word on your computer and type in a sequence of keys, it may result in the selection and underlining of some text. However, if you are running Adobe Photoshop on the same computer, the same sequence of keys may result in a totally different action (the deletion of the text, for example). The behavior of the computer is a function of its current internal state (determined by, among other things, the program being run) and the new information flowing into the system (i.e., the data, including keypresses).

Perhaps the same principles that make computers information processing systems make humans information processing systems. What makes computers sensitive to the content of information is not mysterious. It is the mapping of formal systems onto some physical circuitry, outlined in the appendix of this chapter. As a bonus, these formal systems naturally possess the properties of systematicity, productivity, compositionality, and inferential coherence identified earlier as necessary for human thought and language.

So, the cognitive mind is turned into a computer program. As a computer program, it naturally traffics in sentence-like symbols or propositions and can naturally engage in the semantic and logical relations that form a subset of the more general coherence relations. However, these computational systems cannot deal with conceptual inferences because they involve not just semantic and logical relations between propositions but also knowledge of the world. This is not a trivial matter. Many authors, myself included, have pointed out the various limitations of this computational story; however, it is still a useful working model.[16]

Summarizing the Reasoning Mind along the Five Dimensions

We have been classifying behaviors along the following five dimensions: function, tightness of causal coupling, origins, conceptual mechanisms, and brain structures. Table 6.1 summarizes the characteristics of the four kinds of minds introduced in chapters 3–6.

In the first instance, the function of the reasoning mind is to allow for greater flexibility in individual behavior, and thus more finely tuned responses to environmental stimuli than can be accommodated by the autonomic, instinctive, and associative minds. Coherence relations determine rational actions given beliefs and desires with propositional content. Reasoning is a system for generating new beliefs from observations and/or existing beliefs and maintaining consistency of our beliefs (i.e., mental representations of the world). It allows us to generate possible options for actions and

Table 6.1

Kinds of minds. A summary of how the four behaviors of interest—autonomic, instinctive, associative, and rational—are distinguished along five dimensions. The inclusion of earlier evolved brain structures in more recently evolved behaviors is meant to foreshadow the tethering explicitly discussed in chapter 10.

Reasoning mind	
Example behaviors	• Decision-making
Causal coupling	• Gap between stimulus and response; no specific stimulus is causally necessary or sufficient for any specific response
Origins	• Beliefs and some desires are learned; structure of propositional attitudes and basic coherence relations are innate
Mechanisms	• Propositional attitudes and coherence relations (semantic, logical, conceptual)
Brain structures	• Brain stem + diencephalon + subcortical structures + hippocampus + large neocortex
Function	• Individual-specific • Monitoring and controlling external environment • Tracking environmental changes, even when they are not present; considering counterfactual scenarios; predicting and modeling future events

Associative mind	
Example behaviors	• Driving, writing, balancing on a bike
Causal coupling	• Training: positive and negative outcomes causally necessary and sufficient for learning (within biological constraints) • Execution: stimulus usually causally sufficient but not always
Origins	• Learned (within biological constraints)
Mechanisms	• Tracking co-occurrence relations
Brain structures	• Brain stem + diencephalon + subcortical structures + hippocampus + some cortex
Function	• Individual-specific • Monitoring and controlling external environment • Guiding behavior in response to within-generation environmental fluctuations

Instinctive mind	
Example behaviors	• Baby's suckle response, nest building by birds
Causal coupling	• Stimulus usually necessary and sufficient for response (with noted degrees of freedom)
Origins	• Innate
Mechanisms	• Lorenz-type causal model (action-specific energy reservoir, innate releasing mechanism, fixed action pattern)
Brain structures	• Brain stem + diencephalon + subcortical structures
Function	• Species-specific • Monitoring and controlling external environment • Guiding behavior that is essential (high cost of error); needed prior to learning opportunities; and stable across generations

(continued)

Table 6.1

(continued)

Autonomic mind	
Example behaviors	• Digestion, blood sugar monitoring
Causal coupling	• Stimulus causally sufficient and often necessary for response, within context of specific biology
Origins	• Innate
Mechanisms	• Reflex arcs, biochemical reactions, homeostasis
Brain structures	• Brain stem and diencephalon
Function	• Species-specific • Monitoring and controlling predictable internal environment • Guiding essential processes and behaviors that are needed prior to any learning opportunity

identify those that are consistent or inconsistent with achieving a given goal. Inconsistent action possibilities can be ruled out. Consistent action possibilities can be further broken down into those that are certain or necessary, plausible (but not certain), and indeterminate. Any creature whose actions are a function of representations of the world—in particular, representations that have propositional content—will need some means to perform these dual functions of generating inferences (and action possibilities) from perceptual input and existing beliefs and maintaining consistency of beliefs. Such a system can also be used to consider and develop responses to situations that have not even occurred yet (e.g., through counterfactual scenarios).

A key feature of the reasoning mind is the gap between stimulus and response. As we saw in the Hamlet example earlier, while many reasons can justify an action, and indeed a given reason can justify many actions, no *specific* reason is necessary or sufficient for any *specific* action. This causal relationship between stimulus and response is very different from that found in autonomic, instinctive, and associative systems.

The origins of reasoning behavior have both innate and learned components. I have suggested that basic constructs and relations underlying coherence are innate, though enhanced through learning and practice; the contents of beliefs are clearly learned.

What about the underlying mechanisms? At the conceptual level the mechanisms are propositional attitudes and coherence relations. We believe we can discharge them in computational systems, such as physical symbol systems, though large gaps remain in the story. There is also the issue of realizing physical symbol systems in neural networks. Given that we have already noted that these neural networks can be used to build logic gates, it follows that they can indeed be used to build physical symbol systems. The more interesting variation on this question is how the brain itself organizes these neural networks into physical symbol systems (or systems that exhibit the properties of systematicity, productivity, compositionality, and inferential coherence). The answer to this question is less clear but is being actively researched (Dauphin, Fan, Auli, & Grangier, 2017; Halford, Wilson, & Phillips, 2010; Pater, 2019; Prince & Smolensky, 1997; Shen, Tan, Sordoni, & Courville, 2019; Socher, Lin, Ng, & Manning, 2011).

A more specific question with respect to brain systems is which regions are involved in rational thought processes. When it comes to logical reasoning, we can begin to tell the outlines of a story largely (but not exclusively) involving the neocortex. My lab carried out some of the first studies to explore this question (Goel, 2007; Goel, Gold, Kapur, & Houle, 1997). We are still in the very early stages of understanding the neural basis of rationality, but the emerging picture is that there is no single system of reasoning in the brain. Our ability to engage in rational thought is underwritten by several types of hypothesis generation systems and a common system for detecting conflicts or inconsistencies. These systems are variously located in the occipital, parietal, temporal, and frontal cortices, with some involvement of subcortical structures (Goel, 2019).

Is the Reasoning Mind Enough (or Too Much)?

Now we come to the same question we have asked with respect to the autonomic, instinctive, and associative minds, "Is the reasoning mind enough?," but with a twist. In each previous case, when it came to explaining the range of behaviors that humans exhibit, particularly *rational* behaviors, we found the previous systems wanting and called for additional mechanisms. In questioning the sufficiency of the cognitive mind to explain rational behavior, many of my colleagues will want to answer in the affirmative. As I refocus my attention from laboratory reasoning problems to real-world reasoning situations, I am convinced otherwise. Ironically, my concern is that in many real-world cases the reasoning mind is *too much*. Before considering shortcomings, it is important to emphasize and appreciate what the reasoning mind does explain. We have already considered the Eve example. Let us also look at a real-world example.

On August 28, 2019, British prime minister Boris Johnson, determined to deliver the withdrawal of Britain from the European Union (Brexit) by

October 31, 2019, with or without an agreement, asked Queen Elizabeth II to prorogue Parliament from September 9 to October 14. This caused an uproar in British politics because it meant that Parliament would not have sufficient time to debate and weigh in on any agreement that may or may not be reached by the deadline. It also led to an unprecedented request for judicial review of the decision. No court can challenge the monarch's absolute right to prorogue Parliament, so what was challenged was the advice that the queen received from the prime minister (Scottish Legal News, 2019):

> If the Prime Minister asks the Queen to suspend Parliament she faces an impossible choice. Either she ignores his advice and breaks with convention or she dismisses Parliament so the Prime Minister can use her prerogative to force through No Deal. Both options explode the notion of the UK as a modern, functioning democracy. We will ask the Courts to assist Her Majesty by ruling on that choice.

The petitioners claimed that the reason the prime minister asked to prorogue Parliament was to do an end run around it (i.e., passing a No Deal without parliamentary oversight). Given that the prime minister was determined to leave by October 31, with or without a deal, and that Parliament would not return until October 14, it would not have sufficient time to intervene legislatively. Therefore, the motive for proroguing Parliament was undemocratic and illicit. The prime minister argued that his reason for proroguing Parliament was that the current session had been sitting for more than 340 days, longer than any other session in history, and needed to be brought to a close so that he could "bring forward a new bold and ambitious domestic legislative agenda for the renewal of our country after Brexit" (Sandhu, 2019).

This is the rational mind at work. It is this type of reasoning and decision-making that the cognitive mind explains very well. The prime minister and the dissenters both have *reasons* to hold a particular point of view. These reasons are causally efficacious. In fact, the court is being asked to decide which of the two reasons (doing an end run around Parliament or putting forward "a new bold and ambitious domestic legislative agenda for the renewal of our country") is the *real* (i.e., causally efficacious) reason for the prime minister's actions. Both can justify the action. The former would be illicit, the latter appropriate. The only way the court can decide is to judge the strength of the coherence relation between each reason and the action, given the context (i.e., everything else that is *relevant*). Determining the strength of coherence relations is a job for the reasoning mind.

The cognitive science literature on reasoning largely focuses on instances where the coherence relation seems to be violated, resulting in *irrational* choices. Many such violations have been cataloged in this literature and are routinely explained by invoking "heuristics" rather than logical inference systems (Evans & Over, 1996; Todd & Gigerenzer, 2000; Tversky & Kahneman, 1974). I regard violation of coherence relations as internal issues for cognitive theories of reasoning and will take them up in chapter 7. My larger concern in this book is not with irrationality but rather with *arationality*. By *arationality* I do not mean a violation of the coherence relation. An arational response is one where the action is not selected on the basis of the coherence relation. Either the coherence relation is irrelevant, because propositional attitudes are not involved, or it may be modulated by responses generated by simpler autonomic, instinctive, and associative systems.

To illustrate the intrusion of the autonomic system, I remind the reader of the examples about low blood sugar levels and snapping at my wife, and the data from the parole judges from chapter 3. In both cases, we have instances of the autonomic mind overriding, or certainly protruding into, the reasoning mind. My snapping at my wife and the decisions of the parole judges prior to lunch were not irrational, meaning they did not involve any violation of coherence relations; they were *arational*. They were simply not reason-based. The behavior was triggered by autonomic system processes.

To illustrate the intrusion of instincts into the reasoning mind, I could use any of the four scenarios outlined in chapter 1 (weight management, infidelity, climate change denial, and American aversion to universal healthcare), but the underlying mechanisms have not yet been explained. Therefore, I will introduce a new, more overt and intuitive example of instincts involving interaction with my daughter when she was a teenager. On school days the school bus would pick her up at 8:30 a.m. She would get up around 6 a.m. to prepare for school and then run out of the house at 8:29 a.m. to catch the bus, without having eaten a proper breakfast. She would then come home in the afternoon complaining of being hungry (because she did not have time to eat breakfast). There is a straightforward rational solution to this problem—make time to eat breakfast—so I said to her, "You are awake for two and a half hours before the bus arrives. What is it that you're doing all this time?" She was doing what many teenagers do in the morning before they go out: grooming. I offered her the advice that I had received as a teenager, suggesting that it would be better to prioritize breakfast over grooming so she was not hungry and could focus in class because, after all, "it matters more what is inside your head than what is on top of it." When I was given this advice as a teenager, I think I was naive enough to believe it. My daughter, being smarter, turned around and asked, "What planet are you from?" Initially I dismissed her remarks. However, they stayed with me and became one impetus for this book. The more I thought about them, the more I realized that she was right and I was wrong. There is enormous evidence to indicate that what we look like (i.e., whether we are short, tall, fat, thin, have crooked teeth, bad skin, etc.) makes an enormous difference to every aspect of our lives (Etcoff, 2011). My daughter, of course, did not have access to this body of research. Therefore, her rejection of my argument was not reasoned. She did not have explicit beliefs about the relative importance of grooming versus eating breakfast, but like many teenagers, she was instinctively driven to prioritize grooming activities. Her choice was not rational, but it was evolutionarily adaptive, reminding us that the cognitive system is tethered to evolutionarily older systems, including the instinctive system.

My final example involves an intrusion of the associative mind into the reasoning mind. I return to the story of my suspicion of all advertising from chapter 5. I have come to harbor the belief that all advertising is deceptive. This is not a belief based on consideration of a broad range of evidence. It is based on one-trial operant conditioning. Generalizing it to all advertisers may not be rational, but based on the traumatic negative reinforcement experienced in my childhood, my behavior continues to be guided by this generalization. If it were rational, I would be able to revise the belief if the data warranted. I cannot.

Notice the commonalities and differences across these three examples. What is common to all three cases is that there is no rational explanation for the behavior or the rational choice has been modulated, or even overridden, by an arational choice. What is different across the three examples is that in each case different systems are doing the modulation or overriding. In the blood sugar example, it is the autonomic system. In the grooming example, it is the instinctive system. In the advertising example, it is the associative system.

*   *   *

All systems and mechanisms that are generally postulated to explain organismic behaviors have now been introduced and articulated. Of these, only the cognitive mind can account for our ability to reason. Part III reviews some specific models of reasoning that have been built on top of this cognitive system. They do a reasonable job of explaining certain types of reasoning problems but do less well explaining others. However, they do not even begin to address the types of behaviors highlighted in the preceding three examples (and in chapter 1). The problem here is that, ironically, the cognitive mind—on its own—is in some sense too powerful to account for these behaviors. We need to go back to the autonomic, instinctive, and associative mechanisms discussed in previous chapters and put together an interactional story wherein these simpler systems modulate rational responses (and in turn are modulated by them).

Appendix: Formal Systems and Information Processing in Physical Symbol Systems

Information processing in physical symbol systems requires that certain states of the system carry information about certain aspects of the world (i.e., are representational) and that the reference or content of the states be causally efficacious in the behavior of the system. To understand such a system, we need answers to at least three questions. First, what are informational or semantic properties? Second, how can physical states have semantic properties and preserve them during state changes? Third, how can a thought or content cause another thought or the movement of my body? There is no consensus on the answer to the first question. However, an appeal to formal systems and their mapping onto physical dynamical systems can provide some answers to the second and third questions.

To elucidate certain key aspects of formal systems, let's return to box 6.1 and the example of the propositional calculus. (This simple system does not accommodate the subject-predicate distinction necessary for propositional attitudes and natural languages, but it is sufficient for our limited purposes.) A formal system consists of (1) a finite vocabulary or collection of symbols, (2) a syntax (which determines the grammatical sequences of symbols, i.e., the well-formed formulas), and (3) a finite set of rules that allow for the transformation of well-formed formulas (wffs) into other wffs. The propositional calculus example specifies a vocabulary and the recursive rules that determine the syntax or which patterns constitute wffs. Rules of transformation (in this case "inference rules") were not shown. Here is one transformation rule for the propositional calculus:

$$(A \vee B) \Rightarrow (B \vee A)$$

The way to understand such rules is that if you have a wff that matches the pattern on the left-hand side of the arrow, then you may rewrite it as the pattern on the right-hand side of the arrow. Why? Because that's what the rule says. What does each wff on either side of the arrow *mean*? We don't know. It need not mean anything.

However, to be interesting for our purposes, the wffs need to be *interpreted* or assigned meanings. The interpretation of the connectives is fixed and incorporated into the inference or transformation rules of the system. The intended interpretation of $\vee$ in the preceding rule from the propositional calculus is OR. The variables are usually assigned to some *arbitrary* states of the world that we simply agree on. So, we might interpret A as "John loves Mary" and B as "Mary loves George." Or we could just as well interpret A as "it is sunny today" and B as "dinosaurs are extinct." What is important is that the inference rules of the propositional calculus are *truth preserving*, meaning that if the wff on the left-hand side of the arrow is true on some interpretation, then the wff on the right-hand side of the arrow will also be true on that same interpretation. For example, if the left-hand side is interpreted as "John loves Mary or Mary loves George," then the right-hand side of the arrow becomes "Mary loves George or John loves Mary." If the first is true, the second will also be true. Notice what is happening here. The rule we used to do this transformation knows nothing about John, Mary, George, dinosaurs, or the weather today. It is defined strictly over wffs, that is, over mere syntactic patterns. But despite this, it manages to preserve truth, a semantic property. This is generally expressed by saying that in formal systems "the semantics follows the syntax" or "if you take care of the syntax, the semantics will take care of itself" (Haugeland, 1981).

A physical symbol system can be viewed as a mapping of formal systems onto some physical dynamical system such that the syntactic properties are mapped onto some physical states and the dynamics of these physical state changes mirror, or are isomorphic to, the rule-based transformations of the formal system. If we design the dynamical system such that its operations or state changes are sensitive to the very physical properties instantiating the syntactic properties of the formal system, the symbols and/or their syntactic patterns can be considered causally efficacious in the behavior of the machine. In this way, the explanation for the transformations or state changes of computational systems—that the transformation process is sensitive to the syntactic (meaning physical) properties of the states—also provides an explanation for the puzzle of semantic causation: semantic causation is just physical causation (Fodor, 1975, 1980). While many have questioned the adequacy of this answer to explain human information processing (Goel, 1995; Searle, 1980), it remains the closest we have come to connecting the semantic and causal properties of symbols.

III Reasoning with the Cognitive Mind

Reasons are the pillars of the mind.
—Edward Counsel

We can't avoid reasoning; we can only avoid doing it well.
—Peter Kreeft

With the description of the cognitive mind in chapter 6, we now have in place the infrastructure cognitive psychologists use to build their theories of reasoning. The two are so tightly connected that I have referred to the cognitive mind as the reasoning mind. Separate research efforts have been directed at the semantic, logical, and conceptual inference components of coherence relations. I will examine the basic findings on logical inference in chapter 7, followed by conceptual inference in chapter 8, bypassing the literature on semantic inference.

The logical inference literature leads us to the immensely popular "dual mechanism theories" where successful reasoning is explained via the engagement of an "analytic" reasoning system and failures are explained by the engagement of a "heuristic" system. These accounts have become extremely popular and deeply embedded in the social sciences, but I argue that they harbor some conceptual confusions and are not particularly relevant to explaining the pursuit of food, sex, and politics.

The chapter on conceptual inference gets us closer to the issues of interest. We begin with discussions of the key philosophical puzzles surrounding inductive inference and briefly examine the types of questions and explanations under consideration by psychologists. But the most important part of the chapter is the discussion of the first impeachment of the forty-fifth president of the United States, Donald Trump. Here we see firsthand how the White House public strategy uses reason to intentionally, consciously, activate non-reasoning instinctual systems in the MAGA faithful to drive their behavior.

7 Logical Inference: Heuristic and Analytical Systems

A sane mind should not be guilty of a logical fallacy, yet there are very fine minds incapable of following mathematical demonstrations.
—Henri Poincare

The major strategy utilized by researchers studying logical inference is to identify the many reasoning errors that we all naturally make, notice the nonrandom pattern in these errors, and then propose theories that can account for them. Two important theories that have been proposed are mental logic theory and mental model theory. These theories can explain a number of important facts about human reasoning. But one crucial fact that they cannot explain well is how and why content affects logical reasoning. This has led to the development of "dual mechanism theories." From these theories, we get our vocabulary of heuristic systems and analytical systems. While originally developed to deal with content effects, dual mechanism theories have become immensely popular, to the point that the concept of heuristics is now used to explain *everything*. I will argue that—despite some confusion in the literature—the distinction between heuristic and analytical systems is useful for many purposes. However, heuristics are largely irrelevant for explaining the types of behaviors I'm concerned with. The reader unencumbered with the belief that heuristics adequately explain the behaviors of interest in chapter 1 could skip this chapter.

Cognitive Theories of Logical Reasoning

Psychologists who study reasoning are focused on articulating specific mechanisms—within the cognitive computational framework described in chapter 6—that allow for coherence inferences. Regarding our ability to make logical inferences, two basic ideas have long dominated the field. The first idea is that we have inside our heads something akin to the machinery of formal logic that allows us to draw inferences from given information. This system operates on proposition-like structures, which we encountered in chapter 6. We are, of course, unconscious of these logical rules—just as we are unconscious of the grammatical rules of the natural language that we speak—but they nonetheless guide our inferences. This "mental logic" approach has been articulated over the decades by psychologists such as Mary Henle (1962), Martin Braine (1978), and Lance Rips (1994). It is very congenial to the computational mind framework.

The second idea is similar to the first in that it also relies on a system of logic to account for our inference capabilities. However, it differs from the first in that it postulates that we have access to the semantics of logical terms (such as "all," "some," "none," or "if then"). We use the fixed meanings of these terms to build spatial mental models and then determine whether the conclusion is true across the various models that can be built. If it is true in all the models, then the argument is deemed to be valid; otherwise it is not. This theory has been extensively developed and articulated by Philip Johnson-Laird (2006) and his many students and postdocs. It remains widely influential.

Much of the cognitive reasoning literature for the past 50 years has debated the merits of these two accounts in excruciating detail. What is important for our purposes is that both accounts incorporate aspects of logical theories (and probability theory, where the premises involve numerical probabilities).[1] Psychological inferences that are considered rational on these accounts are the same as those considered valid within formal logic or have the highest probabilities. This is a much narrower application of the term *rational* than the one I am advocating with the intuitive notion of coherence, meaning roughly "making sense."

If we do indeed have some version of a formal logic or probability theory engine inside our heads, we should all be excellent reasoners. We are not. We all make mistakes. We often generate responses other than those predicted by the normative theories of logic. Most researchers in the field accept that humans are largely rational beings. Therefore, reconciling the belief that we are rational with the fact that we make many reasoning errors (at least when compared to the normative models) is the central theme of the cognitive research program on human reasoning. The enterprise is one of analyzing the pattern of errors and then drawing conclusions from this analysis about the specific nature of the cognitive mechanisms underlying human reasoning abilities.

Let us begin by considering a problem similar to the oral contraceptive pill example from chapter 1. Psychologist Gerd Gigerenzer and his colleagues (Gigerenzer, Gaissmaier, Kurz-Milcke, Schwartz, & Woloshin, 2007) asked 160 trained gynecologists the following question:

Assume you conduct breast cancer screening using mammography in a certain region. You know the following information about the women in this region:

- The probability that a woman has breast cancer is 1% (prevalence).
- If a woman has breast cancer, the probability that she tests positive is 90% (sensitivity).
- If a woman does not have breast cancer, the probability that she nevertheless tests positive is 9% (false-positive rate).

A woman tests positive. She wants to know from you whether that means that she has breast cancer for sure or what the chances are. What is the best answer?

A. The probability that she has breast cancer is about 81%.

B. Out of 10 women with a positive mammogram, about 9 have breast cancer.

C. Out of 10 women with a positive mammogram, about 1 has breast cancer.

D. The probability that she has breast cancer is about 1%.

The majority of doctors responded (A) 81% or (B) 90%, significantly overestimating the actual probability of the woman having cancer. Only 21% of the doctors selected the correct answer (C). This certainly looks like a failure of inference. How is it to be explained?

One common way of explaining these types of errors is to assume that while we do have a fully formed reasoning engine, roughly based on the internalization of formal logic and/or probability theory, its operation is subject to what psychologists call "performance constraints." Performance constraints are factors such as how we understand the task, whether we are paying attention, our ability to hold all the information in short-term memory, and other factors that can result in real-time errors. That is, while the basic postulates of logic and probability theory are intuitive and self-evident, our ability to scale up from these intuitions is a function of cognitive capacity, effort, and training. In many cases, such "performance explanations" are adequate to explain errors.

However, in this particular case, Gigerenzer offers a slightly different explanation. He notes that the problem is formulated in terms of conditional probabilities. He reformulates the first part of the problem in terms of "natural frequencies" as follows:

Assume you conduct breast cancer screening using mammography in a certain region. You know the following information about the women in this region:

- Ten out of every 1,000 women have breast cancer.
- Of these 10 women with breast cancer, 9 test positive.
- Of the 990 women without cancer, about 89 nevertheless test positive.

In this formulation of the problem, 87% of the doctors understood that about "1 in 10" (C) was the correct answer. This result suggests that our cognitive architecture is organized such that certain representational formats are much more intuitive and easier to work with than others. Specifically, we are not very good at calculating *conditional* probabilities. We are much better at calculating *natural* frequencies. The fact that different representational formats make explicit (or make implicit) different aspects of representational contents, and that this has enormous consequences for the algorithms operating on them, is well understood within the computer science community. This insight, applied to the structure of the cognitive system, provides a plausible explanation for differential performance on these problems.

This analysis also applies to the contraceptive pill example from chapter 1. If it had been reported that the risk of blood clots increased from 1 to 2 in 7,000, most women who abandoned the pill would have made a different decision. There are approximately a dozen types of errors that we reliably make in deductive and probabilistic reasoning, including the atmosphere effect, confirmation bias, denying the antecedent, and affirming the consequent, among others, and most can be explained by some combination of performance considerations and architectural and representational constraints on the cognitive system. But not all reasoning errors can be so explained.

Content Effects in Reasoning

The error that dominates the field is the content effect. Minna Wilkins (1928) was among the first to report that people reason much more accurately when the logical conclusion of an argument is consistent with their beliefs about the world than when it is inconsistent with their beliefs. This is puzzling because, at least in formal logic, the content of the premises is irrelevant to the validity of the argument. Arguments are valid by virtue of their logical structure or form. For example, you will agree that the following two arguments are both valid:

(1a) Tweety is a robin, no robins are migrants, so therefore Tweety is not a migrant.

(1b) Jerry is a mouse, no mice like cheese, so therefore Jerry does not like cheese.

Suppose I ask you whether you know Tweety the robin or Jerry the mouse. If you do not, how do you know that Tweety is not a migrant or that Jerry does not like cheese? The point, of course, is that this knowledge is irrelevant to determining the validity of these arguments. They are valid simply by virtue of having the following form:

(2)   M has F; nothing with F has G; therefore M does not have G.

Any argument with this form will be valid. Despite this, when people are given the following two arguments with identical logical structures, their responses are very different (Evans, Barston, & Pollard, 1983):

(3a) No cigarettes are inexpensive, some addictive things are inexpensive, so therefore some addictive things are not cigarettes.

(3b) No addictive things are inexpensive, some cigarettes are inexpensive, so therefore some cigarettes are not addictive.

While both arguments are valid, the first has a believable conclusion, whereas the second has an unbelievable conclusion. University students happily rate the first argument as valid 92% of the time but rate the second argument as valid only 46% of the time. They also respond much faster to 3a than to 3b. If the same argument is presented without any belief-triggering content (e.g., no A are B, some C are B, so therefore some C are A), the accuracy falls between these two extremes of 92% and 46%.

Now consider one of the most famous problems in the reasoning and decision-making literature, the Linda problem (Tversky & Kahneman, 1983):

(4)   Linda is 31 years old, single, outspoken, and very bright. She majored in philosophy. As a student, she was deeply concerned with issues of discrimination and social justice and also participated in antinuclear demonstrations.
      Which statement is most likely true of Linda?

      (a) Linda is a bank teller and active in the feminist movement.

      (b) Linda is a bank teller.

Many intelligent individuals will choose (a) over (b) as most likely to be true of Linda. Technically, this is incorrect, because formal logic and probability theory tell us that a conjunction, as in (a), cannot be more likely than one of the component conjuncts, as in (b).

Another example, this time involving explicit probabilities, is provided by the base rate fallacy task (Tversky & Kahneman, 1982):

(5)   I have a jar that contains the names and descriptions of 100 individuals. Ten of these individuals are engineers and 90 of them are lawyers.

I draw a random name from the jar and read the following description about the individual:

Jack is 36 years old. He is not married and is somewhat introverted. He likes to spend his free time reading science fiction and writing computer programs.

Which statement is most likely true of Jack?

(a) Jack is an engineer.

(b) Jack is a lawyer.

Many individuals will often choose (a) over (b). Again, in terms of normative models of rationality, this is problematic because, based on the information given about the number of names of engineers and lawyers in the jar, there is a 90% chance that Jack is a lawyer and only a 10% chance that he is an engineer. How do we explain the discrepancy in responses?

In each of examples (3)–(5), we are confronted with the dilemma that intelligent, educated people are being "misled" by the content of the arguments to make choices that violate the norms of logic and probability theory, which they all otherwise accept. Does this mean they are not rational, or are there other explanations? Earlier, we invoked performance factors and architectural constraints to account for certain types of errors. However, it has proven difficult to convincingly explain content-based errors in problems (3)–(5) as performance factors.

Amos Tversky and Daniel Kahneman (1974, 1982, 1983), based on their analysis of problems like the Linda task and the base rate fallacy task, offered a variation on the architectural constraints account in terms of "heuristics and biases." They proposed that since reasoning tasks can be very complex, our minds are built in such a way that they contain certain shortcuts or *heuristics* that allow us to avoid complex, time-consuming calculations in favor of simpler, faster operations. This can also be construed as an architectural claim about the cognitive system. These heuristic shortcuts often lead to quick, useful responses, but they are not sensitive to the laws of logic and probability theory, and therefore they can also result in severe and systematic errors (i.e., nonnormative responses). The *biases* they mention are those features of the reasoning system that, in certain situations, lead to a selection of a heuristic rather than the calculation of the normative response. This resulted in a research program of identifying the various heuristics humans use. In the case of the Linda and the base rate fallacy problems (4) and (5), Tversky and Kahneman appealed to the heuristic of "representativeness." The idea was that, based on what we have been told about Linda and Jack, Linda being active in the feminist movement is more representative of Linda than just being a bank teller, and Jack being an engineer is more representative of Jack than being a lawyer. Even though formal logic and probability theory indicate otherwise, the biasing triggers the representativeness heuristic, leading to nonnormative responses. Daniel Kahneman was awarded the 2002 Nobel Prize in Economics for this body of work (with Amos Tversky having passed away).[2]

Dual Mechanism Theories of Reasoning

Dual mechanism accounts of reasoning emerged in the 1990s and soon came to dominate the reasoning literature.[3] These accounts have been most prominently championed by Jonathan Evans and David Over (1996), Steven Sloman (1996), and Keith Stanovich (2004). The goal of dual mechanism theory is to preserve normative rationality while accounting for the systematic errors people make without relying on the "performance error" explanation. Because the underlying intuitions are easy to grasp, these theories have become immensely popular, even to the point that Daniel Kahneman (2003) reframed the heuristics-and-biases account as a dual mechanism account in his Nobel Prize lecture.

Dual mechanism theories begin by making an intuitive distinction between formal, deliberate, rule-based reasoning and implicit, unschooled, intuitive, automatic reasoning and postulate two different reasoning systems (or processes) to deal with them. The theories come in three different flavors but largely agree regarding the types of properties that cohere with one system and the types that cohere with the other system. In a very influential article, Stanovich and West (2000) dubbed the two systems System 1 and System 2. System 1 is widely referred to as the "heuristic system," and System 2 is the "analytic system."

In early formulations, System 1 was characterized by properties such as "automatic," "effortless," "associative," "preconscious," and "implicit," while System 2 was characterized by properties such as "explicit," "conscious," "formal," "rule-based," "effortful and slow processing" (Sloman, 1996; Stanovich & West, 2000). These properties were initially meant to be constitutive. In later formations, they have become "suggestive" or "typical" rather than definitive of the two systems (Evans & Stanovich, 2013).

In examples (3)–(5), System 1 would give the intuitive, nonnormative, content-based (heuristic) responses, judging that the argument conclusion in 3b ("some cigarettes are not addictive") is invalid; selecting "Linda is a bank teller and active in the feminist movement" as more probable than "Linda is a bank teller" in the Linda problem; and selecting "Jack is an engineer" over "Jack is a lawyer" in the base rate problem. The reason in each case is the same. These responses are more believable, given what we know about the world and what we have been told in the problems. System 2 would give the normative, logical (analytical) responses in each case, judging the argument conclusion in 3b ("some cigarettes are not addictive") as valid and selecting "Linda is a bank teller" as more probable than "Linda is a bank teller and active in the feminist movement" and "Jack is a lawyer" as more probable than "Jack is an engineer."

The first systematic characterization of these two systems in the reasoning literature was offered by Jonathan Evans and David Over (1996). In their book *Rationality and Reasoning*, they took a broad, evolution-based approach and characterized System 1 as the "old brain" system, which we share with other animals, such as pigeons and mice. They conceived of it as an innate module or instinct, and the world knowledge that it operated on was acquired through learning via associative mechanisms. Keith Stanovich (2004) went so far as to say that "the classic example of a System 1 is the reflex arc. Like the reflex arc they [System 1] provided a causal link between trigger and response, they belonged to the old brain, driven by mechanisms that drive other automatic behaviors across the evolutionary spectrum like foraging and mating." System 2 was conceived of as distinctly human, belonging to more recent brain systems in the neocortex. It permitted abstract and hypothetical reasoning and was subject to the constraints of the computational mind.

Steve Sloman (1996) offered a different but overlapping account. He postulated two different types of reasoning processes rather than different systems. The mechanism corresponding to System 1 was an associative process, while the mechanism corresponding to System 2 was a rule-based process. The former process appeals to the associative mind, while the latter appeals to the cognitive mind, discussed in chapters 5 and 6, respectively.

Daniel Kahneman's dual mechanism theory commitments were different still, but the end result was similar. He began with the intuitions that (1) "thoughts differ in accessibility: some come to mind more easily than others" (i.e., the speed-of-processing issue) and (2) there is a distinction between intuitive and deliberate thought processes. He came to the conclusion that intuitive judgments occupy a position "between the automatic operations of perception and the deliberate operations of reasoning." These intuitions are then mapped onto the dual mechanism view as follows (Kahneman, 2003, p. 699):

> The perceptual system and the intuitive operations of System 1 generate *impressions* of the attributes of objects of perception and thought. These impressions are

neither voluntary nor verbally explicit. In contrast, *judgments* are always intentional and explicit even when they are not overtly expressed. Thus, System 2 is involved in all judgments, whether they originate in impressions or in deliberate reasoning. The label *intuitive* is applied to judgments that directly reflect impressions—they are not modified by System 2.

If you're going to have two systems of reasoning, you need a way for them to interact and determine which one is going to respond in any given situation. The literature suggests two possibilities. One, known as the default interventionist account, is that System 1 processes are dominant. In order to generate the alternative (normative) response, System 2 must inhibit System 1 and produce an alternative response (Evans & Over, 1996). The other selection strategy is known as the parallel competitive model. It requires that both systems compete and the one that completes the task first cues the response (Sloman, 1996). Control structures are further discussed in chapter 12.

Data Supporting Dual Mechanism Theories

Four sources of data are cited to support these theories. First, and perhaps most influential, are the experimental manipulations involving reaction times (De Neys, 2006a; Evans & Curtis-Holmes, 2005). The belief-bias (System 1) responses are faster than formal logical (System 2) responses. For example, reaction times for valid responses to the cigarette argument 3a with the believable conclusion (belief-biased, System 1) are faster than for the cigarette argument 3b with the unbelievable conclusion. Invalid responses to the cigarette argument 3b (belief-biased, System 1) are faster than valid responses (System 2).

"Dual-task paradigm" studies (De Neys, 2006a, 2006b) are a second source of supporting data. In these studies, individuals are asked to solve problems like the cigarette arguments 3a and 3b and at the same time do an additional task, such as counting backward. The finding is that this additional task will impair the accuracy of reasoned responses in problem 3b with the unbelievable conclusion "some cigarettes are not addictive" (requiring System 2) much more than in 3a with the believable conclusion "some cigarettes are not addictive" (engaging System 1).

Third, psychometric studies by Stanovich and West (2000) noted that the normative formal (System 2) responses to logical problems correlated with an individual's IQ but the intuitive (System 1) responses did not, suggesting underlying differences.

A fourth source of data was provided by our lab (Goel, 2007; Goel, Buchel, Frith, & Dolan, 2000; Goel & Dolan, 2003). In a series of brain imaging studies, we demonstrated that different brain systems were involved in processing heuristic belief-biased responses (System 1) and formal or normative (System 2) responses.

Critique of Dual Mechanism Theories

After becoming insanely popular, dual mechanism accounts, at least as characterized here, have begun to unravel. The characterization of System 2 has been relatively uncontroversial. It is just an appeal to a model that internalizes some theory of logic or probability. But the characterization of System 1 has been a source of considerable confusion. I believe two types of errors have led many researchers astray. The first error can be characterized as the "duck problem," and the second is a failure to keep in mind the machinery necessary for inference, outlined in chapter 6.

The duck problem is the problem of individuation of systems based purely on behavioral data: if it quacks like a duck, then it must be a duck. Well, it might be a duck, or it could be a hunter looking to shoot ducks. Dual mechanism theories assume that if a response is slow, it belongs to one system; if it is fast, it belongs to the other system. Speed of response is one of the few behaviors that can be directly measured. But on their own, such behavioral categories are superficial and largely uninteresting for scientific purposes. For instance, suppose I want to understand modes of locomotion. I can make a category of "all things that move fast" and another category of "all things that move slow." In the fast category, I could put things such as cars, planes, comets, and electromagnetic waves. In the slow category, I could put such things as bicycles, motor scooters, bears, fish, and nuclear submarines. Notice the several problems here. First, it is unclear how fast or slow something has to move for membership in the respective category. Do cars move fast enough to be in the fast category? They move fast when compared to bicycles and bears but not when compared to electromagnetic waves. Second, and more importantly, while these categories may be of interest for some purposes, they are of little scientific interest, even for the purpose of understanding locomotion, because the members of each category do not share underlying structural and causal principles of locomotion. For example, cars and scooters, the former belonging to the fast category and the latter belonging to the slow category, actually share the internal combustion engine as a means of locomotion. Airplanes and comets, both belonging to the fast category, have very different mechanisms of locomotion. Scientifically interesting categories individuate along structural and causal lines. Behavioral categories may be a reasonable place to start, but they are not the place to stop. I believe all the dual mechanism accounts that we have reviewed are guilty of overemphasizing differences in processing speed, without any attempts to look deeper into underlying mechanisms. In fact, the mistake is even more egregious. There is actually very little published data supporting speed differences between System 1 and System 2. I was only able to find two studies to cite above. In our own studies, we get trends but few significant differences (Goel et al., 2000; Goel, Makale, & Grafman, 2004; Goel & Dolan, 2003).

The second error is forgetting the machinery necessary for inference. The specifics of this error are different in each of the three accounts. Let's begin with Evans and Stanovich. I like that they start with innate, automatic, mandatory processes belonging to the "old brain" system. They also point out that System 1 is not a single process but rather a collection of low-level processes. I have differentiated several of these processes in terms of causal coupling, origins, conceptual mechanisms, brain structures, and function in part II of the volume. So I can readily agree up to this point. The problem arises when these researchers characterize belief-biased reasoning responses to problems such as (3)–(5) as "innate, automatic, and mandatory," belonging to the "old brain," and explicitly compare them to reflex arcs (Stanovich, 2004). On the contrary, these responses are very high-level, conceptual inferences involving propositional attitudes and knowledge of the world. I have spent several chapters discussing why they cannot be accommodated by reflex arcs, instincts, and associative systems. They require the machinery of the cognitive brain. This was an unfortunate misstep in the development of dual mechanism theories.

At the expense of some repetition, let us take up the issue of "innate, automatic, and mandatory" in the context of the eye blink reflex and the non-normative responses to the three problems here. As discussed in chapter 3, an eye blink reflex is *truly* innate, automatic, and mandatory. If I suddenly snap my fingers in the vicinity of your eyes, you will blink. If I forewarn you of my intention prior to snapping my fingers, you will still blink. Even if I forewarn you and assure you that I will not touch your eyes (and you trust me and believe me), you will still blink. Even if I forewarn you, assure you that I will not touch your eyes, and offer you a large monetary reward for not blinking, you will still blink when I snap my fingers in the vicinity of your eyes. The snapping of my fingers in the proximity of your eyes is causally sufficient for you to blink, for reasons having to do with the underlying neural machinery discussed in chapter 3.

Compare this to what is happening in the Linda problem. There are several important dissimilarities between reflexes and the nonnormative response in such problems. First, individuals can give sensible reasons for their response; for instance, "I was focusing on x instead of y." Second, when the correct answer is pointed out to them (and the underlying reason), they can acknowledge that they have made a mistake and apply the analytic system next time around and generate the normative response. The phenomenology of the belief-bias effect—to say nothing of the neurobiology—is very different from that of an eye blink. Reasoning fallacies are simply not automatic and mandatory in the same sense as reflex arcs. Reflex arcs belong to the autonomic mind. They cannot traffic in propositional attitudes. The reader is encouraged to revisit table 6.1 as a reminder that the reasoning mind differs from the autonomic mind on each of the five criteria that we have been considering. To think otherwise is to invite conceptual confusion.

To be clear, I'm not criticizing the appeal to "reflex arc type" noncognitive factors in the explanation of human behavior. Indeed, this book is about how critical and ubiquitous the contribution of noncognitive factors is in our behavior. In previous chapters, I have given several examples of noncognitive factors belonging to simpler, "old brain" systems: low blood sugar level, instinctual biases, and associative learning. There is an interesting story to be told about how they modulate reasoning processes, but that is not the story being told by Jonathan Evans, David Over, Keith Stanovich, and this branch of the dual mechanism literature.

Steven Sloman's characterization of System 1 is in terms of associative inferences. If this account requires constructing thoughts and inferences just from co-occurrence associations, we've already discussed some of the challenges involved. But that may not be required. It may be that our system of propositional attitudes can, in addition to logical relations, also participate in co-occurrence relations. We saw an example of this with my story of the racing car set. But it should be noted that the Eve example in chapter 6 shows that one certainly does not need to appeal to associations to explain the nonnormative response to the Linda problem. For example, we know that Linda was concerned with social justice. We may have some sort of causal model whereby the traits that lead to concerns about social justice also lead to participation in feminist movements. Linda has these properties by virtue of being concerned about social justice, and therefore Linda is likely to be active in the feminist movement.

Daniel Kahneman's account of heuristics or System 1 appeals to perceptual system processes. System 1 operates on "impressions" rather than intentional, explicit representations, which are the province of System 2. It is clear that System 2 uses the conceptual or propositional representations, and the accompanying machinery, identified in chapter 6. I confess to not understanding what "impressions" are and how one might draw inferences from them.

It is possible that by "impressions" Kahneman means some form of non-propositional or "nonconceptual" representation. We discussed propositional representations in chapter 6. Sixty years of research in cognitive science (to say nothing of the more than a century of philosophical consideration) largely converge on the idea that inferences require something akin to propositional representations. The notion of nonpropositional or nonconceptual representations is much less understood, to the point that many philosophers are skeptical of the coherence of the concept. There are a few philosophers, such as Ron Chrisley of Sussex University, trying to develop notions of nonconceptual content, though they are far from telling a story of how these representations might be involved in inferences. In essence, if by reasoning with "impressions" Kahneman is appealing to some sort of nonconceptual or nonpropositional inference, we currently do not understand what that might be (Frixione and Lieto, 2014). This means that the resulting notion of heuristics is unclear. But perhaps the more immediate point is the one made in the discussion of Evans and Stanovich's model, that the inferences in examples (3)–(5) are all high-level, logical, and conceptual inferences, requiring the cognitive machinery outlined in chapter 6. If there is a coherent notion of heuristic inference based on "impressions," it does not apply to these types of problems. If we want to apply the label of "heuristic" to the inferences in these problems, then it must be construed as belonging to the cognitive mind.

In addition to these major conceptual problems, which question the coherence of dual mechanism theories, there is also a growing body of data that is inconsistent with these theories. First, the neuroimaging data contributed by our lab support the idea of distinct neural systems involved in formal and belief-biased responses to problems (3)–(5). The formal responses typically engage the parietal systems, while the belief-biased responses engage high-level language systems in left hemisphere frontal-temporal systems (Goel, 2007). These are not the types of systems that we share with rats and pigeons; nor are they the structures that would be involved in perception-based inferences on "impressions." Also, recent behavioral data indicate that logical inferences (System 2) can sometimes be faster than belief-biased inferences (System 1) (Handley, Newstead, & Trippas, 2011; Trippas, Thompson, & Handley, 2017). Based on the original logic, this would suggest that in some cases logical reasoning is accomplished by System 1 rather than System 2, upending the whole framework or at least requiring further reorganization (De Neys, 2017; De Neys & Pennycook, 2019).

For completeness, it should be noted that at least Jonathan Evans and Keith Stanovich (2013) have recognized the problems with their accounts and walked back the theory. In its latest iteration, it has been reduced to little more than the claim that the critical distinction between the two systems is the utilization of working memory. System 2 utilizes working memory, while System 1 does not. The problem here is that *all* computational processes require working memory. If dual mechanism theories are going to explain inferences in terms of the standard computational model, then the distinction between System 1 and System 2 cannot be one of using and not using working memory. It will need to be one of using more or less working memory.

Having critiqued this body of work, it is incumbent on me to say how the content effect is to be explained. To begin with, I find myself less impressed than many of my colleagues by examples (3)–(5).[4] These problems artificially induce a conflict between logical coherence and conceptual coherence, and researchers fret when individuals choose the conceptual over the logical inference.

A number of years ago, Wim De Neys from the French National Center for Scientific Research and I undertook some neuroimaging studies on the base rate fallacy task (problem (5)). As we were developing the stimuli, I tested them on my son, who was 14 years old at the time. To my surprise, he asked me: "Do you want the response based on the numbers or the response based on the descriptions?" This caught me totally off guard, as there was nothing in the literature to suggest that people are consciously aware of the conflict that has been set up and are prepared to give either response. When we ran the actual study, many people who went into the MRI scanner to do the task asked the same question. The response we gave them was, "Give us your best answer." So the subjects had an understanding of the base rates. They fully understood that if an occupation occurred 90% of the time, it would be a safer response than one that occurred only 10% of the time. However, they were making a judgment about the saliency of the description. How closely did it fit their prototypes of (for instance) an engineer or lawyer? If there was a tight fit, then they were making the judgment that on this specific trial, the description was sufficiently compelling to override the base rates. I personally do not see the irrationality of this response, at least if we view it more broadly. It may not be the normative response, but it is a perfectly coherent response. Given that base rates address overall probabilities, it is perfectly coherent to supplement the base rate information, or even override it, and select on the basis of the strength of the description on any *specific* trial. Note that if the description is neutral, as in the following problem, individuals will select on the basis of the base rates.

(6) I have a jar that contains the names and descriptions of 100 individuals. Ten of these individuals are pool players and 90 of them are basketball players. I draw a random name from the jar and read the following description about the individual:

John is 29 years old and has lived his whole life in New York. He has green eyes and black hair. He drives a light gray car.

Which of the following is most likely?

(a) John is a pool player.
(b) John is a basketball player.

Here, no conceptual inference is cued and the "fallacy" disappears.

A similar, but not identical, analysis applies to the Linda problem. Again, a conflict has been set up, this time between the description of Linda and the choice offered by the single conjunct (Linda is a bank teller). Being a bank teller is somewhat insufficient, if not inconsistent, with what we have been told about Linda. Being active in the feminist movement, on the other hand, rings truer or more consistent with her past. A plausible explanation for the nonnormative selection is that, rather than simply being a logical inference task, the task invites Gricean pragmatic maxims (Grice, 1975) for effective communications in social situations. In particular, participants assume that all the information provided will be truthful, relevant, informative, and will avoid obscurity and ambiguity, so they are not looking for a trick and may not even notice the conjunction, or at least discount it. The fallacy, unsurprisingly, disappears if we use "Linda is active in the feminist movement" as the single conjunct and "Linda is a bank teller" as part of the conjunction. Here people will overwhelmingly choose the single feminist conjunct. Also, when the conflicting conceptual inference is replaced with a neutral one, as in the following problem, people will also choose the single conjunct (b).

(7) Kelly is 12 years old and lives in a town in Illinois. Kelly likes watching sports on TV and is a big fan of the local football team. Which of the statements is most likely true of Kelly?

(a) Kelly is a boy and sometimes visits his grandparents.
(b) Kelly is a boy.

Regarding the syllogisms in problem (3), the issue is one of prioritizing consistency of the conclusion with our belief network. If the conclusion is believable, we may stop processing out of laziness, or to conserve resources, and respond on the basis of the belief bias. If we notice an inconsistency, we will suppress the belief-biased response and calculate the normative response. This can be described without postulating any "low-level" or noncognitive systems. A simple distinction between logical-inference and conceptual-inference systems is perfectly adequate. Both systems belong to the cognitive mind.

The only reason to see irrationality in examples (3)–(5) is if we assume that people explicitly construe these problems as deductive reasoning tasks. The preceding discussion suggests that they may not. In the real world, deductive and inductive inferences are intertwined. Even though, technically, deduction should override induction, real-world deductive inferences with untrue premises are worthless, but conceptual inferences which make errors in any embedded deductive inference may nonetheless be useful. For example, given my belief that if it rains then the grass will be wet, and my morning observation that the grass is indeed wet, I might draw the invalid but potentially useful inference that it rained last night. Conceptual inference is discussed in chapter 8.

In concluding this chapter, I want to leave the reader with two important points. First, examples (3)–(5) only result in an existential crisis if we limit our notion of rationality (or give a privileged status) to the normative accounts provided by the formal models. These models are, of course, extrapolations of our basic logical and probabilistic intuitions. We intuitively understand and accept their basic axioms, though we may well struggle with complex derivations. But we also engage in conceptual inferences. While conceptual inferences lack the necessity associated with formal logical inferences, they are perhaps more important for real-world functioning, and should not be treated as second-class citizens.

The second, and perhaps more important, point I want to convey is that, however you construe these examples, your analysis will necessarily be confined to the cognitive reasoning system. There are no noncognitive factors in play here. To think otherwise is to needlessly invite confusion. That is, it may well make sense to postulate two different systems of reasoning—a formal logic system and a conceptual inference system—but both belong to the cognitive mind.[5] For this reason the heuristics/analytic (System 1/System 2; "fast and slow") distinction is not particularly relevant to explaining the behaviors under consideration in this volume.[6]

*   *   *

The types of problems reviewed in this chapter do not require an appeal to noncognitive factors. Indeed, they might well leave us with the impression that the cognitive mind is necessary and sufficient to account for all rational behaviors. This is because they are artificial, contrived problems. Such constrained problems are often used in psychological research and can be very useful for exploring specific issues under controlled conditions. However, generalizations based exclusively on such problems can mislead. If we step beyond these textbook problems and consider the type of reasoning that preoccupies us on a daily basis, we will find that it contains these issues and many more. In chapter 8, I consider conceptual inferences, beginning with textbook examples, followed by real-world examples from science and politics. In the political example, we will clearly encounter the intrusion of genuine noncognitive factors in the reasoning process.

8 Conceptual Inference in the Real World:
From Science to Politics

Inferences of science and common sense differ from those of deductive logic and mathematics in a very important respect, namely, when the premises are true and the reasoning correct, the conclusion is only probable.

—Bertrand Russell

Most readers will be familiar with Arthur Conan Doyle's sleuth Sherlock Holmes, renowned for his skills of "deduction." Here he is impressing his friend Watson (Doyle, [1892] 2019, pp. 2–4):

> Then he stood before the fire, and looked me over in his singular introspective fashion. "Wedlock suits you," he remarked . . . "And in practice again, I observe. You did not tell me that you intended to go into harness."
>
> "Then how do you know?"
>
> "I see it, I deduce it. How do I know that you have been getting yourself very wet lately, and that you have a most clumsy and careless servant girl?" . . .
>
> "It is simplicity itself," said he; "my eyes tell me that on the inside of your left shoe, just where the firelight strikes it, the leather is scored by six almost parallel cuts. Obviously they have been caused by someone who has very carelessly scraped round the edges of the sole in order to remove crusted mud from it. Hence, you see, my double deduction that you had been out in vile weather, and that you had a particularly malignant boot-slitting specimen of the London slavery. As to your practice, if a gentleman walks into my room, smelling of iodoform, with a black mark of nitrate of silver upon his right forefinger, and a bulge on the side of his top-hat to show where he has secreted his stethoscope, I must be dull, indeed, if I do not pronounce him to be an active member of the medical profession."

Entertaining though these inferences may be, they have nothing to do with deductive reasoning. They are, however, excellent examples of inductive reasoning. They are plausible conceptual inferences. But, as in all conceptual inferences, there are many equally plausible alternative explanations for Holmes's observations.

Most real-world reasoning involves conceptual inference, with logical inference embedded throughout. Conceptual inferences differ from logical inferences in that they are not a function of logical structure but rather involve the integration or evaluation of the given propositions in the context of all other propositions that we might believe or entertain. Hence, formal theories such as those discussed in chapter 7 provide no insight into the basis of conceptual coherence relations.

This chapter is divided into three sections. We begin by reviewing the two deep epistemological puzzles surrounding inductive inference. The first is David Hume's problem of justifying generalization of properties based on a limited number of observations; the second is Nelson Goodman's problem of knowing which properties are generalizable. The second section reviews some of the basic research on inductive inference by cognitive psychologists. Unlike our understanding of deductive reasoning, our understanding of inductive reasoning is extremely superficial. Very little is known about what underwrites conceptual coherence relations. We then move from the consideration of laboratory problems to see how induction works in the real world. Two examples, one from science, the other from politics, are discussed. These examples highlight some of the errors, pitfalls, and fallacies that induction is subject to. When we come to the political example (the first impeachment of Donald Trump), we will encounter instances of coherence relations being intentionally backgrounded and replaced by systems belonging to the instinctive mind.

Conceptual Coherence from Hume to Goodman

Arguments involving conceptual coherence relations are broadly called inductive arguments,[1] defined as arguments where the premises provide only limited grounds for accepting any given conclusion. The classic form studied is generalization from particulars. Consider the report of a unique, well-preserved fossil of an armored dinosaur, *Borealopelta markmitchelli*—including skin, scale, and very fine-grained detail of stomach contents—recently found in the tar sands of Alberta, Canada. After extensive examination of the gut contents, scientists drew conclusions about stomach physiology and dietary habits, not just of the one specimen but of all *Borealopelta* (Brown et al., 2020). They reasoned as follows:

(A) The only fossilized remains of *Borealopelta markmitchelli* stomach contents ever discovered reveal fern, cycad, and conifer remains.

∴ The diets of all *Borealopelta markmitchelli* included fern, cycad, and conifer plants.

Clearly, this is not a valid logical argument. The premises involve the observation of one *Borealopelta* (though in most cases we would have multiple observations). The truth of these limited observations cannot guarantee the truth of the conclusion, which involves *all Borealopelta*. However, most of us would be prepared to accept argument (A) as plausible, reasonable, or coherent.

One question that immediately presents itself has to do with the justification of this inference. David Hume ([1739] 1888) famously considered such problems and argued that the conclusion is neither a report of direct experience nor a logical consequence of it. It cannot be the former because we have viewed a limited number of *Borealopelta* remains, and it cannot be the latter because an inference from the premises (or observations) would require an appeal to a Principle of the Uniformity of Nature, according to which *"instances, of which we have had no experience, must resemble those, of which we have had experience, and that the course of nature continues always uniformly the same"* (p. 89). But such a principle cannot, of course, be established by observation or deductive inference. It can only be established by inductive inference, which presupposes the principle, thus leading to a vicious circle. If Hume is correct, this negative argument rules out the possibility of justifying induction. In other words, the epistemological problem of induction may be unsolvable.

Despite Hume's observation, it is a fact that human beings are almost compelled to draw inferences from limited information, as in argument (A). Why? Hume's positive contribution to the problem of induction is the suggestion that the experience of constant conjunction results in a "habit of mind" that leads us to anticipate the same conclusion whenever we encounter another instance of the premises. For example, seeing one (or even several) *Borealopelta* fossils and identifying gut contents results in a "habit of mind" leading to the expectation that any future *Borealopelta* fossil will reveal the same dietary contents. Hume is appealing to the associative structure of our minds. That is, he provides a *psychological solution* to the epistemological problem by appealing to the mechanism of association discussed in chapter 5.

It is also the case that the same unique fossil revealed broken scales (Brown et al., 2020), leading to the following argument:

(B) The only fossilized remains of *Borealopelta markmitchelli* skin and scales ever discovered reveal broken and missing scales.

∴ All *Borealopelta markmitchelli* had broken and missing scales.

Most of us would not be prepared to accept the conclusion of (B). Frankly, it sounds crazy. So, what is the difference between (A) and (B) that causes us to accept (A) as plausible but reject (B) as implausible? Unlike in the case of deduction, we cannot appeal to logical form to differentiate between the plausibility of (A) and (B) because both of them have the same invalid logical form:

(C) X has the property alpha and X has the property beta

∴ Everything with the property alpha has the property beta.

To state the problem in Hume's vocabulary, why does observing the fossilized remains of stomach contents and skin and scales of one dinosaur result in the formation of a "habit of mind" or expectation that all dinosaurs of the same species would have the same diet but not the expectation that all dinosaurs of the same species would have broken and missing scales? Given that the evidence for both is identical, why is the mind prepared to generalize the former but not the latter?

This is the New Riddle of Induction illustrated more rigorously by Nelson Goodman (1955) with the famous grue example. Consider the following observations and plausible inference:

(D) Emerald x is green

Emerald y is green

etc.

∴ All emeralds are green.

Goodman then introduces the predicate "grue" and defines it as applying to all things that are green and observed before a certain time, let's say January 1, 2135, and to things that are blue and examined on or after January 1, 2135. This leads to the following observations and plausible inference:

(E) Emerald x is grue

Emerald y is grue

etc.

∴ All emeralds are grue.

All available evidence to date supports both conclusions, "all emeralds are green" and "all emeralds are grue." However, on January 1, 2135, the conclusion "all emeralds are grue" will be false, resulting in the dilemma that the very same observations that support the conclusion that all future observed emeralds will be green also support the incompatible conclusion that all future observed emeralds will be grue. How do we select which of these predicates to project or generalize?

Goodman is pointing out that while Hume was correct in appealing to "habits of mind," he failed to notice that the mind is only prepared to generalize or project certain regularities but not others. Why is this? How do we differentiate projectable properties from nonprojectable ones? It is sometimes said that properties that project or generalize in the required manner (such as having a mother or all members of a species having similar dietary habits) are lawlike, while those that do not project or generalize (such as having a sister or having broken scales) are a matter of individual accident, but this is not particularly helpful, because lawlike properties are just defined as those that project or generalize.

Cognitive Approaches to Inductive Reasoning

These philosophical analyses of induction provide psychology with two basic empirical questions that need to be answered with respect to conceptual inference: (1) what is the psychological mechanism responsible for forming Hume's "habits of mind" from previously observed regularities (i.e., what structures of mind allow us to generalize from past experience to the future?); and (2) Goodman's question: what are the cognitive structures and mechanisms involved in determining whether a particular property is generalizable (or projectable)? While there may be no epistemological solutions to the problem of induction, we know that psychological and biological solutions do exist (because we do induction every day). The psychological task is one of discovering these solutions. Cognitive scientists have made some progress with respect to the first issue, but the second remains elusive.

In fact, much of the research program of the behaviorists addressed the issue of how minds make connections between past and future events. The examples we discussed in chapter 5 ranged from imprinting in some birds, to taste aversion in many animals, to my negative reinforcement experience leading to the formation of my early belief that advertisers are out to deceive me. Each of these examples exhibits Hume's "habit of mind" and can be explained in terms of associations formed through co-occurrence. However, we also saw that co-occurrence associations on their own are not sufficient to explain the logical, semantic, and conceptual relations that hold between propositional attitudes. Considerable work needs to be done to understand how these more complex relations can be built from neural networks sensitive to simple co-occurrence relations.

Psychologists study the Goodmanian question of what the relevant or projectable properties of events and entities are by showing participants inductive arguments and asking them to make judgments about the strength of coherence or plausibility relations between the premises and conclusion. These ratings are then used to gain insight into what factors may be driving coherence judgments. The two main ideas that have emerged are that the key factors in our ability to differentiate more projectable (generalizable) properties from less projectable ones may be similarity and causation. Let's consider similarity first.

The basic claim is that similarities between the subjects of observations increase confidence in the generalizability of applied predicates (and hence conclusions). In the following examples (Heit, 2007)

(F) Dogs have hearts;

∴ Wolves have hearts.

is judged to be a more plausible argument than

(G) Dogs have hearts;

∴ Bees have hearts.

because wolves are much more similar to dogs than bees are. So far, so good. But it is also the case that (Sloman & Lagnado, 2005)

(H) Robins require magnesium to live; ostriches require magnesium to live;

∴ All birds require magnesium to live.

is considered more plausible than

(I) Robins require magnesium to live; sparrows require magnesium to live;

∴ All birds require magnesium to live.

In both cases, we are generalizing to all birds. We consider robins to be more similar to sparrows than they are to ostriches. In (H), the dissimilarity between robins and ostriches (or diversity of observations) is strengthening the plausibility of the conclusion, while in (I) the similarity between robins and sparrows is weakening the plausibility of the generalization.

The other means of differentiating between regularities that humans are prepared to project and those that they are not is an appeal to causality. It is claimed that generalization or projection from one instance to another instance is warranted if the same causal laws or mechanisms underwrite both instances. Accordingly, subjects prefer the inference (Heit & Rubinstein, 1994)

(J) Hawks prefer to feed at night;

∴ Tigers prefer to feed at night.

to the inference

(K) Hawks prefer to feed at night;

∴ Chickens prefer to feed at night.

The explanation is that, despite the greater anatomical similarity between hawks and chickens, subjects are focusing on the underlying causal story that hawks and tigers are hunters and carnivores, while chickens are not. However, people also prefer the inference

(L) Hawks have a liver with two chambers;

∴ Chickens have a liver with two chambers.

to the inference

(M) Hawks have a liver with two chambers;

∴ Tigers have a liver with two chambers.

This is consistent with the similarity account. Subjects are back to focusing on the biological properties of chickens, hawks, and tigers and conclude that, in terms of anatomy, chickens and hawks are more closely related than hawks and tigers. Thus, what emerges from this line of research are a series of "principles" that presumably constrain the observed regularities that humans are and are not willing to project or generalize. However, the application of these principles seems to be ad hoc and subject to counterexamples.

More generally, reviewing the state of our understanding of inductive inference is a humbling experience. First, while many psychologists believe that similarity is a useful explanatory concept, philosophers and some psychologists recognize that it largely begs the question (Goodman, 1955, 1976; Sloman, 1996). Any two objects can share an infinite number of properties.[2] What matters for purposes of inductive reasoning is the identification of the *relevant* properties. The appeal to similarity in this literature is meant to explain the notion of relevance, but it seems that an independent notion of relevance is required to explain similarity.

There is something intuitively correct about appealing to causality. However, the problem is that causality seems to be a projection of the mind onto the world rather than an objective property of the world. This returns us to the circularity that Hume pointed out. The justification of causation requires a principle of uniformity that itself can only be established via an inductive inference.[3] Therefore, an appeal to causation in order to explain induction may be less than satisfactory.

Against this intellectual background, the cognitive research program on induction can be characterized as trying to understand the factors that contribute to the strength of the coherence relation in conceptual inferences. But interestingly, as we do not have a normative account of conceptual inference, we cannot have a normative standard for conceptual coherence. So the ability to recognize coherence relationships is not questioned (or studied). But nonetheless we do make many fallacious inferences in real-world reasoning. In the next section, we discuss some examples of such failures and the various underlying reasons for them.

Inductive Reasoning in the Real World

Rhetoricians and philosophers have identified a number of fallacies commonly found in everyday reasoning. For our purposes, it may be useful to organize them into three broad groups: (1) logical fallacies committed while drawing conceptual inferences; (2) fallacies resulting from complexities of collecting and interpreting data; and (3) "fallacies" of falling prey to a speaker's use of rhetorical devices to bypass coherence relations between premises and conclusion and trigger noncognitive systems.

Common logical fallacies include affirming the consequent, denying the antecedent, and circular reasoning. An instance of affirming the consequent is when, given the background belief that if it rains in the night, then the grass will be wet in the morning, and based on the observation that the grass is indeed wet in the morning, I conclude that it rained in the night. An instance of denying the antecedent is when, given the same background belief and observing that it did not rain in the night, I infer that the grass will not be wet in the morning. In both cases, there can be other reasons for the grass being wet. Both are logical fallacies, but both can sometimes result in reasonable and useful inductive inferences.

The fallacy of circular reasoning was illustrated in Hume's discussion of justifying inductive inference and also by the relationship between similarity and relevance. In the former case, the Principle of the Uniformity of Nature is necessary to justify inductive inference, but the principle itself can only be justified by inductive inference. In the case of similarity, it is usually explained in terms of relevance, but relevance in turn is discharged in terms of similarity. Strictly speaking, these are fallacies of formal logical inference, but they are often embedded in real-world conceptual inferences.

Many logical fallacies have been identified and studied in the heuristics and biases literature (Tversky & Kahneman, 1974) introduced in chapter 7. I will briefly mention three to give the uninitiated reader a sense for them. One popular one is known as the Gambler's Fallacy. After observing a long run of black on a roulette wheel, most people will believe that a red is now "due" and is more likely to occur. Actually, irrespective of the number of preceding stops on black, there's approximately a 50–50 chance that the next stop will be black (since the ratio of reds to blacks on the roulette wheel has not changed and previous rolls do not influence future rolls). A second fallacy results from the Availability Heuristic. If asked, "What is the risk of a heart attack among middle-aged men?," people will offer a much higher percentage if they have recently heard of someone middle-aged having had a heart attack. The Adjustment and Anchoring Heuristic provides the third example. If asked a question such as, "How many African countries are in the UN?" and given the probes

Is it more than 5? How many?

Is it more than 25? How many?

the cues provided (5 and 25, respectively) will be used to anchor the response. As a result, the answer provided in response to the second probe will be much higher than the number provided in response to the first probe.

Data collection strategies and judgments regarding interpretation are subject to numerous errors. For example, any prediction about the winner of a national election on the basis of interviewing 10 people will be unreliable. So-called sampling errors are another instance of such fallacies. Any prediction based on an unrepresentative (biased) sample (e.g., asking only liberal voters) will be unreliable. Furthermore, data always need to be interpreted. One common error is to confuse (potentially spurious) co-occurrence relations with causal relations, as in the earlier example of a 99% correlation between the consumption of margarine and divorce rates in the state of Maine. Much graduate school education in the natural and social sciences involves teaching students about experimental design strategies, statistical techniques to separate noise from data, subtleties of data interpretation, and the role of argumentation in drawing sound inferences.

There is a wonderful example from the history of science worth retelling to drive home the importance of argumentation in science. At some point in school, we all learned that Galileo simultaneously dropped two cannonballs from the Leaning Tower of Pisa, one of which weighed twice as much as the other. The prevailing wisdom, based on the writings of Aristotle, predicted that the heavier ball would fall twice as fast as the lighter one. Galileo, however, based on his studies of motion, predicted that they would reach the ground at the same time. Historians of science tell us that the incident of the Tower of Pisa probably never happened. Galileo did, however, build a number of inclined tracks along which he rolled metal balls of various sizes and weights, taking very careful measurements of speeds of descent. We are also taught that the two balls arrived at the ground (or the end of the track) at the same time, proving Galileo right and Aristotle wrong. Historians of science can confirm that this is also not true. The heavier ball actually arrives a little bit ahead of the lighter ball. So, on what basis can Galileo claim that he is right and Aristotle is wrong? He offers the following argument in the dialogue between Simplicio and Salviati in the *Two New Sciences* (Galilei, [1638] 1954, pp. 64–65):

Simplicio: Your discussion is really admirable; yet I do not find it easy to believe that a bird-shot falls as swiftly as a cannon ball.

Salviati: Why not say a grain of sand as rapidly as a grindstone? But, Simplicio, I trust you will not follow the example of many others who divert the discussion from its main intent and fasten upon some statement of mine which lacks a hairsbreadth of the truth and, under this hair, hide the fault of another which is as big as a ship's cable. Aristotle says that 'an iron ball of one hundred pounds falling from a height of one hundred cubits reaches the ground before a one-pound ball has fallen a single cubit.' I say that they arrive at the same time. You find, on making the experiment, that the larger outstrips the smaller by two finger-breadths, that is, when the larger has reached the ground, the other is short of it by two finger-breadths; now you would not hide behind these two fingers the ninety-nine cubits of Aristotle, nor would you mention my small error and at the same time pass over in silence his very large one.

This is a wonderful, but fallacious, argument. One might term it the Lesser of Two Evils Fallacy. Neither Galileo's nor Aristotle's predictions were fully supported by his experimental data. Galileo offers some reasons to prefer his account over Aristotle's. He is essentially saying, "I'm less wrong than Aristotle." Perhaps this is a reason, but it is not a great reason. Given that the data do not fit either account, both could be equally incorrect. It is possible that there is a third account that the data would fit. (Galileo was of course ultimately correct. He just didn't know about friction. Subsequently, scientists learned that the small discrepancy in Galileo's data was the result of friction. When it is controlled for, the two balls do arrive simultaneously, as he predicted.)[4]

Let's now extend this analysis of fallacies in conceptual inference to the very different realm of politics, particularly the first impeachment of the forty-fifth president of the United States, Donald Trump. On September 24, 2019, House Speaker Nancy Pelosi announced the launch of a formal impeachment inquiry against President Trump, focusing on whether Trump abused his presidential powers in soliciting help from a foreign government for his 2020 reelection (Przybla & Edelman, 2019). Specifically, the allegation was that Trump pressured President Zelenskyy of Ukraine, by withholding financial military aid authorized by Congress, to pressure him to build a case of criminal misconduct against his expected political rival in the 2020 presidential election, former vice president Joe Biden (and his son). The charge was based on a whistleblower's complaint. The next day, September 25, the White House released a partial transcript of a July 25, 2019, telephone conversation between President Trump and President Zelenskyy of Ukraine. Here are some key extracts from the transcript:

The President: . . . I will say that we do a lot for Ukraine. We spend a lot of effort and a lot of time. Much more than the European countries are doing. . . . I wouldn't say that it's reciprocal necessarily because things are happening that are not good but the United States has been very very good to Ukraine.

President Zelenskyy: . . . I'm very grateful to you for that because the United States is doing quite a lot for Ukraine. . . . I would also like to thank you for your great support in the area of defense. We are ready to continue to cooperate for the next steps specifically we are almost ready to buy more Javelins [antitank missiles] from the United States for defense purposes.

The President: I would like you to do us a favor though because our country has been through a lot and Ukraine knows a lot about it. I would like you to find out what happened with this whole situation with Ukraine, they say Crowdstrike [a cyber security company]. . . . The server, they say Ukraine has it [the Democratic National Committee's computer servers]. . . . I would like to have the Attorney General call you or your people and I would like you to get to the bottom of it. . . . Whatever you can do, it's very important that you do it, if that's possible.

President Zelenskyy: Yes, it is very important for me and everything that you just mentioned earlier. . . .

The President: . . . The other thing, there's a lot of talk about Biden's son, that Biden stopped the prosecution and a lot of people want to find out about that so whatever you can do with the Attorney General would be great. Biden went around bragging that he stopped the prosecution so if you can look into it. . . . It sounds horrible to me.

President Zelenskyy: . . . Since we have won the absolute majority in our Parliament, the next prosecutor general will be 100% my person, my candidate, who will be approved by the parliament and will start as a new prosecutor in September. He or she will look into the situation, specifically to the company that you mentioned in this issue. . . .

The President: . . . I will have Mr. Giuliani give you a call and I am also going to have Attorney General Barr call and we will get to the bottom of it. . . .

On September 26, 2019, the whistleblower complaint was publicly released. The complaint provided additional context and details of events and noted that White House officials were so concerned that they quickly intervened to "lock down" all records of the phone call.

Given the context and the telephone call, the question of interest, in terms of an impeachment inquiry, is whether the president abused the power of his office by asking a foreign government to intervene in the 2020 US presidential elections, by digging up or manufacturing dirt on his political opponents (the Democrats and former vice president Biden), and even more egregiously by pressuring the Ukrainian president to do so by withholding authorized military aid. This is a question for the rational mind. The evidence details a series of events and actions. We do not explicitly know the rationale for the actions. They must be inferred. How do the actions, including the telephone call, connect or cohere with the president's beliefs and desires?

The interpretation drawn by Democrats and major press outlets (Cochrane, Lipton, & Cameron, 2020) was roughly that the president withheld much needed military aid from Ukraine and, in effect, said to the president of Ukraine, "We have been good to you, but you have not reciprocated. . . . I want you to do me a favor. . . . Investigate the Democrats and Biden and his son . . . in cooperation with my people." But this is, of course, an inference based on the available evidence, and like all conceptual inferences, it may be incomplete or incorrect. There are many ways of connecting any given set of dots. In the legal world, the widely accepted standard of correctness (or, more accurately, plausibility) is what inference disinterested, reasonable men and women would draw from the facts.

A number of issues can be raised in such an inquiry: (1) one could question whether the list of reported events is complete and accurate (e.g., did a particular meeting occur, what was the specific language that was used?); (2) one could point out that the president did not explicitly say "if you don't do this I will suspend aid" or "I will release the aid only when you have done this"; and (3) one could raise the concern of co-occurrence versus causation. Just because withholding of the aid preceded the asking of the favor does not necessarily mean that it is causally related to the asking of the favor. Perhaps the aid was withheld for some other reason (e.g., Ukraine was not doing enough to root out corruption, or European countries were not giving enough aid) (Rupar, 2019). In this case, there would be no threat or quid pro quo that the release of the aid was contingent on the execution of the favor. These are all reason-based issues to consider and resolve in drawing conclusions from the preceding series of events.

With respect to the first issue, the accuracy of the data or evidence can be verified by documents and witnesses. With respect to the second, is it necessary to make a threat explicitly, or is a contextually implied threat equally culpable? For example, if someone holding a gun smiles and asks for a donation, and I hand over my wallet, was I robbed or did I simply gift my wallet? Judges and juries routinely make such determinations in criminal trials involving organized crime. With respect to the third issue, are there any other reasons to believe that the president would accept, much less request, election assistance from foreign governments? Is there any evidence that he has done this in the past? The findings of the Mueller report suggest he has. Is there any evidence that he would do so in the future? He has publicly stated there is nothing wrong in accepting "dirt on one's opponents" from foreign governments and said he would do so (ABC News, 2019). This would all be relevant context in evaluating claims and counterclaims.

A legal argument was eventually made during the Senate trial to the effect that if the president does something that he believes will get him reelected, and he believes that his reelection is in the public interest, then "that cannot be the kind of quid pro quo that results in impeachment" (Sherman, 2020). This strikes my legally untrained mind as absurd. Presumably, it would allow a president to deploy the military to California on election day to ensure that everyone voted for him, because he genuinely believes that is in the public interest. This is motivated reasoning but reasoning nonetheless (see chapter 13). It offers *reasons* for the president's actions and *reasons* why they are not impeachable. One can use the reasoning mind to question the coherency and merit of these reasons, as did many legal scholars.

The Reasoning Mind Recruits the Instinctive Mind

Interestingly, this is not the route the White House took to make the public case against any wrongdoing by the president. Rather, it intentionally and consciously utilized well-known reasoning fallacies to reach *below* the reasoning mind to the instinctive mind, betting that many, or even most, members of the electorate would not notice the absence of coherence in the arguments or indeed the absence of arguments. This was a calculated, rational decision. The rhetorical devices used included ad hominem attacks, deflection, arguments from consequences, false dilemmas, arguments from fear, and "fake news" arguments. Each is described here. The purpose in each case was to use framing and contextual devices—that do not actually affect the coherency of the argument (or lack thereof)—to activate instinctual biases, which will then trigger associated action tendencies.

Ad hominem arguments focus on properties (desirable or undesirable) of the individual making the argument rather than the argument itself. We saw a classic example in chapter 1, when Tania rejected the science of climate change by imputing malicious, mercenary motives to the scientists rather than examining the coherency relation between theory and data. In the impeachment case, the attacks were directed at Democrats, specific members of Congress, and the whistleblower. They ranged from (1) derision ("His [whistleblower's] term-paper report is laden with anonymously sourced rumors. . . . Scary references abound to the supposed laws that the legal-eagle whistleblower believes were violated" (Hanson, 2019)); to (2) charges of ulterior motives ("IG found whistleblower had political bias in favor of Trump 2020 rivals" (Pollak, 2019)); (3) to outright contempt ("Can you imagine if these Do Nothing Democrat Savages, people like Nadler, Schiff, AOC Plus 3, and many more, had a Republican Party who would have done to Obama what the Do Nothings are doing to me. Oh well, maybe next time!" (@realDonaldTrump, September 28, 2019)). Even if all these assertions are true, attacking the individual presenting the evidence or argument for impeachment has no bearing on the accuracy of the data and the coherency of the argument.

Deflection is another common strategy. Rather than addressing the evidence and argument, the conversation is directed elsewhere; for example, "Joe Biden and the corrupt Democrats are getting away with murder. . . . This is the real story." If true, it requires a separate investigation, but it has no bearing on the allegations against the president. Many Republicans argued that the real problem was that Trump's conversation with the Ukrainian president leaked. Judiciary Committee Chairman Lindsey Graham said it was imperative to find out which White House staffers talked to the whistleblower and why. Representative Devin Nunes expressed alarm that "a cabal of leakers are ginning up a fake story with no regard to the monumental damage they're causing to our public institutions and trust in government" (McLeod, 2019). Again, if laws were broken here, they need to be investigated, but this has no bearing on the evidence and the coherency of the argument for impeachment.

Another strategy is to focus on the consequences. For example, "If Trump is impeached, his voters will have been disenfranchised." Vice President Mike Pence noted that Democrats "keep trying to overturn the will of the American people" (Malloy, 2019). Such consequences may be extremely undesirable in a democracy but have little bearing on the coherency of the argument for impeachment, a constitutional remedy for "treason, bribery, or other high crimes and misdemeanors." The issue is whether the actions of the president meet the criteria for impeachment set down in the Constitution.

Setting up false dilemmas is another widely used strategy: "If you do not support Trump you are supporting the Democrats." This argument confines you to a false dilemma. One can fail to support Trump without supporting Democrats. There are other conservative contenders one can vote for. Even impeachment would have resulted in Vice President Mike Pence, a Republican, becoming president. The dilemma is not only false; it has nothing to do with the coherence relation.

A fifth strategy is the argument from fear: "If Trump is impeached and Democrats come into power, they will open the borders to rapists and drug dealers, bankrupt the country by giving handouts to immigrants, and of course come for your guns." Again, this has nothing to do with the coherence relationship between the evidence and conclusions under consideration.

A sixth strategy is simply to deny the given evidence and even make up "alternative facts." Many Trump supporters interviewed about the transcript and surrounding events simply said, "It didn't happen. It is fake news," or there was the headline "BREAKING: Trump Releases Transcript of Ukrainian Phone Call, Major Bust for Democrats" (Sabia, 2019). Why this technique should be so successful in an age of ubiquitous recording and instant fact checking is puzzling. It has nonetheless become an effective go-to strategy.

None of these responses address the coherence relation between evidence and conclusion, that is, the soundness of the argument for impeachment. They all commit common reasoning fallacies, but they do so intentionally, consciously, rationally! The official impeachment counteroffensive relied on the calculation that most of the MAGA faithful would fail to accept any evidence of wrongdoing by the President, if universal in-group/out-group instinctual systems could be activated. The group we belong to is always good, pure, innocent, and of course beloved of God; the out-group consists of elites, socialists, Muslims, and others trying to destroy us and our way of life for nefarious purposes:

> They're [Democrats] evil people. They want to abort babies up to the 9-month gestation age. That's sick and twisted, okay? They want to tell me that if I disagree with homosexuality because of my Christian views, that I hate gays. I don't hate gay people. I'm not out there trying to harm gay people. I don't hate gay people any more than I hate a murderer. (Hensley-Clancy, 2019)

If we can get people to strongly identify with one or the other group, we can get them to engage in an existential battle for the soul of the republic (or whatever is at stake) rather than examine the merits of the data and soundness of the arguments (which is not only more mundane but also more intellectually challenging).

*   *   *

There are significant limits in our understanding of conceptual inference. To begin, the epistemological or philosophical problem of conceptual inference may be unsolvable. Psychologists continue to struggle to understand the empirical basis of predicate generalization, with limited success. The main candidates considered thus far have been similarity and causality. The results have been mixed, both experimentally and conceptually. Experiments have shown that it is reasonably easy to find counterexamples to the principles of similarity and causality. On a conceptual level, there seems to be a degree of circularity in the utilization of these concepts. The third point of the chapter was to review and emphasize that conceptual inference is subject to a range of fallacies attributable to education, training, and heuristics.

But the most important issues for our purposes are raised in consideration of the impeachment debate. Here, behavior of the MAGA faithful was not driven by coherence relations. Rather, I claimed that *rational* strategies were used to activate *nonrational* instinctual systems, such as in-group/out-group bias, and that these instincts directly drove many people's behavior. Such a story cannot be told within the context of standard cognitive theories of reasoning because they do not recognize such noncognitive systems. We now turn to tethered rationality to provide a framework to accommodate such a story and return to complete the explanation of the impeachment response in part V (chapter 13).

IV The Tethered Mind

Many choose to ignore the likelihood that raw affective experiences—primal manifestations of "mind"—are natural functions of mammalian brains which could serve as key empirical entry points for understanding the experienced reward and punishment functions of the human mind. To proceed on this track, investigators would need to accept one grand but empirically robust premise—that higher aspects of the human mind are still strongly linked to the basic neuropsychological processes of "lower" animal minds.

—Jaak Panksepp

We now have some understanding of the different kinds of minds that underwrite behavior of human and nonhuman animals. With respect to human behavior, the reasoning mind is particularly important; little human behavior can be fully explained without it. I have associated reason with the cognitive mind and reviewed its scope, power, and limits. We have examined the types of behavior that it explains very well and those that it stumbles on. The latter cases have been presented as a series of anecdotal stories ranging from the behavior of teenage daughters to climate change denial and the Trump impeachment defense. In these "failed" cases, the problem is not irrationality but the absence of rationality. It is not that additional, more powerful machinery is required; ironically, it is that the reasoning mind is tethered to *simpler* machinery, and it is this tethering that we must appeal to for a more accurate accounting of behavior. Accordingly, I have proposed that most human behaviors incomprehensible just in terms of the reasoning mind are being modulated by the autonomic mind, the instinctive mind, and the associative mind. This leads to the postulation of the *blended response hypothesis*: that most human behaviors result from a blending of inputs from these various systems.

This part of the volume makes the positive case for the tethered mind. This requires actual data for the blended response hypothesis. Chapter 9 examines data from the economic decision-making literature and argues that it is consistent with the blended response hypothesis. Once we have behavioral data to support a tethered mind, chapter 10 makes the case for tethering as a natural byproduct of brain evolution: hierarchically organized and tethered behaviors are supported by hierarchically organized and tethered brain structures. Chapter 11 considers the puzzle of communications across these various systems. It introduces the idea that feelings—generated in evolutionarily old, widely conserved brain stem structures—are evolution's solution to initiating and selecting all behaviors, and provide the common currency for the four different systems to interact. All four systems contribute to behavior and the overall control structure is one that maximizes pleasure and minimizes displeasure (chapter 12).

9 The Instinctive Mind Resurrected: Modularity, Reciprocity, and Blended Response

It is Nature that causes all movement. Deluded by the ego, the fool harbors the perception that says "I did it."
—Bhagavad Gita (3:27)

He who understands baboons would do more towards metaphysics than Locke.
—Charles Darwin

Before we dive into the tethered mind, we need to address the small elephant in the room. As noted in chapter 1, there is a model of human behavior in evolutionary psychology called "massive modularity," which accepts the reality of instincts (referring to them as "modules") and argues that they are sufficient to explain all or most human behaviors. It is not widely accepted beyond evolutionary psychology, but if I'm to advocate for the inclusion of lower-level systems in models of human behavior, it is necessary to address this literature. It is a little bit of a detour because this model does not recognize a blended response; it argues that all human behavior—including reasoning—is based only or largely on instincts. This is a bold, counterintuitive hypothesis worth considering and evaluating, if only to set it aside.

I will begin by introducing and reviewing the now infamous Wason card selection reasoning task and Leda Cosmides's novel "cheater detection" instinct explanation for the pattern of results. The task and her interpretation of the results do offer a clear, simple illustration of how evolutionary psychologists think we "reason" with instincts. While I'm sympathetic to her interpretation, this task has become so embroiled in controversy that it can yield few uncontentious conclusions. It is also the case that no evidence was offered by Cosmides that "cheater detection" is actually an instinct. Therefore, I set aside this task (and massive modularity) to seek evidence for the existence of "cheater detection" instincts and their use in a broader range of reasoning and decision-making tasks.

Data from comparative animal research and experiments on young children provide evidence that cheater detection and the related traits of self-maximization, fairness, cheating, and punishment are indeed reasonable candidates for instincts, though all but self-maximization are largely confined to humans. I then turn to the work of a small group of behavioral economists and mathematical biologists who study cooperative monetary decision-making and explain their results as an interplay between these very traits of self-maximization, fairness, cheating, cheater detection, and punishment. On closer examination, it is apparent that learning, beliefs, and reason are also a critical part of their explanation. These data take us beyond anecdotal stories and provide experimental evidence for a model of tethered rationality, whereby reason interacts with instincts to guide human behavior.

Massive Modularity: Instincts All the Way Up

The term *massive modularity* is most closely associated with Leda Cosmides and John Tooby, of the University of California, Santa Barbara. Leda Cosmides (1989) began her academic career by giving an instinct-based account of a famous reasoning task, the Wason card selection task, named after its inventor, Peter Wason. The Wason card selection task involves four cards that are placed on a table in front of the participant (figure 9.1). In the standard version, the so-called no-content version, a letter is written on one side of the card and a number is written on the other side. The cards are placed such that two letters and two numbers are visible (figure 9.1a). The participants are also given a rule; for example, "if the letter on one side of the card is a vowel, then the number on the other side must be even." The four cards are placed on the table such that one vowel, one consonant, one even number, and one odd number are visible. The participant is asked to indicate which cards need to be turned over for verification of the rule. The task is a disguised form of conditional reasoning (if P then Q), with the choices of the cards corresponding to "P," "not P," "Q," and "not Q." It can be trivially completed by turning over all the cards. However, the instructions are to complete it by turning over as few cards as possible. Surprisingly, the accuracy rate on this task ranges from 10% to 25% (Goel, Shuren, Sheesley, & Grafman, 2004; Griggs & Cox, 1982). Everyone will select E to see if there is an even number on the other side (confirmation bias effect). Many people will stop at this point. Others will additionally select K or 4,

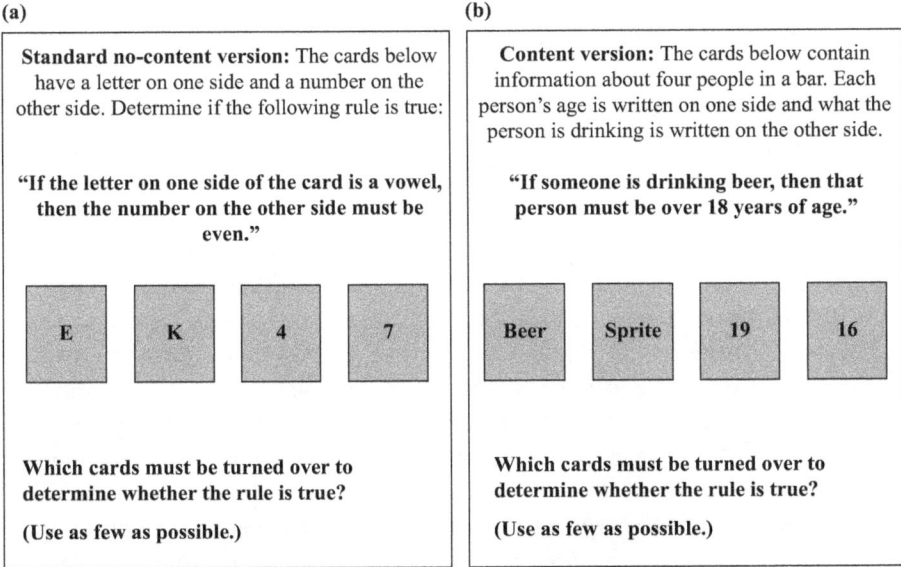

(a)

Standard no-content version: The cards below have a letter on one side and a number on the other side. Determine if the following rule is true:

"If the letter on one side of the card is a vowel, then the number on the other side must be even."

E	K	4	7

Which cards must be turned over to determine whether the rule is true?

(Use as few as possible.)

(b)

Content version: The cards below contain information about four people in a bar. Each person's age is written on one side and what the person is drinking is written on the other side.

"If someone is drinking beer, then that person must be over 18 years of age."

Beer	Sprite	19	16

Which cards must be turned over to determine whether the rule is true?

(Use as few as possible.)

Figure 9.1
Wason card selection task.

which are both irrelevant, but very rarely is 7 selected. The selection of E (corresponding to P) and 7 (corresponding to not Q) is the correct response.

Now consider the second version of this task (figure 9.1b). Four cards are again laid out on the table. On one side of each card is written the name of a beverage and on the other side a number indicating the age of the person drinking the beverage. The layout is such that the beverages "beer" and "Sprite" are visible, along with the ages "19" and "16." The rule that must be confirmed is, "if someone is drinking beer, then that person must be over 18 years of age." Again, the task is to determine whether the rule is being violated by turning over as few cards as possible. In this logically identical version of the task, accuracy rates jump to between 65% and 90% (Goel et al., 2004; Ragni, Kola, & Johnson-Laird, 2017). Under this condition, many participants correctly select "beer" and "16." Turning over the "beer" card allows them to confirm that the individual drinking beer is over 18 years of age. Turning over the "16" card allows them to confirm that no one under the age of 18 is drinking beer. Participants rarely turn over the "Sprite" and "19" cards. The rule places no restrictions on the age for drinking Sprite, so the selection of "Sprite" provides no relevant information. Similarly, turning over "19" provides no relevant information because the rule allows for people older than 18 to drink whatever they want. This task has spawned many hundreds of studies and at least a dozen different theories to become one of the most studied and contested tasks in the cognitive reasoning literature (Ragni et al., 2017).

The manipulation in this task falls under the content effect (chapter 7), but not all content has the same effect on all participants. For example, a rule involving content about different classes of mail and the price of postage stamps increased accuracy in participants in the United Kingdom but not in the United States (Griggs & Cox, 1982; Johnson-Laird, Legrenzi, & Legrenzi, 1972). Much of the literature on this task has been concerned with what aspects of content allow for dramatic increase in performance accuracy. Cosmides (1989) made the novel suggestion that what was in play here was not just a content effect but rather the activation of an instinct for detecting cheaters. People are aware that we have laws that establish a minimum drinking age for alcohol. If someone drinking alcohol is underage, they are breaking the law. Like many social, cooperative species, we have evolved adaptive "cheater detection" mechanisms that allow us to quickly identify anyone who might be breaking the law or violating a "fairness" norm (i.e., cheating). Notice that it is indeed the identification of the underage individual (the cheater) that accounts for the large accuracy swing. The situation simply triggers the cheater detection module—just as the swollen abdomen and the posture of the female stickleback unlocks the innate release mechanism of the male's mating behavior—and that generates the correct answer. It has nothing to do with coherence relations. Many cognitive psychologists have responded to Cosmides by questioning her data and interpretation (Atran, 2001; Carlisle & Shafir, 2005) or ignoring it (Ragni et al., 2017). I came to find the results and interpretation plausible only after observing my children interact when they were younger.

Based on such data, and steeped in evolutionary and computational ideas, Cosmides and Tooby (1994a, 1994b) went on to develop an account of cognition and reasoning different from that reported in chapter 6. Rather than a general-purpose reasoning system, based on coherence relations between proposition-like representational structures, they postulated a system of numerous domain-specific instincts or modules triggered by specific environmental cues. Each module is a specific adaptation. Adaptations are specific traits of organisms that arise based on how they improved the reproductive success of the organisms' ancestors. An organism is broken down into a collection of individual traits or adaptive solutions capable of solving specific problems, such as solicitation of assistance from parents, detection of safe and unsafe foods, coalition formation, cooperation, cheater detection, in-group/out-group formation, inference of intentions from facial expressions, incest avoidance, mate selection, object recognition, and spatial distribution of objects in the local environment, among others. Instincts embody these adaptive solutions. There is little else to the mind. It is instincts all the way up from the bottom to the top. Importantly, in the context of these modules, information is not thought of as beliefs with propositional contents, introduced in chapter 6.

The model of the mind that massive modularity presents is one where all behavior (including reasoning) emerges from the interaction of these various individual modules or instincts. Somehow, these modules, which were developed in response to the challenges faced by our hunter-gatherer ancestors in the Pleistocene environment, interact not only to allow us to avoid incest, select suitable mates, and detect cheaters but also to reason and assess how gravity affects the fabric of space and time. Such an account may be possible but remains unrealized.[1]

I believe evolutionary psychologists are correct in reminding us that instincts play as important a role in our behavior as they do in the behaviors of bats, beavers, and baboons. This obvious insight has great value and should be embraced. It needs to be part of any model that purports to explain real-world human behavior. However, the massive modularity model itself is a nonstarter for me because it cannot accommodate propositional attitudes and coherence relations. Without accounting for these, there is no accounting for reason. The initiated reader will know that there are deep conceptual issues in play here that have been widely discussed in the literature (Buller, 2006; Fodor, 2000). I register my own conceptual critique in the appendix of this chapter. The reader more interested in tethered rationality than in the conceptual issues surrounding massive modularity can usefully skip this appendix.

The balance of this chapter delves into the following questions: (1) What is the evidence that cheater detection and related traits are instincts? (2) Are there data from tasks less controversial than the Wason card selection task to support the role of cheater detection in human decision-making? (3) What sort of model of reasoning and decision-making do the data portend? I conclude that there are indeed good reasons to regard cheater detection (along with related traits of self-maximization, fairness, cheating, and punishment) as instincts and that data from financial decision-making tasks suggest that they are clearly involved in decision-making, but importantly they are modulated by reasoning systems. This allows us to start building a model of tethered rationality.

Reciprocity and Cheater Detection in Nonhuman Animals

Darwinian selection is based on the inherently selfish mechanism of "survival of the fittest" or, as Tennyson versified, nature being "red in tooth and claw." This implies that organisms will maximize resources (food, mates, shelter) for themselves rather than for others. It is easy to find examples of this on any branch of the phylogenetic tree. However, it is also the case that humans and many nonhuman animals live in socially organized groups and will help other members of the group even at a net cost to themselves. Classic examples are food sharing among vampire bats and sentinel duty and alarm calling (presumably at greater risk to oneself) among some birds and mammals. Prima facie such altruism is problematic for the theory of evolution.

In a seminal paper, "The Evolution of Reciprocal Altruism," Robert Trivers (1971) proposed a solution to this problem. He argued that though *seemingly* altruistic, cooperative behavior actually confers fitness benefits to the donor (altruist) because it is rendered with the expectation that the recipient will reciprocate in the future. For the practice to flourish, a number of conditions need to be met, including (1) that individuals be nontransient and live in stable social groups (to maximize the number of opportunities for donors to be reciprocated); (2) that the social groups have flat dominance relations (dominant individuals can take what they want without reciprocating); (3) that individuals be able to recognize other individuals and retain memory of past interactions; (4) that the recipient must reciprocate at least once; (5) that the benefits to the recipient must be greater than the cost to the donor; and (6) that donors must be able to recognize cheaters and expel them from the system. Without this last condition, the recipient of the aid could enjoy the benefits without having to reciprocate in the future and would be at a net advantage and the donor at a net loss. Reciprocators would disappear in such a system. Hence, the evolution of an ability to recognize and punish cheaters is necessary to maintain the stability of reciprocity (Wade & Breden, 1980).

Trivers's account resulted in a nice evolutionary story of a behavior rooted in biology and widely available across the phylogenetic tree. It provided Cosmides with the cheater detection explanation of people's behavior in the Wason card selection task. But subsequent research suggests that the story is not so simple. The data indicate that helping behavior in nonhuman animals is not reciprocal altruism. It is either kin based (i.e., the assistance is being offered to genetically related individuals) or does not meet Trivers's criteria for reciprocity (de Waal & Brosnan, 2006; Riehl

& Frederickson, 2016). In these cases, no cheater detection mechanism is required. Does such a mechanism actually exist?

Certain mammals, such as meerkats and Belding's ground squirrels, perform sentinel duty as members of a group. One would think that such activities are altruistic in that, to help the group, the individual is increasing its own chance of predation. A careful examination suggests otherwise. In the case of meerkats, once an individual is well fed, sentinel duty allows the animal to benefit from early detection of danger and actually reduces its risk of predation compared to other members of the group (Clutton-Brock, 1999). For Belding's ground squirrels, the picture is a little bit more complex. When they detect an airborne predator, such as a hawk, they "whistle." When they detect a mammalian predator (e.g., a weasel or coyote), they emit a "trill." When they emit a whistle to warn against airborne predators, they are attacked 2% of the time, compared to 28% of the time for nonwhistlers, again greatly reducing their chance of predation. So, both sentinel duty and whistle calls, despite appearances, are actually selfish acts that increase individual fitness, consistent with the Darwinian account. However, when Belding's ground squirrels emit a trill to warn against mammalian predators, they are attacked 8% of the time, compared to 4% of the time for nontrillers. In this case, the individual trilling squirrel is putting itself in harm's way to help conspecifics. But this behavior is actually driven by kinship relations (Sherman, 1985).

Even kin-based altruism was problematic for the notion of direct (or individual) fitness in the Darwinian theory of evolution, but it was nicely reconciled by several important developments in the neo-Darwinian update that incorporated knowledge of the gene and in the 1960s led to the insight that the unit of selection in evolution was not the individual, group, or species but rather the gene (Hamilton, 1963, pp. 354–355):[2] "Despite the principle of 'survival of the fittest' the ultimate criterion which determines whether [a gene] G will spread is not whether the behavior is to the benefit of the behaver, but whether it is to the benefit of the gene G."

The individual was simply the container for the gene. Based on this reformulation, William Hamilton (1964a, 1964b) redefined the notion of fitness as "inclusive fitness" and proposed kinship theory. Inclusive fitness equals direct fitness (individual reproductive success) plus indirect fitness (reproductive success of relatives the altruist has aided). Hamilton argued that inclusive fitness can be maximized by helping genetically related individuals, even at a cost to oneself, and proposed the following rule: altruistic behavior will spread if $rB > C$, where r = coefficient of relatedness, B = fitness benefit to the beneficiary, and C = fitness cost to the altruist.[3]

An example of reciprocal altruism that meets some of Trivers's criteria but fails others is provided by food sharing among vampire bats. Vampire bats feed only on blood. Approximately 8% of adults will fail to feed successfully on any given night. If a bat fails to feed on two or three consecutive nights, it will starve to death. Food sharing through regurgitation of blood meals from a successful individual to an unsuccessful individual is commonly observed in vampire bats (Wilkinson, 1984). Much of the sharing is between mothers and pups and other related female individuals (Carter & Wilkinson, 2013). However, sharing also occurs in the case of familiar but unrelated individuals (Denault & McFarlane, 1995). This practice seems to meet a number of criteria for reciprocal altruism: vampire bats live in social groups, the females are nontransient and have weak dominance hierarchies, and the benefit to the recipient bat is greater than the cost to the donor bat (because the former could otherwise die). However, there is no evidence that vampire bats can distinguish kin from associates, nor is there any evidence that they can track cooperative returns (i.e., detect cheaters) and base future donations on these returns (Carter & Wilkinson, 2013).

But what about more encephalized animals such as nonhuman primates, particularly chimpanzees, our closest living relatives? Chimpanzees live in social groups and cooperate with unrelated partners. Males cooperate in patrolling territory, hunting, sharing food, grooming, and joint mate guarding, and even form within-group coalitions for aggressive actions against other members. Despite all these prosocial behaviors, in experimental settings they will not volunteer to help another familiar but unrelated individual obtain food, even at no cost to themselves (Silk et al., 2005). In one study, they show a small increase (5.7%) in sharing food with a familiar individual who has groomed them within the past two hours compared to an individual who has not groomed them within this time period (de Waal, 1997). In other studies, they fail to show that their altruism is conditional on reciprocity.

Sarah Brosnan and her colleagues (2009) carried out an experiment on captive chimpanzees to determine whether they would more readily share food with a partner from their home group who had shared food with them on previous trials versus partners who had not shared food with them. Individuals familiar with each other were tested in pairs. One individual was offered a choice between two options: (a) deliver a food reward to themselves and another equal one to the other individual (prosocial behavior) or (b) deliver a food reward to themselves and nothing to the other individual (selfish behavior). On the next trial, the other individual was offered the same choice. The trials were repeated a number of times. Interestingly, the choices individuals made were not affected by the choices that their partners made in previous trials. That is, any food cooperation was not contingent on previous interactions. Several leading primatologists now agree that reciprocal altruism (and hence cheater detection) is nonexistent among nonhuman animals, even including nonhuman primates (de Waal & Brosnan, 2006; Stevens & Hauser, 2004). It may be something unique to humans.

Frans de Waal and Sarah Brosnan (2006) propose three levels of reciprocity, of which the first two can be found among nonhuman animals and the third seems exclusive to humans: symmetry-based, attitudinal, and calculated or contingent reciprocity. The simplest form, symmetry-based reciprocity (i.e., "we are friends"), requires that both parties behave similarly with each other; it is based on existing relationships such as kinship, group membership, alliances, and similarity in age. It does not require scorekeeping. There is a very low degree of contingency. The altruistic behavior of meerkats, Belding's ground squirrels, and vampire bats would fall into this category. By contrast, attitudinal reciprocity requires that an individual's willingness to cooperate covary with the recent attitude of the partner ("if you are nice, I will be nice"). Both parties may not benefit simultaneously, but the requirement of scorekeeping is minimal. The contingency is immediate. The exchange is based on "general social disposition rather than specific costs and benefits" (Brosnan, de Waal, & Proctor, 2014, p. 24). The altruistic behavior of chimpanzees reported here would fall in this category. Finally, in calculated or contingent reciprocity (Trivers's reciprocal altruism), individuals expect reciprocation of at least equal value, though allow for significant time lags. If expectations are violated, cheaters will be punished.

In addition to the prerequisites identified by Trivers—the benefit to the recipient greater than the cost to the donor, opportunity for repeated interaction, reasonably flat dominance hierarchies, and the cheater detection mechanism—full-fledged contingent or calculated reciprocity requires quite sophisticated cognitive abilities, such as recognition of individuals, memory of previous events, scorekeeping, numerical discrimination, and even temporal discounting. The only robust examples of it occur in humans.

The task used to test for calculated reciprocity in chimpanzees can also be used on young children. It is reported that children 3–4 years of age choose like the chimpanzees (House, Henrich, Sarnecka, & Silk, 2013). They usually choose not to share, and this choice is independent of whether the other child shared with them on previous trials. Children 5–7 years of age, on the other hand, make the prosocial choice (70% of the time) when their partners have made the prosocial choice on previous trials. Children of this age are beginning to show signs of contingent or calculated reciprocity.

According to these data, any claims that calculated reciprocity and cheater detection are evolutionary traits widely dispersed along the phylogenetic tree are false. Contingent reciprocity does have simpler precursors (i.e., symmetry-based and attitudinal reciprocity)—that do not require cheater detection—but fails to present itself in unambiguous form even in our closest living relatives, where we might well expect it. It is possible that it arose only on the hominina or even homo branch of the phylogenetic tree, piggybacking on increased cognitive capacities, perhaps even propositional attitudes. Indeed, Trivers's original description is replete with appeals to propositional attitudes and other sophisticated cognitive and emotional systems largely confined to humans. The failure to find robust calculated reciprocity in nonhuman animals does not preclude it as a candidate for an instinct. However, the fact that it does not arise in humans until five years of age suggests a period of maturation and/or socialization. There may also be some more basic instinctual systems that feed into it.

Reciprocity in Humans: Self-Maximization, Fairness, Cheating, and Punishment

Any Darwinian model of human behavior must begin with the selfish maximization of resources. However, resources can sometimes be multiplied exponentially through mutual cooperation. A single individual may be able to hunt a rabbit or build a hut. A cooperating group of individuals can bring down a mammoth and build Chartres Cathedral; the group result may be greater than the sum of individual efforts. The starting point for any model of human cooperation needs to be based on sharing of effort and rewards. This assumes and requires a sense of *fairness*. While *self-maximization* of resources is widely present along the phylogenetic tree, fairness may be unique to the hominina or even homo branch. Human cooperative behavior is an interplay between fairness and self-maximization. Unchecked self-maximization will lead to a violation of fairness (i.e., cheating). Unabated cheating would result in a breakdown of cooperation. Fairness (hence cooperation) is maintained by not only *detecting* cheaters but also actively *punishing* them (even at a cost to self). Computational models suggest that such interacting systems can result in stable cooperation (Axelrod & Hamilton, 1981; Boyd, Gintis, Bowles, & Richerson, 2003; Fowler, 2005).

Are these traits learned or are they instincts? It has long been a tenet of Western society that these concepts, particularly fairness, are cultural and social, even religious, constructs. If this is the case, they must be learned, will emerge late with socialization, and will correlate with individual and societal variations in beliefs. The instinct-based view advocated by evolutionary psychologists is that these notions are innate constructs and a common heritage of at least *Homo sapiens*. There is indeed empirical evidence to suggest that self-maximization, fairness, cheating, and punishment are all adaptations or instincts. They have not been arrived at through socialization (i.e., learning and reasoning).

The most interesting data in support of the innateness view of fairness are emerging from the study of young infants. Infants as young as 19 months old expect resources and rewards to be divided equally between two individuals. In a game-playing scenario with pairs of infants, 21-month-olds expect the experimenter to distribute rewards equally when both infants worked to complete the task but not when only one worked at the task and the other played another game (Sloane, Baillargeon, & Premack, 2012). It is very difficult to argue that cultural and social influences are driving the behavior at this early stage.

Three-year-olds share more equally with a collaborating partner than with a freeloader (Melis, Altrichter, & Tomasello, 2013). Three- and four-year-old children engaged in collaborative tasks objected to inequitable reward distribution, even when it favored themselves, and in such cases equalized rewards by transferring some of theirs to the partner. Chimpanzees performing the same task are insensitive to the inequity and are only concerned with maximizing their own resources (Ulber, Hamann, & Tomasello, 2017). These data indicate that the concept of fairness emerges very early, prior to extensive cultural socialization, and is thus best considered innate or instinctive. Furthermore, if the concept of fairness emerges so early, and fair-minded individuals can be taken advantage of by cheaters, emergence of cheating should follow. This is indeed the case.

While children seem to understand the concept of fairness at a very early age, they do not always follow the principle when their own resources are at stake (i.e., they often cheat). When children three to eight years old were given stickers and asked to share with children who did not receive any, they all said giving them half the stickers would be fair. When it actually came time to share, the seven-to-eight-year-olds did share half their stickers, but the younger children gave fewer than half their stickers (Blake, McAuliffe, & Warneken, 2014). The result with the older, more socialized children shows that instinctive biases can in certain situations be modulated (to varying extents) by social, belief-based factors.

In another study, children ranging in age from 5 to 15 years were asked to toss a fair coin and privately record the results (black or white). The children were to be rewarded based on the number of white trials they reported.

They were told that the experimenters would not check the actual tosses but rather just take their word for it. Statistically, one would expect approximately 50% white trials. The children reported on average 85% white trials, well above the expected 50% but also below 100%. There were no age or sex differences. The experiment was then repeated with a prior admonition not to cheat. Here the overall white responses were reduced by 13% in boys and 36% in girls (Bucciol & Piovesan, 2011). Interestingly, the admonition dampened but did not eliminate the cheating. This speaks to a role for socialization (learning and reason-based beliefs) in shaping cheating behavior and also to the limits of socialization. The fact that cheating behavior cannot be eliminated by socialization speaks to some innate components.

Given that the notion of fairness and the propensity to cheat develop very early and universally, and seem to be only modestly affected by socialization, the development of cheater detection and punishment should not be far behind. Consistent with this expectation, it is reported that children two to three years of age can understand normative rules, as in the structure of simple games, and detect violations (i.e., detect cheaters) (Rakoczy, Warneken, & Tomasello, 2008). In the context of moral transgressions by a third party, three-year-old children can detect such violations and even intervene by tattling on the transgressor (punishment) and behaving more prosocially toward the injured party (Vaish, Missana, & Tomasello, 2011). These data are consistent with the evolutionary psychologists' claim that these traits are instincts. The evolutionary psychologists would further argue that we should be able to explain human cooperative decision-making as the interaction of these various traits or instincts. To evaluate this claim, we now consider how these traits actually come into play in cooperative decision-making. That is, is the massive modularity model—of just interacting instincts—sufficient to account for the cooperative decision-making data, or do we need to introduce learning and reason to explain the data?

Tethered Rationality: Blend of Instincts and Reason in Cooperative Economic Decision-Making

A small number of economists and mathematical biologists have recently traded in the exclusively self-maximizing *Homo economicus* model of decision-making for models based on the actual study of human nature. Prominent among this group are Martin Nowak of Harvard University, and Ernst Fehr of the University of Zürich, and their colleagues. They accept the innateness of the mechanisms of self-maximization, fairness, cheater detection, and punishment, with some even including the notion of trust

(Berg, Dickhaut, & McCabe, 1995; Ortmann, Fitzgerald, & Boeing, 2000), and explore their interaction in cooperative monetary decision-making. In fact, some of these economists expand Trivers's notion of contingent or calculated reciprocity, where we reward or punish if it is in our long-term self-interest, to a notion of "strong reciprocity," where we will bear the cost of rewarding and punishing even in the absence of any long-term benefits to ourselves (Fehr & Fischbacher, 2003). While it remains unclear how this extension reconciles with the theory of evolution, it is an interesting conjecture.[4]

The question of whether human reciprocity is adequately characterized as contingent reciprocity or strong reciprocity is interesting but not particularly germane for our purpose. Either way, there are in place a set of adaptations or instincts guiding our cooperative behavior. Numerous studies characterizing human cooperative monetary decisions as a function of these traits have been undertaken. I will suggest that the valuable data and insights that they have generated are best accommodated by a model of tethered rationality where human cooperative choices are a blend of beliefs, coherence relations, and instincts.

Human decision-making, specifically choice in cooperative resource-allocation situations, is studied by economists through the use of a handful of simple games. Four such games are the Dictator Game, the Ultimatum Game, the Trust Game, and Social Dilemma Games. In the Dictator Game, there are two players. One player (called the donor) is endowed with a sum of money and instructed that he can keep it all or share a portion with the other player (called the recipient), who has received nothing from the experimenter. The recipient must accept whatever (if anything) is offered by the donor. Because the donor did nothing to earn the reward, fairness would dictate that he or she share half the money, whereas the self-maximizing choice would be to give the recipient player nothing. When the game is played as a single shot (i.e., no repetitions) and with actual money, there are significant individual differences among players: 40% of donors will choose to keep all the funds and only 20% will share equally with the recipient player, with others sharing smaller amounts (Forsythe, Horowitz, Savin, & Sefton, 1994). These results show similar trade-offs between fairness and self-maximization, with a preference for the latter, as noted in the children's data.

The results can be shifted dramatically and reliably under certain conditions. When the game is played with imaginary money, 80% of the donors will share 40%–50% of the funds with the recipient (it costs them nothing) (Forsythe et al., 1994). If there are repeat trials of the game, with the donor and recipient alternating, then donors become even more "generous," because they know they will be at the receiving end in the next trial and will have to deal with the consequences of their reputation for defecting or cooperating (Berg et al., 1995). *Reputation* is a critically important *belief-based* factor, discussed shortly. Generosity can also be manipulated by instilling certain beliefs in the donor about the recipient (for example, the recipient is dying of cancer or has recently insulted the donor) (Eimontaite, Nicolle, Schindler, & Goel, 2013; Eimontaite et al., 2019). In these manipulations the instilled beliefs—albeit with important emotional components—modulated the outcome of the choice. Beliefs do not even have to be explicitly instilled. A manipulation whereby the two players spend a few minutes silently looking at each other increases generosity in single-shot games compared to totally anonymous interactions (Bohnet & Frey, 1999). This process allows not only for the humanization of the other player but also for identification for future interactions (i.e., it raises concerns about reputation).

The Ultimatum Game also involves two players (donor and recipient), with only one (the donor) receiving an initial sum of money. However, there is an interesting twist. The donor must offer some of the money to the nonreceiving (recipient) player. If the recipient accepts the offer, then they both get to keep the allocated funds. If the recipient rejects the offer, both walk away with nothing. The self-maximizing choice for the donor is to offer as little as possible to the recipient (as in the Dictator Game). The self-maximizing choice for the recipient is to accept any nonzero amount offered, because even if only 1¢ is offered from $10, that 1¢ is greater than the alternative of 0¢.

In actuality, any offer less than 25% of the original amount is roundly rebuffed. Only offers around the 40%–50% mark are routinely accepted. This is an instance of fairness being enforced by cheater detection and punishment. The recipient detects a violation of fairness (i.e., cheating) in low offers and punishes the donor player at a cost to themselves. They would rather have nothing—and have the donor receive nothing—than accept an unfair offer. In anticipation of this response, donors usually offer something in the 40%–50% range. Interestingly, in the case of the Ultimatum Game, it seems to make no difference whether the game is played with real money or imaginary money, presumably because of the presence of the real threat of punishment by the recipient (Forsythe et al., 1994). This is an example of a self-maximizing choice being rejected in favor of punishing the cheating behavior.

Punishment is costly. In the preceding example, in order to punish the donor, the recipient has to forgo whatever amount the donor offers. What is even more interesting is that we will expend resources to punish not just those who cheat or harm us (or could potentially do so in the future) but also those who cheat others. This speaks to our strong sense of fairness. This can be illustrated by a modified version of the Dictator Game that includes three players: a donor endowed with the money, a potential recipient, and a third party. The donor is endowed with $10, the potential recipient with nothing, and the third party with $5. The donor may give whatever he wishes to the recipient player. Once the transfer is made, the third party is informed that they can spend money to punish the donor if they so wish. Every dollar spent by the third party reduces the income of the donor by $3. Where the donor has violated fairness norms, the third party will use some of their own funds to punish them (Fehr & Fischbacher, 2004). Again, this is not immediately self-maximizing for the third party but speaks to the important role of cheater punishment in our behavior. Empirical data (Dreber, Rand, Fudenberg, & Nowak, 2008; Fehr & Gächter, 2000) and computational models (Boyd et al., 2003; Fowler, 2005) suggest that punishment (even at a cost to oneself) is critical for maintaining cooperation based on reciprocity. But there may also be a reason-based calculation involved, specifically that long-term punishment may lead to formation of a reputation as a "punisher," thereby reducing the probability of being cheated in future interactions (Hilbe & Traulsen, 2012). Other evidence of reason-based modulation of punishment behavior includes calculation of cost-benefit trade-offs such that people are more likely to punish if the cost of the punishment is less than its consequences (Egas & Riedl, 2008).

Cheating and punishment instincts emerge very early and are ubiquitous in human social affairs. In every newspaper around the world, we will find stories such as the following (Ingalls, 2011): "Woman in Washington State living in a million dollar waterfront mansion with her Jaguar driving chiropractor husband, receives monthly welfare assistance of $1272 for housing, federal and state payments for a disability, and food stamps." We pay attention to such stories. We become outraged. We demand punishment for the cheaters. This is all in line with the adaptation story. When the cheaters belong to an out-group, such as immigrants, we are extra incensed, demand greater punishment, and generalize more broadly (e.g., "all immigrants are welfare cheats"). This may result from interaction of the cheater detection adaptation and the out-group aversion adaptation (chapter 13). Reactions to instances of cheating may be instinctive but can clearly be modulated by beliefs, as predicted by the model of tethered rationality.

Beliefs about the trustworthiness of other members—known as their *reputation*—are a critical factor in cooperation and punishment (Milinski,

2016). An individual's reputation is established by their history of choices in previous interactions. It is transmitted either directly (through firsthand knowledge of previous choices) or indirectly, via language or some other system of symbols. Either way, potential donors are less likely to help (i.e., more likely to punish) those who have previously violated fairness norms. In repeated game interactions, where both individuals know that the other will have knowledge of their past interactions, cooperation rates rise dramatically. In fact, individuals are aware of the value of a good reputation and will expend resources to gain one. In an experimental situation where an individual has the possibility of developing a positive reputation, the cooperation rate rises from 37% to 74%. Reciprocity and a good reputation reinforce each other (Gächter & Falk, 2002).

Reputation also modulates punishment. Consider the differences in how corporate tax avoidance is viewed and punished in the United States compared to single mothers collecting housing vouchers and food stamps while holding an unreported secondary job. A recent study by Oxfam America (2020) reports that between 2008 and 2014, the 50 biggest US companies received $27 in federal loans, loan guarantees, and bailouts for every dollar they paid in taxes. Each dollar that the biggest companies spend on lobbying is associated with $130 in tax breaks and more than $4,000 in federal loans, loan guarantees, and bailouts. Another study by economists estimates that tax avoidance by major corporations costs the US Treasury up to $111 billion a year (Clausing, 2016). Interestingly, these facts rarely make the front page of most newspapers or incense most of us.

Corporations are run by humans. *Homo sapiens* will cheat, to some extent, if they can get away with it, irrespective of whether they are corporate CEOs or welfare recipients. The question is, why doesn't corporate cheating activate our cheater detection and punishment modules to the same extent as a single mother welfare recipient working an unreported side job? Americans have been raised to have different beliefs about corporations and single mothers on welfare, which either attenuate or accentuate the triggering of the relevant instincts. For example, we are taught that corporations are the backbone of society. They provide jobs. They grow food, build cars, and provide health insurance, among other things. Corporations are good. They spend billions of dollars shaping our beliefs, and it works (Wu, Balliet, & Van Lange, 2016). Single welfare mothers do not have the lobbyists to explain why it might be necessary to hold down a couple of side jobs while claiming welfare assistance to pay the rent and buy food (Kohler-Hausmann, 2007), so they are stuck with the following reputation (Feagin, 1972): "About welfare? What do I think about welfare? It ought to be cut back. The goddamn people sit around when they should be working and then they're having illegitimate kids to get more money. You know, their morals are different. They don't give a damn."

In both cases, whether it is corporate CEOs lying to get around regulations in order to increase profits or single mothers on welfare lying about holding a second job, it is cheating and should activate our detection and punishment systems equally. However, because these systems are modulated by reputations, which are often in the form of beliefs—and corporate CEOs have much better reputations than welfare mothers—welfare cheats are much more likely than corporate CEOs to go to jail.[5] These modulations can both amplify and attenuate cheater punishment, *but they cannot eliminate it.*

We will not only punish cheaters but also reward those who play fair. Consider the Trust Game. There are again two players. In this case, both players are given an equal amount of money. The first player has the option of transferring some arbitrary portion of his money to the second player, with the understanding that the experimenter will triple any amount that is transferred. The second player can then decide whether to keep all the funds or send a portion back to the first player. If the first player decides not to transfer any funds to the second player, each player keeps the initial funds. However, given the tripling rule, the self-maximizing choice for the first player is to transfer all their funds to the second player, as long as the second player then transfers half the tripled amount (or at least more than they received) back to the first player. This way, both players come out ahead. But there is a danger. What if the second player violates fairness and keeps everything for himself? If funds are transferred, it is immediately self-maximizing for the second player to keep all the proceeds and not send anything back. In this situation, the self-maximizing outcome is distant for the first player and relies on fairness, while the self-maximizing outcome for the second player is immediate and relies on cheating. Despite this, even in single-shot games, most players choose to make a substantial transfer, and the transfer back (i.e., reward) made by the second player correlates with the amount of the initial transfer (Eimontaite et al., 2013; Fehr & Fischbacher, 2003). Any knowledge, beliefs, and perceptions about the trustworthiness of the other player (i.e., their reputation) modulates the initial transfer amount as well as the returned amount (Berg et al., 1995). Repeat trial games automatically generate such knowledge and, of course, affect trust.

Tragedy of the Commons

Some of the most intractable societal problems involve allocation of public goods and take the form of a social dilemma that Garrett Hardin (1968) famously labeled the "tragedy of the commons." These are all problems of exploiting common resources for selfish ends. The problem of initiating action to combat global warming is a classic example. The problem of maintaining universal healthcare schemes is another example. Both these problems were introduced in chapter 1. However, I will undertake the discussion of the tragedy of the commons with an analogous historical example where we have the advantage of 20/20 hindsight: the collapse of the Canadian Maritimes fisheries.

One of the greatest fish resources in the world was found on the Grand Banks, off the coast of Newfoundland, Canada. For 500 years, European vessels plied these waters to exploit the resource. It seemed endless. Based on self-maximization, each fisherman should maximize his take. Every extra fish means extra income. According to the logic of Adam Smith's "invisible hand" doctrine, individual self-maximization would be "led by an invisible hand to promote . . . the public interest" (quoted in Hardin, 1968, p. 1244). What could go wrong?

In a world of infinite resources (or where the amount removed from the resource is always less than or equal to the replenished amount), this might be reasonable advice. In the actual world, every resource is finite. As technological advances in fishing dramatically increased the catch of individual fishermen, individual trawlers, and individual corporations, they all did become dramatically wealthier, as did the Canadian Maritimes as a whole (along with communities in Iceland and Portugal). In 1968, the cod catch from the Grand Banks was 810,000 tons. In 1974, it dropped precipitously to 34,000 tons! Seemingly overnight, the cod population totally collapsed because of overfishing. This left 40,000 fishermen and related workers unemployed and financially decimated the Canadian Maritime Provinces (Pilkey & Pilkey-Jarvis, 2007). If every individual had limited their catch in line with the available resource, they and their children would still be fishing and prospering today, as would the Maritimes as a whole.

This is the tragedy of the commons. It constitutes a social dilemma where an individual receives a higher (self-maximizing) benefit for the socially noncooperating choice (e.g., overfishing, using excess energy, polluting, accessing healthcare without paying enough into the system), irrespective of what others do; however, everyone is better off if everyone cooperates. If not enough people cooperate (i.e., they take out too much or don't pay enough into the system), the resource will be depleted. In this case, everyone loses. Individual interest is at odds with the group interest.[6] The dilemma requires that the payoff matrix be as follows:

1. Payoff to the defectors > payoff to cooperators

2. Payoff for universal defection < payoff for universal cooperation

The situation can be modeled (poorly) in simple experiments such as the following. Take 10 players and give each $10. They are then given the opportunity to invest some or all their funds and are told that the return on investment will be distributed equally among the group. They privately place their contribution in an envelope. The experimenter collects the envelopes, multiplies the total of all the envelopes by five, and distributes the new total equally among all the players. If everyone contributes their $10, for a total of $100, everyone will take home $50 (after the $100 is multiplied by five and the $500 is distributed equally among the 10 players). However, if one player contributes nothing and the others contribute $10, for a total of $90, which is then multiplied by five to become $450 and redistributed, the noncontributing player will take home $55 ($10 plus $45) and the others will take home $45 each. Therefore, *Homo economicus* should contribute nothing in this situation. In every scenario in which the multiplier is less than the number of participants, the noncontributing player will come out ahead by contributing nothing or less than others. But if no one contributes, everyone will lose, as did the individual fishermen and the Maritime Provinces when the cod fisheries collapsed.

These experiments demonstrate a great deal of individual variation. More people than might be expected by the self-maximizing principle actually cooperate, but approximately 30% start as freeloaders, and this percentage increases to 80% or 90% by the tenth round of the game, leading to a rapid decline in cooperation and a depletion of the common resource (Isaac & Walker, 1988). Unsurprisingly, rates of cooperation increase with the introduction of punishment (Fehr & Gächter, 2000). Reputation also helps (Milinski, 2016). Our knowledge and beliefs regarding what others are contributing increases our own contribution, perhaps by triggering the fairness instinct. Interestingly, it does not affect the number of free riders (approximately 30%), but 50% of participants match their contributions to what they believe others are contributing, while 14% match contributions up to a certain point and then decline. The net overall result is a positive contribution, and the common resource is sustained (Dawes, 1980; Fischbacher, Gächter, & Fehr, 2001). Another important factor is the determination of the payoff matrix. One needs to understand the situation to understand the payoff matrix. Is it really advantageous to defect? How severe is the cost of group failure? Is it really the case that more is being taken out than put back in? For instance, if the multiplier in the preceding example is greater than 10 (i.e., greater than the number of participants), it is more advantageous to cooperate. These modulating factors speak to the contribution of knowledge and the rational mind in decision-making.

If cooperation can be sustained by these reason-based modulations (at least in artificial scenarios), how did a wealthy, technologically advanced country like Canada succumb to the tragedy of the commons? Reason was employed. When questions of sustainability of the harvest arose, steps were taken to eliminate foreign participants from the fishing grounds by extending the coastal boundary line from 3 miles to 200 miles, and the best available science and technology was used to find the sustainable limit; that is, to find the actual payoff matrix. Fisheries experts from the Department of Fisheries and Oceans were consulted. They used sampling techniques to estimate the current cod population and used mathematical models to project future population levels. Based on these models, the Department of Fisheries and Oceans advised the government that imposition of proper catch limits would allow a recovery of the population within 10 years, and thereafter it would be sustainable to annually harvest 16% of the population (estimated at 500,000 tons). For 1989, they recommended a total catch of 125,000 tons. That is, the models indicated a very small individual gain and very high individual and group costs for defecting (noncooperating).

In this situation, it is irrational to continue unrestricted fishing for immediate marginal individual gains, given the inevitable dire consequences (destruction of livelihood). A rational choice would be to sustain the resource so it can continue to provide current and future benefits, albeit at a more moderate level. This choice could be implemented by cooperating with the government to initiate steps known to avert the tragedy, specifically (1) coercion or punishment of noncooperators and rewarding of cooperative behavior and (2) making sure people understand the long-term consequences of cooperating versus defecting (Dawes, 1980). These strategies essentially change the payoff matrix so it becomes less of a dilemma.

What did the fishermen actually do? The imposition of catch limits resulted in an outcry from fishermen, corporations, their communities, and the Maritime Provinces. They claimed, without any direct evidence to the contrary, that the population estimates of the Department of Fisheries and Oceans were inaccurately *low*. In response to the outcry and political pressure, the Ministry of Fisheries arbitrarily increased the quota to 235,000 tons. This saga played out annually for several years. In actuality, because of inaccuracies in sampling and the little-understood complexities of ecological systems, the Department of Fisheries and Oceans' estimates were much too *high*, resulting in an annual harvest of 60% of the total population in the last few years instead of the predicted sustainable rate of 16%! In January 1992, the Department of Fisheries and Oceans recommended a harvest of 185,000 tons, but by June 1992 they had revised their recommendation to a complete halt to cod fishing. The fisheries were gone.

What can be learned about human decision-making from this example? Two obvious points can be highlighted for current purposes. Self-maximizing is a powerful force in every aspect of life. Like all evolutionary adaptations, it is local and shortsighted. Its concerns are immediate. But beliefs and reason were also an integral part of the tragedy. The dispute between the Department of Fisheries and Oceans and the fishing community played out as a disagreement about the payoff matrix. The fishing community did not overtly state, "We want to be selfish rather than altruistic." They did not state, "We do not care if the fisheries collapse and we all lose our livelihood a few years down the road." The fishing community argued (without evidence) that the department's methods for estimating fish populations were inaccurately low. The fish were still plentiful. *The fishing community refused to believe that they were harvesting more than was being replenished.* Why?

It is reasonable to question the accuracy of any model, based on reasons and evidence. If a model is incorrect, it can err in either direction (underestimating and overestimating). In questioning the department's estimates and mathematical models, the fishing communities were not privy to any special or additional information. Nonetheless, the fishermen argued that the model was underestimating the number of remaining fish. They refused to entertain the possibility that it might be overestimating the number of fish. They had few evidence-based reasons for their belief. Why didn't they reason as follows? "Even if it is underestimating the number of fish and recovery rates, and we nonetheless curtail our harvest, we will still benefit by having a greater future yield, whereas if it is overestimating, then curtailing the harvest is essential for our survival." Isn't "better safe than sorry" the rational choice here? One plausible explanation for the failure of reason in this case is the predominance of the principle of immediate self-maximization. Self-maximization would be one factor that biased the reasoning system, leading to a faulty conclusion, self-harmful behavior, and the tragic destruction of the fishermen's livelihood.[7]

But there was another important factor in play: the failure to consider, acknowledge, and accept the severity of the consequences of being wrong, of refusing to believe the facts as presented by the best available science at the time. We will consider two types of explanations for this refusal in chapters 13 and 14. The first will involve the introduction of another instinct, in-group/out-group bias, and the second will involve the conjecture that worldviews are very difficult to revise once neural systems have matured. While this discussion has been undertaken with the historical example of the Canadian Maritimes fisheries, the same scenario is tragically playing out in the climate change debate, where the stakes are even higher.

This is also the appropriate time to revisit my American friend from chapter 1 and evaluate his aversion to the concept of universal healthcare. Maintaining a universal healthcare system is also a classic social dilemma situation, but the concern of my friend was not with maintaining the system but rather not wanting to opt into it. Many Americans have good health insurance coverage through their job, and if they are over 65, have subsidized Medicare coverage through the government. For them, universal healthcare offers no personal benefits. In fact, it may be in their self-interest to oppose it if they believe that the expansion of coverage will add freeloaders to the system and dilute care for them. (Ironically, the people on Medicare themselves are not fully paying into the system—and may or may not have contributed a fair share during their working years—but are equally concerned about *other* freeloaders.) Any sense of fairness or altruism is strongly subdued by self-maximization and self-righteously reinforced by reason fueled by beliefs about the "other." Remember the single welfare mothers? They are not deserving like us: "You know, their morals are different. They don't give a damn." While this may begin as a belief, it quickly activates the in-group/out-group system that will be discussed in chapter 13.

There is also a surprisingly large percentage of Americans who do not have good health insurance but also object to universal coverage. They would actually be better off with universal healthcare, despite the existence of freeloaders. It would be their rational choice. This is the category my friend falls into, but his cheater punishment instinct is so powerful that he is unable to dampen it and tolerate some cheaters in order to be personally better off. But there may also be another strand to the explanation. The universal healthcare plans being proposed are all by the *other* political party (Democrats), the un-American socialist party. They involve death panels (Gonyea, 2017). When my friend's anointed political representatives ascend to power, they are going to deliver a patriotic, American solution that will "cost much, much less and deliver much, much more" (Costa & Goldstein, 2017). We will also return to complete this discussion in chapter 13.

The data reviewed so far relate to variability in individual choices. If this variability is at least in part a function of beliefs and reason-based modulations, one would expect to see societal-level variations where there are large differences in beliefs. This is indeed the case. For example, all industrialized societies provide some social assistance programs to their citizens. However, the variability in the amount of assistance as a percentage of GDP (for 2018) ranges from 11.1% in South Korea, to 18.7% in the United States, to 31.2% in France (OECD, 2020). Consistent with these differences, studies of 15 small-scale preindustrialized societies from around the world, including societies that engage in foraging, slash and burn horticulture, nomadic herding, and small-scale agriculture, revealed that the cooperative economic decisions of all groups were a function of self-maximization, fairness, cheating, and cheater punishment, but with considerable group variation attributable to societal factors (Buchan, Croson, & Dawes, 2002; Henrich et al., 2001). For example, in the Ultimatum Game, the mean offers in Western societies (as represented by undergraduate students) are approximately 44%. In the 15 societies in this study, they ranged from 26% to 58%. Rejection rates also varied widely. Western undergraduate students reject offers below 25% with high probability. In some of the preindustrialized societies, low offers were rarely rejected. In others, offers in the vicinity of 50% are frequently rejected. In a Social Dilemma Game, there's a 30% freeloading rate among Western undergraduate students. In one of the preindustrialized societies studied, not a single subject cooperated fully. As all these differences emerge across societal groups, it is plausible that they are a function of learned social norms or beliefs rather than instincts. This again indicates some modulation of instincts by learning and belief systems.

*   *   *

This discussion of reciprocity has highlighted several interesting features of instincts. Some instincts, such as the suckling response in mammals, are widely available; others, such as fairness and cheating, are largely restricted to the hominina or even homo branch of the tree. This has two obvious consequences. First, just because an instinct appears in a common ancestor does not necessarily mean that humans will (or will not) also possess it. Second, traits that are unique to humans need not be lesser candidates for instincts. Data and details matter.

The discussion has also highlighted the possibility of complex interactions between instincts. For example, when I approach a robin's nest containing chicks, the mother robin takes flight. She soon turns around to fly back to protect the chicks, but my presence near the nest again frightens her such that she stops in midflight and retreats, only to approach and retreat again and again. The behavior of the bird from moment to moment is a function of the relative strengths of the different signals from fear and maternal instincts. Similarly, in accounting for contingent reciprocity, the "simpler" traits of self-maximization and cheating are consistent, but fairness tugs in the opposite direction. This type of relationship among instincts will complicate the prediction of behavior of individuals (as one will need knowledge of individual differences in the strength or intensity of various instincts) but will not affect the nature of the causal relationship between stimulus and response. This is the type of interaction envisioned by the massive modularity model.

But the data also clearly show the modulation of economic decision-making choice by beliefs and coherence relations (i.e., by reason).[8] Reason is a double-edged sword. It can be used to either attenuate or accentuate instincts. Conversely, instincts can also either reinforce or overcome reason (to a certain extent). In the face of this overwhelming evidence (to say nothing of common sense) for the interaction of instincts, beliefs, and coherence relations, any model restricted to instincts will be insufficient. Trying to explain these data without the postulation of reason is as futile as trying to explain them without the postulation of instincts.

What is the appropriate model to accommodate the data on cooperative decision-making that provide evidence for the involvement of both instincts and reason in economic decision-making? Part of the story is undoubtedly the existence and interaction of instincts as envisioned by the massive modularity model. Individual differences in the "setting" of these instincts lead to individual differences in choice, but the other, equally important part of the story is reason. The proposed model of tethered rationality—characterized by different kinds of minds, ranging from autonomic, instinctive, associative, and reasoning minds, with evolutionarily newer levels tethered to evolutionarily older levels—acknowledges all these critical components. Chapter 10 delves into the comparative neuroanatomy literature and makes the case for the evolution of hierarchically organized neural infrastructure to support tethered rationality.

Appendix: A Conceptual Critique of Massive Modularity

I find great value in the basic insights of evolutionary psychology that reiterate the importance of instincts in human behavior, but I reject the claim that all human behavior is to be explained in terms of instincts. This rejection is based on common sense and the empirical data reviewed in this chapter, but there are also some deep conceptual reasons to reject the specific massive modularity instantiation of the insights of evolutionary psychology (Buller, 2006; Fodor, 2000). I register my objections here and conclude by reiterating the difference between reason and instincts. The reader more concerned about my positive account of tethered rationality can skip this appendix without loss of continuity.

The case for massive modularity is typically made at the level of computational architecture. Massive modularity is associated with a specific type of computational architecture very different from the physical symbol system architecture we met in chapter 6. Allen Newell and Herbert Simon celebrated the fact that their single GPS computer program could solve problems from different domains simply by switching data sets and could also solve the same problem in different ways by switching the algorithm, demonstrating both generality and flexibility. Cosmides and Tooby see a host of problems with this approach. They celebrate the fact that their computational architecture consists of a collection of independent, specialized programs that each solve very simple specific problems. As the number of behaviors an organism is capable of increases, so will the number of necessary modules. There could be hundreds, thousands, perhaps even hundreds of thousands of these modules, depending on the complexity of the organism. That is why this theory is referred to as "massive modularity." The human brain consists of a large number of these independent, task-specific computer programs or modules. As Cosmides and Tooby (1994a, p. 330) note:

> The human mind is powerful and intelligent not because it contains general-purpose rational methods (although it may include some), but primarily because it comes equipped with a large array of what we might call reasoning instincts. Although instincts are often thought of as the polar opposite of reasoning, a growing body of evidence indicates that humans have many reasoning, learning, and preference circuits that (i) are complexly specialized for solving the specific adaptive problems our hominoid ancestors regularly encountered; (ii) reliably develop in all normal humans; (iii) develop without any conscious effort; (iv) develop without any formal instruction; (v) are applied without any awareness of their underlying logic; (vi) are distinct from more general abilities to process information or behave intelligently. In other words, these reasoning, learning, and preference circuits have all the hallmarks of what people usually think of as "instincts."

In this passage, Cosmides and Tooby are unimpressed with the idea of a general-purpose reasoning system (i.e., the reasoning mind introduced in chapter 6). They suggest that reason and instincts are not polar opposites— even referring to "reasoning instincts"—and that we can understand the former in terms of the latter. Let's take up their concerns about generality and then revisit the relationship between reason and instincts.

Objections to General-Purpose Reasoning Systems

Cosmides and Tooby (1994a, 1994b) raise three main objections to a general-purpose reasoning system. They argue that generality is (1) overrated and unnecessary, (2) inconsistent with evolutionary theory, and (3) leads to certain intractable computational problems. I worry that much of this discussion in the literature is conflating computational and conceptual issues. My own characterization of the reasoning mind was at the conceptual level, with the computational instantiation as a means of capturing the conceptual machinery. The computational issues only come into play after the conceptual issues have been sorted out. Accordingly, I will address the conceptual issues surrounding generality.

Objection 1: Generality is overrated   The first objection is essentially that specialists (specific modules for solving specific problems) will do a better job of solving any given problem than a generalist (i.e., a general-purpose program for solving arbitrary problems). This may be true, but it misses the mark on the need for generality. Let's review how and why generality and flexibility enter into the reasoning mind. Conceptually, the reasoning mind is committed to a system whereby any specific stimulus is neither necessary nor sufficient for a specific response. That is, given any specific input, a reasoning mind is not predisposed to any specific response. This was the "gap" between stimulus and response in reasoning systems identified by Ernst Cassirer. It was discussed in the context of Hamlet killing his uncle Claudius. As noted in chapter 6, the various reasons proposed were all capable of justifying the act, but none was necessary or sufficient to cause the act. Furthermore, it is widely believed that the key to realizing such a system is the conceptual machinery of propositional attitudes and coherence relations. This same apparatus allows us to find novel ways of getting to work in the mornings and allows us to land a rover on Mars. Additionally, it has been proposed—with numerous caveats—that such a system can be mechanistically realized using a particular general-purpose computer architecture (Fodor, 1975; Fodor & Pylyshyn, 1988; Newell, 1980). Cosmides and Tooby seem to focus largely on criticizing this particular computational architecture rather than the conceptual system the architecture is meant to realize. If the criticism is that a particular computational architecture may not be the best way to realize this conceptual system, that is fine, but it fails to address why generality and flexibility at the conceptual level are unimportant.

Objection 2: Generality could not have evolved The claim that generality is inconsistent with evolutionary theory is predicated on a very narrow view of evolution that emphasizes the stability of the evolutionary environment of our Pleistocene ancestors, natural selection, gradualism, specific adaptations, and an increase in the complexity of organisms through a linear, uniform addition of adaptations. This leads to the conclusion that only situation-specific adaptations are possible and that general-purpose reasoning systems could not have evolved. I want to make the obvious suggestion that what evolved were mental representations with propositional content responsive to coherence relations. The problem they solved was that of maintaining *veridicality* between mental representations and the world and *consistency* of mental representations (chapters 6 and 11). Generality is just a consequence of this system. If a particular formulation of the theory of evolution cannot account for propositional attitudes, then based on the same rationale used to reject behaviorism in chapter 5, I would say so much the worse for that theory; it needs to be updated and enriched. One cannot pretend that the phenomenon does not exist. Chapter 11 illustrates how an evolutionary account based on comparative neuroanatomy does have the potential to naturally accommodate both reason and instincts.

Objection 3: Generality leads to the intractable Goodman relevance problem; massive modularity solves it The main objection that Cosmides and Tooby raise about general-purpose reasoning systems is that they are susceptible to the intractable "frame problem." This is actually a loose collection of problems, and it is not clear that they are all identical (Shanahan, 2016), but discussions with my evolutionary psychology colleagues suggest that the problem they are referring to is the Goodman relevance problem of selecting properties suitable for generalization (i.e., projectable predicates).[9] This problem was introduced and discussed with the dinosaur and grue examples in chapter 8. Recall that on finding one *Borealopelta* with fossilized stomach contents, we happily generalized the dietary habits of all *Borealopelta* dinosaurs, but finding the *identical* evidence for broken and missing scales, we were unwilling to generalize that all *Borealopelta* had broken and missing scales. The former property was relevant for generalization, the latter not. The issue was formulated more precisely by Goodman with the grue example.

Cosmides and Tooby are correct in noting the seriousness of this problem and the fact that the conceptual model of the reasoning mind presented in chapter 6 has no solution for it. Without a solution to this problem, there is no science-based psychology. Any candidate solution needs to be considered carefully and, if it solves the problem, embraced. If I understand correctly, the solution massive modularity is offering for the relevance problem is to replace the large general-purpose database and single reasoning engine of physical symbol systems with many specialized smaller databases, each with its own "reasoning" engine. This reduces the problem search space that any particular module needs to traverse for a solution. By reducing the search space, any potential combinatorial explosions are supposedly avoided. By avoiding combinatorial explosions, Goodman's relevance problem is solved.

It is possible that I have misunderstood both the problem they're trying to solve and the solution they are offering, but if they are dealing with the Goodman problem of projectable predicates, then the size of the database that needs to be searched is irrelevant for determining the relevance of any particular predicate. Whether one has three predicates to consider or three billion, it is equally difficult to determine relevance. In the dinosaur example, there are only two predicates ("dietary contents" and "broken and missing scales"); one is generalizable, the other not. But the evidence provides no basis for differentiating between them. Goodman made the same point more rigorously with the predicates "green" and "grue."

The example that is often given in the literature is that of learning a grammar for natural language. Any given fragment of a natural language can be trivially described by an infinite number of grammars, so one might think reducing this infinite number to 20 or even 2 is a big step forward. Not as far as Goodman's selection problem is concerned. Even if there are only two possible grammars that the module has access to and the sampled data are equally consistent with *both* of them, how does the system decide? Notice that the relevance problem is not a computational problem; it is a conceptual problem (of specifying necessary and sufficient conditions for relevance). It needs a conceptual solution. The size of the database may become an interesting computational factor once the conceptual problem is solved, but it is not a factor in the solution to the problem itself. It is a red herring.

It is important to understand that not all minds have to confront the relevance problem. It is a problem specific to minds reasoning with propositional contents and coherence relations. Instinctual minds like Lorenz's (figure 4.1) don't have to deal with it. There is a causal connection between a specific stimulus and the animal's fixed action pattern (response), as in the example where the swollen abdomen and the posture of the female stickleback unlock the innate release mechanism of the male's mating behavior. The stimulus is causally necessary and sufficient for the response (with the noted degrees of freedom allowed by the model). The male stickleback is not confronted with Goodman's problem. Neither are our homeostatic systems regulating various bodily functions nor our low-level visual system confronted with it. In the latter case, there is a topographic mapping from the retina, to the lateral geniculate nucleus (LGN), to the primary visual cortex. There are specific mechanisms for detecting edges, light and dark areas, line orientation, and other features. Hold up a certain pattern of lines at certain angles, and certain specific neurons in the primary visual cortex will respond. Hold up a pattern of lines in a different orientation, and another set of neurons will be activated. This is a causal story. No propositions are involved until the very end, when you formulate the belief that the sun is setting on the horizon. The Goodman relevance problem emerges with the emergence of propositional attitudes and conceptual coherence relations.

"Solving" the Goodman problem the way the stickleback does it—sidestepping it by replacing conceptual relations with direct causal relations—is a genuine workaround. But it is important to appreciate that it restricts you to a certain kind of mind, the kind that the stickleback has. If massive modularity is signing up for the stickleback solution, then it indeed sidesteps Goodman's relevance problem. But the proposed "cure" may be worse than the proverbial disease itself. It means doing without the "gap" between stimulus and response, which is the *conditio sine qua non* of the reasoning mind. My own view is that this is too high a price to pay. I want to keep my reasoning mind (though tethered to the stickleback's mind), with the understanding that at some future date the relevance problem must be discharged.

There are occasions on which Cosmides and Tooby seem to recognize the shortcomings of the stickleback's mind, and the need for the types of behaviors made possible by propositional attitudes, but are unsure how to get there (Cosmides & Tooby, 2013, p. 182):

> Large amounts of knowledge are embodied in intelligent, domain-specific inference systems, but these systems were designed to be triggered by stimuli in the world. This knowledge could be unlocked and used for many purposes, however, if a way could be found to activate these systems in the absence of the triggering stimuli; that is, if the inference system could be activated by *imagining* a stimulus situation that is not actually occurring: a counterfactual.

When they write "if a way could be found to activate these systems in the absence of the triggering stimuli," I'm assuming they mean a causal trigger. If this is the case, then they seem aware of the dilemma that massive modularity presents: either stick with the stickleback's mind and ignore such situations or accept the solution provided by propositional attitudes and coherence relations (and put up with Goodman's relevance problem). I have chosen the latter.

Reiterating the Distinction between Reason and Instincts

Finally, Cosmides and Tooby insist that reason and instincts are not polar opposites. By contrast, I have claimed that instinct and reason constitute different kinds of minds. The reader is encouraged to return to table 6.1 and review each of the five dimensions along which kinds of minds were differentiated. Reason and instinct differ on each dimension. At the expense of some repetition, I will summarize: instincts are the preferred solution to guiding behavior that is essential, does not need to change across generations, and may be needed prior to any opportunity for learning. Where within-generation environmental fluctuations are in play, instincts on their own will not be sufficient. The least expensive and most widespread solution for this is a mechanism that learns through associations by tracking co-occurrence relations. An even more complex interaction with the environment involves the ability to track stimuli and changes that may not even be present at the time, to consider counterfactual scenarios, and to make flexible individualized responses. For example, I can easily imagine the consequences of rising oceans on coastal cities before it actually happens and choose to respond very differently than my neighbor. This requires still more sophisticated machinery.

In the cognitive account, this more sophisticated machinery consists of psychological intentional states, such as beliefs and desires with proposition-like representational contents, referred to as propositional attitudes. Propositional attitudes possess the properties of productivity, compositionality, systematicity, and inferential coherence; they relate to the world via a reference relation and to each other via semantic, logical, and conceptual coherence relations (chapter 6). Along with this machinery comes Goodman's relevance problem. As far as I can see, the massive modularity solution for doing away with the relevance problem entails doing away with this machinery. So, even if Cosmides and Tooby are correct about instincts and reason not being polar opposites, it should be clear that they are solutions to different types of problems and postulate different machinery. One cannot be successfully substituted for the other. Both are necessary to explain human behavior.

10 Kinds of Brains: Of Mice, Monkeys, and Men

At that time, the generally accepted idea that the differences between the brain of [nonhuman] mammals (cat, dog, monkey, etc.) and that of man are only quantitative seemed to me unlikely and even a little offensive to human dignity. . . . [L]anguage, the capability of abstraction, the ability to create concepts and finally, the art of inventing ingenious instruments . . . , do [these facets] not seem to indicate (even admitting fundamental structural correspondences with the animals) the existence of original resources, of something qualitatively new which justifies the psychological nobility of *Homo sapiens*? Microscope at the ready, I then launched with my usual ardor to conquer the supposed anatomical characteristic of the king of Creation, to reveal these enigmatic strictly human neurons upon which our zoological superiority is founded.

—Santiago Ramon y Cajal (1917)

Chimpanzees are one of our closest extant relatives. We diverged from them only five million years ago and still share 98.6% of our genes with them (Waterson, Lander, Wilson, & the Chimpanzee Sequencing and Analysis Consortium, 2005). In fact, chimpanzees share more of their genetic material with us than they do with gorillas! The similarity in anatomy and physiology is readily apparent. We both have arms, legs, eyes, hearts, livers, and other body parts. We both breathe, eat, fornicate, and defecate. But we are not identical. Chimpanzees have more protruding faces and larger and more powerful jaws. Their arms are longer than their legs. They do not walk upright. Their bodies are coated with hair. But there is sufficient similarity that we can learn about our own physiology by studying theirs.

It is even the case that the first human heart transplant, in 1964, performed by James Hardy, used a chimpanzee heart (Margreiter, 2006). While human and chimpanzee hearts are similar, there are also some interesting differences in terms of endurance capabilities having to do with pressure and volume trade-offs (Shave et al., 2019). Heart disease in the two species has different underlying causes (Varki et al., 2009). As we move further down the phylogenetic tree, the differences become more dramatic. Some reptiles have a three-and-a-half chambered heart, amphibians have a three-chambered heart, fish have a two-chambered heart, and earthworms have an open circulatory system that we might not even recognize as a heart. All this speaks to the evolutionary *continuity* and *change* across species envisioned by Darwin and Wallace.

But unlike the visible similarity in anatomy and physiology, there seems to be an enormous chasm in intellectual abilities between humans and nonhumans, including chimpanzees, at least to the extent that humans have hypothesized, then discovered, and are now investigating black holes, while our nearest cousins on the evolutionary tree have learned to fish for termites using sticks. At the cognitive level, we have attributed this difference to the fact that we have minds with propositional attitudes characterized by productivity, systematicity, generativity, and inferential coherence and the necessary machinery to determine various types of coherence relations among them. Colloquially, we refer to the behavior entailed by this machinery as "special." Special or not, it is qualitatively different from autonomic behaviors, instinctive behaviors, and behaviors resulting from associative learning. Indeed, each of these behaviors is equally distinct from each other and can be organized into a hierarchy based both on complexity and order of appearance on the phylogenetic tree, with simpler behaviors appearing before more complex ones (see table 6.1). The autonomic and instinctive behaviors appear very early in the evolutionary tree, followed by associative behaviors, which are also widely available, and then followed by rational behaviors, which seem to have emerged very recently on the hominina or even homo branch of the tree. It is even possible that they are specific to *Homo sapiens*.

This chapter addresses the corresponding neural bases of these behaviors by considering brain evolution in vertebrates and, more specifically, mammals, to illustrate that human brains have the structure and organization to support a model of tethered rationality. This will be done by considering the following questions: (1) What brain structures account for the different kinds of minds? (2) What is the evolutionary and anatomical evidence for interconnection between these structures? (3) What brain structures account for the rational behavior unique to us? These are unsettled questions, with no universally accepted answers. I will weave together the most plausible account currently warranted by the data.

Comparative neuroanatomical evidence points to hierarchically evolved brain structures, with interestingly different functional and neuronal properties, that map onto the behavioral hierarchy we have been discussing. Later evolving brain structures do not simply float on top of earlier evolved structures; they are anatomically and physiologically tethered to them. With respect to the uniqueness of human reason, the data seem to point to a combination of brain size and organization, along with some neurophysiological differences and innate constraints, particularly in the neocortex. These evolutionary and anatomical considerations provide a sound platform for an account of tethered rationality. The more speculative question of why human brains evolved in the way they did will not be addressed.

Overview of Brain Evolution

All the behaviors we are considering are initiated by nervous systems.[1] The function of the brain is to track and respond advantageously (i.e., in fitness enhancing ways) to environmental changes, be they in the internal environment (e.g., rising body temperature) or the external environment (e.g., decreasing daylight hours). That is, the brain is for controlling behavior. With few exceptions,[2] all multicellular animals have a nervous system, whether it be simple diffused "nerve nets," with various degrees of consolidation, as found in jellyfish (Satterlie, 2011), or the highly developed, integrated brains found in primates. If an organism has a nervous system, it consists *largely* of the same building blocks and principles of operation irrespective of whether that organism is a bat, Botticelli, or Brahmagupta.[3] These building blocks consist of neurons and glial cells, briefly described in the appendix of chapter 5. But any set of building blocks can be used to build many different things. Given the same paints and brushes, Botticelli can create "The Birth of Venus" while I would create something embarrassingly more modest. This opens up the possibility for significant differences among brains.

Comparative neurobiologists, beginning with Ludwig Edinger some hundred years ago, advanced a view of brain development as a ladder-like progression from fish to amphibians, reptiles, birds, mammals, and primates, ending up with humans (Jarvis et al., 2005; Naumann et al., 2015). Brains on higher rungs of the ladder retain ancestral structures from lower rungs with additions and modifications. These distinct physiological layers correspond to distinct behavioral layers. If this is the case, we can trace human brain evolution by examining brains of extant nonhuman vertebrates that appeared earlier on the phylogenetic tree.

The *details* of the accounts of classical comparative neurobiologists have not withstood the last hundred years of research in comparative neuroanatomy. There are two valid criticisms of these models. First is of the (often implicit) assumption of unilinear, ladder-like development. Second are the details of the emergence and transformation of specific brain regions. The fact that the phylogenetic tree is not a ladder is emphasized in box 10.1. For instance, our current understanding is that sauropsid reptiles and therapsids shared a common ancestor (stem amniotes) approximately 300 million years ago. Modern mammals branched from therapsids some 200 million years ago, while birds branched from reptiles some 50 million years later (Jarvis et al., 2005). This is inconsistent with the unilinear development idea that bird brains fit somewhere between reptile brains and mammal brains (figure 10.2). With respect to the second criticism, neurobiologists continue to update origins and functions of different brain structures with increasingly more sophisticated tools.

But both of these "errors" are errors of detail. The basic insight, that at every branch of the phylogenetic tree certain *brain structures* are conserved and propagated while others are added, modified, and expanded in new directions, remains valid. It is also the case that certain *behaviors* are preserved and propagated at branching points, while others are added, modified, and expanded in new directions. The correspondence is not an accident. Hierarchically organized behaviors are underwritten by hierarchically organized brain structures. Taken together, these observations may be our best route to understanding the emergence of different kinds of minds, including the reasoning mind.

Following the comparative neurobiologists, I will characterize brain evolution as an ordered emergence and differentiation of various neural structures at different points in the phylogenetic tree (see figure 10.2). All vertebrate brains consist of a spinal cord, myelencephalon, metencephalon, mesencephalon, diencephalon, and telencephalon (figure 10.2a). The myelencephalon, metencephalon, and mesencephalon are all part of the brain stem. The diencephalon consists of the thalamus and hypothalamus. These brain stem and diencephalon structures differentiated early and are reasonably well established and conserved in vertebrates from fish, to reptiles, to birds, to mammals, and even to primates (Jarvis et al., 2005; Naumann et al., 2015; Wilczynski, 2009). There is considerable variation in the telencephalon.

The telencephalon of fish and amphibians is largely undifferentiated pallium (Wilczynski, 2009). In reptiles, birds, and mammals, the telencephalon develops two major subdivisions, called the pallium and the subpallium.

Box 10.1

Phylogenetic tree

A phylogenetic tree is a treelike diagram that represents evolutionary relationships between organisms (figure 10.1). It has a single trunk, indicating a common origin for all life-forms, and then sprouts many branches and subbranches, indicating the proliferation of different groups (e.g., mammals and amphibians) and species (mice and men). A branching point indicates a common origin for all groups or species emanating from that point. These relationships were initially postulated based on external morphology, internal anatomy, and behaviors. Today, DNA analysis (where available) plays an important role. It is also possible to make time estimates for these branching points. Species that have more recent common ancestors are more closely related. Figure 10.1 indicates a rooting point (i.e., a common ancestor) for humans and chimpanzees approximately five to six million years ago. Humans and chimpanzees shared a common ancestor with elephants approximately 90 million years ago and with chickens some 300 million years ago. Behaviors (and underlying mechanisms) that appeared earlier in the phylogenetic tree may be conserved and shared by a larger number of branches of the tree and thus be more broadly available. To what extent the behaviors are propagated in the original form or modified en route is an important open question.

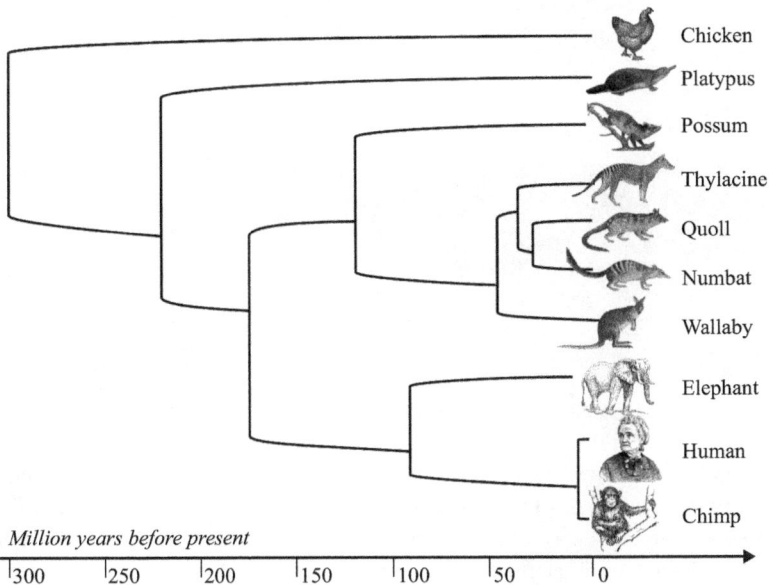

Figure 10.1

Example of a phylogenetic tree. Some animal illustrations are reproduced from Angel Cabrera (1919).

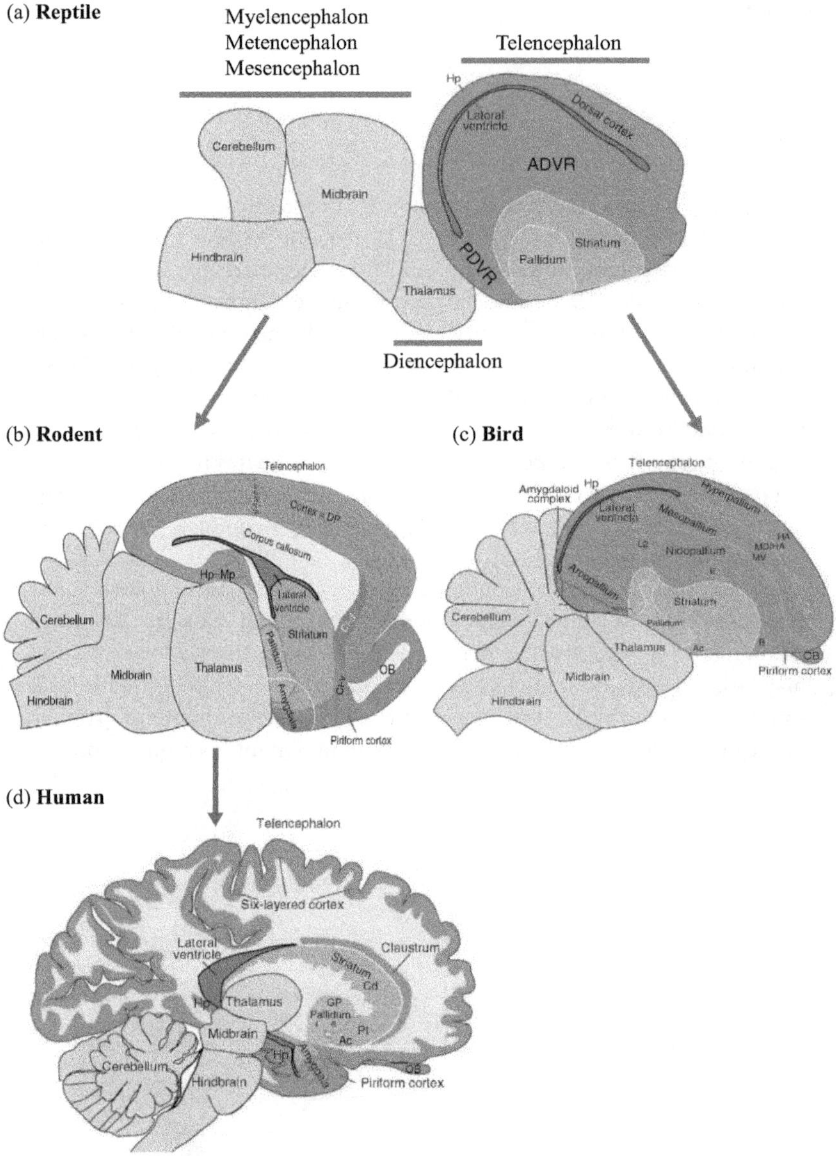

Figure 10.2

Simplified phylogenetic tree of brain development across reptiles, birds, and mammals. There are five things to note: (1) the brain stem, cerebellum, and thalamic structures appear early and are conserved across the evolutionary tree; (2) the telencephalon begins to differentiate into the pallium and subpallium with reptiles; (3) the subpallium evolves into the basal ganglia structure of mammalian brains; (4) the pallium starts as a largely undifferentiated structure in reptiles and becomes more differentiated in birds and even more fully differentiated into subcortical and cortical structures in mammals; and (5) in mammals, the pallium becomes the six-layered neocortex.

The pallium is generally thought to have evolved slowly and differentiated into subcortical and cortical structures, though the details remain unsettled (Goodson & Kingsbury, 2013; Reiner et al., 2004). The subpallium is differentiated into regions called the striatum and pallidum in reptiles (figure 10.2a) and eventually becomes fully articulated as the various basal ganglia structures in mammals (figures 10.2b, 10.2d, and 10.3c).

The pallium varies considerably. In reptiles, it remains largely undifferentiated and is currently divided into the dorsal ventricular ridge (anterior and posterior aspects), olfactory, hippocampal, and perhaps amygdala-like regions (figure 10.2a) (Naumann et al., 2015). The bird pallium warrants finer-grade differentiation into four major regions—hyperpallium, mesopallium, nidopallium, and arcopallium—and olfactory, hippocampal, and amygdala-like regions (figure 10.2c) (Jarvis et al., 2005). The pallium in mammalian brains is fully articulated into the hippocampus, amygdala, olfactory system, and cerebral cortex, in particular a part known as the neocortex, a six-layered structure considered the newest evolutionary addition to the brain (figures 10.2b and 10.2d). There is considerable variation among different mammalian orders as to the size, development, and suborganization of the neocortex.

This brief sketch reveals a basic plan that has been conserved and differentially detailed and upgraded across vertebrates, with perhaps the only exception being the absence of the cerebellum in hagfish and lampreys (Northcutt, 2002). The hierarchical nature of brain evolution is reinforced by the fact that there are no branches of the phylogenetic tree where we find brain stem structures in the absence of the spinal cord, diencephalon structures in the absence of a brain stem, subcortical structures in the absence of diencephalon and brain stem structures, or cortical structures in the absence of subcortical, diencephalon, and brain stem structures.

Based on this evolutionary story mammalian brains can be usefully subdivided into the brain stem, diencephalon, subcortical structures, and cerebral cortex (see figure 10.3).[4] The brain stem is located at the anterior of the spinal cord and consists of the medulla, pons, midbrain, and cerebellum

---

ADVR=anterior dorsal ventricular ridge; Ac=accumbens; B=basorostralis; Cd=caudate; E=entopallium; GP=globus pallidus, internal (i) and external (e) segments; HA=hyperpallium apicale; Hp=hippocampus; IHA=interstitial hyperpallium apicale; L2=field L2; MD=dorsal mesopallium; MV=ventral mesopallium; OB=olfactory bulb; PDVR=posterior dorsal ventricular ridge; Pt=putamen. For birds, there is some uncertainty as to whether the MD is hyperpallium densocellulare (HD) or a separate structure. Reproduced with permission (with slight modifications) from Jarvis (2009).

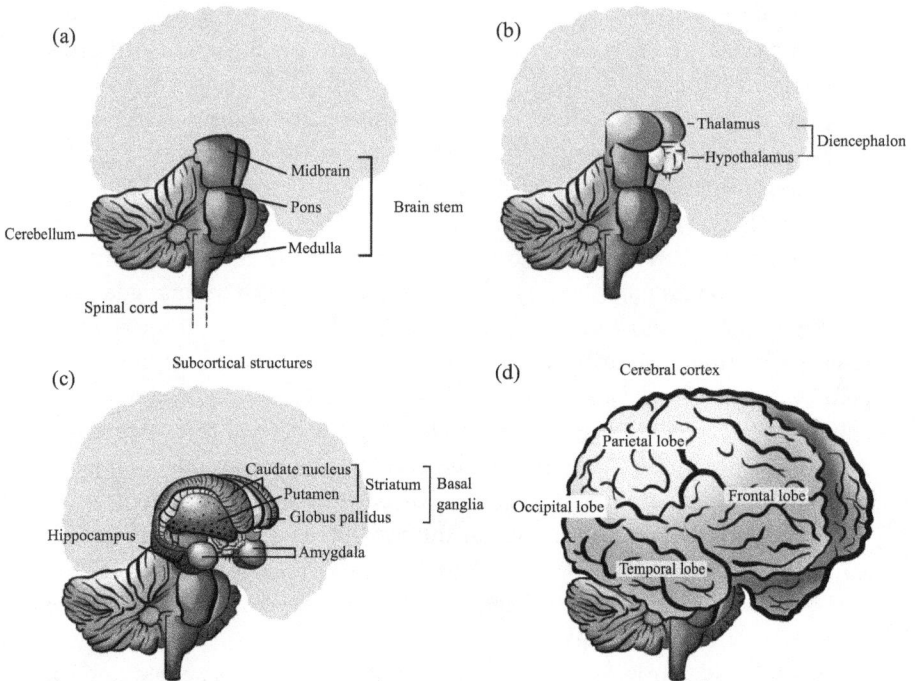

Figure 10.3
The major components of the human brain starting with (a) brain stem and cerebellum, (b) diencephalon (thalamus and hypothalamus), (c) subcortical structures, including basal ganglia nuclei, hippocampus, and amygdala, (d) and cerebral cortex. Figure drawn by Brooklyn McKinley.

(figure 10.3a). On top of the brain stem sits the diencephalon (thalamus, epithalamus, subthalamus, and hypothalamus) (figure 10.3b). Surrounding the thalamic structures are subcortical structures, including the basal ganglia system consisting of the caudate nucleus, putamen, nucleus accumbens, and olfactory tubercle (together called the striatum), globus pallidus, ventral pallidum, substantia nigra, and subthalamic nucleus. Other subcortical structures include the pituitary gland, hippocampus, and amygdala (figure 10.3c). The cerebral cortex is the surface "bark" structure enveloping the subcortical structures (figure 10.3d). It can be further subdivided in multiple ways.

In the next several sections, I will differentiate these brain structures in terms of their functional and neuronal properties and associate them with the different behaviors we are interested in. An important part of this story is the issue of "hardwiring" versus "softwiring" of neural systems and its functional consequences. While these terms are often misused, they do have a well-defined meaning in terms of differential developmental trajectories that will be discussed here and again in chapter 14. We will also encounter the old adage "ontogeny recapitulates phylogeny," as we observe the developmental timeline of brain maturation displaying a similar temporal unfolding as the evolutionary timeline.

Brain Stem, Diencephalon, and Subcortical Systems: Essential and Mostly "Hardwired"

Neural Basis for Autonomic and Instinctive Systems

It should not be surprising that the phylogenetically earliest emerging neural systems should also be those essential to sustain basic life processes, and, once developed, they should be largely conserved, with local modifications. The brain stem systems, consisting of the medulla, the pons, and the midbrain, serve to control many autonomic functions, such as cardiovascular, respiratory, digestive, pain sensitivity, alertness, and awareness. They also function as pathways for conducting sensory information, including pain and pleasure, to and from peripheral nerves, including cranial nerves, to relevant brain regions.

The diencephalon is sandwiched between the basal ganglia and the brain stem. It is the major relay center for sensory information (hearing, vision, smell, touch, and taste) and motor control information traveling between the spinal cord, medulla, and cerebrum. The hypothalamic nucleus in the diencephalon controls several homeostasis functions and essential survival behaviors, such as feeding and reproduction, that we will examine in greater detail in subsequent chapters. The hypothalamus also controls the endocrine system, synthesizing and controlling the secretion of hormones. It is intrinsically connected to the pituitary gland, which serves as the master control of the other glandular systems.

The basal ganglia are in part concerned with motion and modulate connections between the thalamus and motor cortex (see figure 10.4). Several brain stem nuclei, along with several basal ganglia structures, such as the nucleus accumbens, are an integral part of the hedonic reward system, which will be considered in more detail in chapter 11. The hippocampal system is critical for laying down new memories and spatial navigation. The amygdala is associated with processing emotions, particularly fear. The particulars of these functions are complex and will differ across branches of the phylogenetic tree, but they are all variations on the same basic plan.

Unsurprisingly, the brain stem, diencephalon, and subcortical systems found in the oldest parts of the phylogenetic tree play a critical role in the

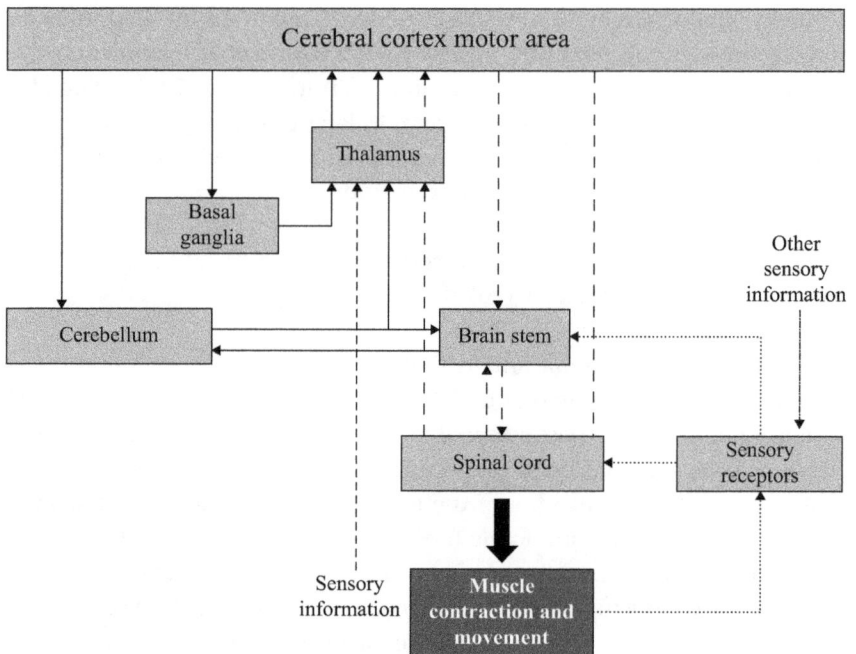

Figure 10.4
Brain systems are hierarchically organized and tethered. This example from the motor system highlights interconnections and coordination across four levels of brain structures (brain stem and cerebellum, thalamus, subcortical structures, and cerebral cortex). Most brain functions require such system integration and coordination. Notice that information flows both ways, top-down and bottom-up. Figure modeled after information in Sherwood and Kell (2009).

genetically encoded, instinctive behaviors of all animals. Like autonomic, sensory, and motor systems, instinctive systems also need to be available prior to extensive learning. Recent reviews of instinctive or innate behaviors in mammals—such as aggression, dominance, social attachment, defense, mating, and parental care—highlight the role of brain stem, diencephalon, and subcortical nuclei, including the periaqueductal gray nucleus, nucleus accumbens, ventral tegmental area, hypothalamus, and amygdala, among others (Ko, 2017; Zha & Xu, 2015). Figure 10.5 indicates some of the subcomponents and pathways that have been identified in rodents for three of these behaviors: aggression, dominance, and social attachment.

This is not to say that only subcortical structures are involved in innate behaviors (Beach, 1937; Febo, Felix-Ortiz, & Johnson, 2010). Figure 10.5

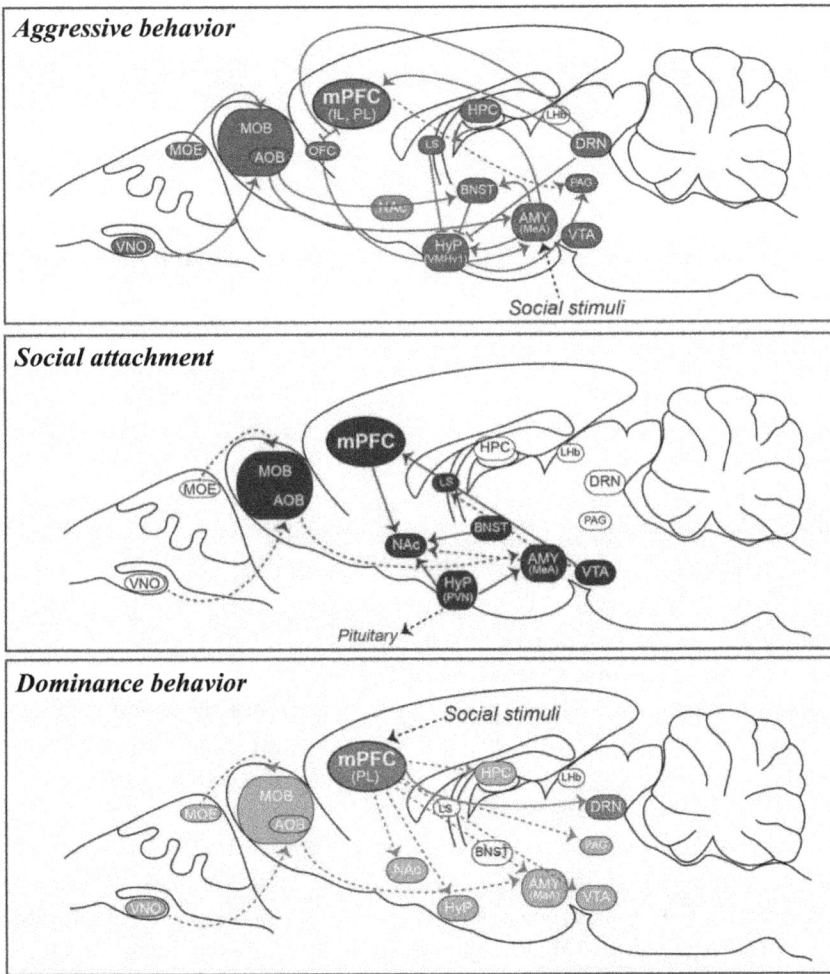

Figure 10.5

Brain stem, diencephalon, and subcortical systems involved in aggression, dominance, and social attachment behaviors in rodents. AMY = amygdala; AOB = accessory olfactory bulb; BNST = bed nucleus of the stria terminalis; DRN = dorsal raphe nucleus; HPC = hippocampus; HyP = hypothalamus; IL = infralimbic division of the mPFC; LHb = lateral habenula; LS = lateral septum; MeA = medial amygdala; MOB = main olfactory bulb; MOE = main olfactory epithelium; mPFC = medial prefrontal cortex; NAc = nucleus accumbens; OFC = orbitofrontal cortex; PAG = periaqueductal gray; PL = prelimbic division of the mPFC; VMHv1 = ventrolateral subdivision of the ventromedial hypothalamus; VNO = vomeronasal organ; VTA = ventral tegmental area. Reproduced with permission from Ko (2017).

also identifies some cortical structures. A recent study on nest-building behavior in zebra finches broke down nest building into simpler activities such as picking up twigs, putting down twigs, time spent in nest, preening, and time singing. These activities were associated with three networks: a motor pathway network, a "social behavior" network, and a dopaminergic reward circuit network. All these networks involved subcortical structures (thalamus, hypothalamus, striatum, septum, and tegmental areas) and an area referred to as the nidopallium, thought to correspond to the mammalian cortex (Hall, Bertin, Bailey, Meddle, & Healy, 2014).

The brain stem, diencephalon, and subcortical structures not only emerged early in evolutionary development but—with the exception of the hippocampus—are also said to be "hardwired." Hardwired neurons are not different types of neurons. They are best thought of as neurons that undergo an *experience-expectant* (Greenough, Black, & Wallace, 1987) developmental trajectory characterized by very early, mostly prenatal maturation; greater involvement of genetically predetermined neural connections; some requirement of parameter settings through environmental interactions; considerable malleability to disruption prior to maturation; and little or no malleability to disruption after maturation. These systems tend to compute specific outputs from specific inputs. They are usually associated with primal life-sustaining processes that need to come online prenatally (e.g., controlling heart rate) or systems that need to be available very quickly after birth (e.g., suckling response), prior to extensive opportunity for environmental sculpting.

Consistent with the first requirement for "hardwiring," data from the rhesus monkey (gestation period of 165 days) in figure 10.6 reveals that brain stem monoamine neurons have an earlier origin and differentiation date than neurons in subcortical and cortical structures (Levitt & Rakic, 1982). The generation of neurons in the lateral geniculate nucleus (part of the thalamus) is complete by 30 to 40 days (Rakic, 1977). The neurons composing the neostriatum are in place by the eightieth day (Brand & Rakic, 1979). The formation of neurons in the visual cortex begins at approximately day 45 of gestation but does not complete until day 102 (Ghosh, Antonini, McConnell, & Shatz, 1990; Rakic, 1974).

There is an obvious reason why neural connections in the brain stem, diencephalon, and subcortical structures need to be largely genetically predetermined: they need to mature and come online prenatally or soon after birth, prior to opportunities for extensive interaction with the external environment. Early maturation requirements necessitate a largely innate, genetically controlled developmental program that tightly prescribes a set

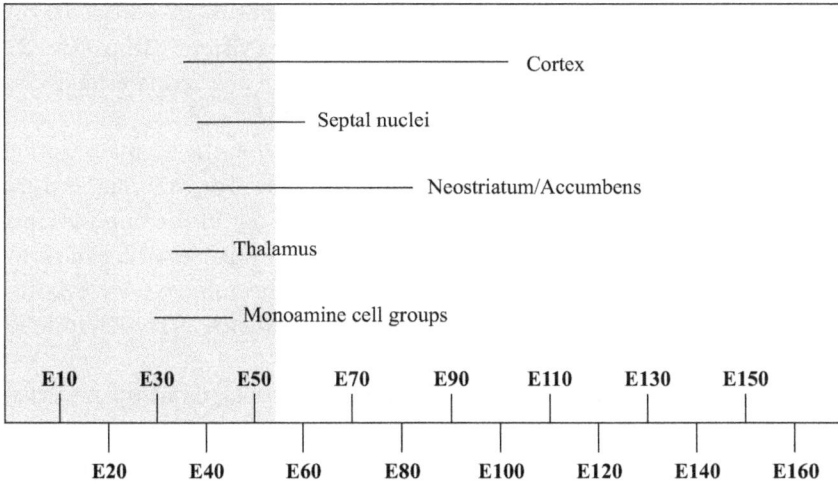

Figure 10.6
Developmental timeline of various brain structures in rhesus monkeys is consistent with the old adage "ontogeny recapitulates phylogeny." Neurogenesis and maturation occur from the inside out, beginning with the brain stem (monoamine cells), diencephalon, and subcortical structures and then cortical structures. "E" indicates embryonic days. Gestation period is 165 days. Data compiled and graphed by Selemon and Zecevic (2015) from studies by Rakic and colleagues. Reproduced with permission.

of connections from one neural region to another neural region or between neural regions and sensory regions, with *minimal impact from external factors*—because these systems are largely beyond the reach of the external environment during the maturation window. This genetically prescribed (or predetermined) wiring is known as neurospecificity.

Vision provides an example of neurospecificity in both types of connections, between neural regions and between neural regions and sensory regions. In normal brain development, the optic nerve connections from the retina arrive at the lateral geniculate nucleus (in the thalamus) and then are mapped onto the primary visual cortex in the occipital lobes. These connections preserve a very specific topographic mapping from the retina to the lateral geniculate nucleus and from the lateral geniculate nucleus to the primary visual cortex.

This neurospecificity was demonstrated in a famous experiment by Roger Sperry. Sperry (1944) severed the optic nerve ("pulled and teased apart in a rough manner") in several species of amphibians, including salamanders, newts, and frogs. In some cases, the eyes were also operationally rotated 180°. Amphibians have the ability to regenerate body parts, so most animals regenerated their optic nerve and recovered vision within 11 to 23 days. Behavioral testing of the animals with the rotated retina established that the regenerated optic axons reconnected to their original positions in the tectum (optic lobe) of the animals. This was completely maladaptive for the animals in terms of navigation and feeding. For example, if a fly appeared behind the frog, it would dart its tongue out in the opposite forward direction, unable to catch the fly. The behavior persisted for months without any modification through learning until the animals were eventually sacrificed. This experiment provides strong evidence of predetermined neural mappings, or neurospecificity.

The sensory systems (somatosensory, auditory, visual, olfactory, and gustatory), while typically not required prenatally, must be ready to function at birth or shortly thereafter. These factors limit, but do not preclude, opportunities for external environmental inputs to complete normal development. In fact, neurospecificity is predicated on certain expectations about the external environment of the organism, such that environmental sampling is actually necessary at certain critical developmental stages for parameter settings. Where environmental expectations are violated prior to maturity of the system, there is often a small window of opportunity for neural rewiring.

This is nicely illustrated by several experiments on the visual systems of cats and monkeys. Normal visual input from both eyes is the default expectation of the visual system. David Hubel and Torsten Wiesel (1970) found that disrupting this expectation by suturing shut one eye of young kittens around the fourth and fifth weeks after birth, even for a few days, results in a sharp decline in neural connections in the lateral geniculate nucleus, with limited prospects for recovery after the eye is opened. The same procedure carried out on adult cats has no effect on the underlying neural organization or behavior. Normal visual input from two eyes also forms a distinctive banded pattern in the primary visual cortex, called ocular dominance columns. When the input to one eye (of young rhesus monkeys two days to three weeks of age) is blocked either through suturing or removal of one eye, disrupting the competitive balance, the bands corresponding to the open or remaining eye expand and encroach on the regions that would normally have been innervated by neurons from the closed eye. The bands corresponding to the closed eye shrink into thin stripes (Hubel, Wiesel, LeVay, Barlow, & Gaze, 1977). When the procedure is carried out on adult monkeys, there is no neural reorganization. This demonstrates a "critical period" in the development of the visual system where it is necessary to have the expected environmental input for the normal completion of the maturation process.

The effect of environmental manipulation on visual development was demonstrated even more dramatically in a behavioral experiment by Colin Blakemore and Grahame Cooper. The mammalian visual system has specialized cells for detecting lines of various orientations in the visual field. Certain cells preferentially respond to lines of certain orientations. Utilizing this fact, Blakemore and Cooper (1970) housed kittens in a completely dark room from birth. After two weeks, they were placed in a special apparatus for five hours per day, where they were completely surrounded by either black and white vertical stripes or horizontal stripes. The manipulation continued well beyond the critical period of visual development and resulted in some permanent neural rewiring. The kittens exposed only to vertical stripes were unable to see or respond to horizontal stripes. The kittens exposed only to horizontal stripes were unable to see or respond to vertical stripes. The effect was permanent and measurable both behaviorally and with single-cell recordings. Equally important, the manipulation was not effective on adult cats.

These experiments illustrate that the impact of environmental input on neurons in the visual system is modulated by the level of neural maturity. The same is true of physical disruptions in neural pathways. If the primary visual cortex (in mice) is ablated prenatally (i.e., prior to maturation), adjacent pieces of cortices will be recruited and reorganized into the primary visual cortex (Sur & Rubenstein, 2005). However, if the primary visual cortex is ablated postnatally, after maturation, there will be blind spots in the corresponding parts of the visual field. That is, postmaturation ablation in hardwired systems displays no plasticity—the ability to recover from injury or rewire for a different function—and leads to very specific and permanent deficits.

Hardwiring is an effective strategy where the problem is fixed and environments are known and stable across generations, and systems have to be working from the get-go, prior to any extensive opportunity for learning. These hardwired systems have been conserved over evolutionary time because they provide highly efficient solutions for autonomic, instinctive, and sensorimotor systems. While these are all critical systems, they are not sufficient to account for the behaviors of most animals. Associative and reasoning behaviors must also be accounted for.

Neural Basis for Associative Learning
Associative learning behavior is also widely distributed among vertebrates (and beyond), but unlike autonomic functions and instincts, its availability is highly variable across species. For example, dolphins and primates have a much more extensive repertoire of learned behaviors than frogs. There

are at least two reasons for this. First, the associations any organism can learn are not arbitrary. They are constrained by its biology and evolutionary history. Second, learning associations requires some flexibility in neuronal wiring. One must be able to learn at any time. This requires neural networks that are *not* hardwired and can modify neural connections (discussed later) in response to environmental experiences such as in spatial orientation, navigation, and memory to and from a homesite or food source. Such neural resources start appearing with subcortical memory systems in the hippocampus and related structures and expand with cortical and neocortical systems (O'Reilly & Norman, 2002; Rolls, 1989).

Species that are better at spatial navigation tasks tend to have a larger hippocampus (Krebs, Sherry, Healy, Perry, & Vaccarino, 1989; Lucas, Brodin, de Kort, & Clayton, 2004). Larger neural volume is expensive to maintain, so it is even the case that the size of the hippocampus varies seasonally in certain species, depending on the utilization of spatial navigation and memory abilities. For example, black-capped chickadees exhibit maximum food hoarding activity in October. It is reported that the hippocampal volume relative to the rest of the brain is greater in October than at any other time of the year (Smulders, Sasson, & DeVoogd, 1995), though these claims have also been contested (Bolhuis & Macphail, 2001).

More direct evidence of hippocampal involvement in associative learning comes from data from single-cell recording studies. One study in which monkeys were rewarded for learning associations between visual "scenes" stimuli and certain target locations showed robust correlation between activity in certain cell groups in the hippocampus and associative learning (Wirth et al., 2003). Lesion studies, also in monkeys, show that cutting the fornix (a major output tract for the hippocampus) impairs the learning of new associations, even in nonspatial paradigms (Brasted, Bussey, Murray, & Wise, 2003).

With autonomic, instinctive, and associative learning, we may have exhausted the behaviors of most, if not all, nonhuman animals, so how do we explain human reasoning behavior, specifically our ability to draw coherence relations among propositions? This behavior is as distinct from learning behavior as learning behavior is from instinctive behavior and autonomic behavior. It cannot be accounted for by hardwired systems. Like associative learning, belief formation and belief revision require neural systems that are not precommitted and can modify their connections in response to new information. In addition, these systems must be pre-structured with innate constraints such that they allow for representations that meet the criteria of productivity, compositionality, systematicity, and inferential coherence. There must also be an innate (hardwired) notion of coherence to guide inference. As noted earlier, hippocampal regions are not hardwired and play an important role in memory and learning, which can potentially support reasoning (Goel, Makale, & Grafman, 2004), yet these systems alone cannot be sufficient for reasoning behavior because they are widely available among mammals, while reasoning is not. This leaves the cerebral cortex as the final evolutionary development and potential neural substrate for reasoning. Interestingly, while the differences in the brain stem and diencephalon and subcortical structures among mammals are slight variations on the same plan, the amount of cerebral cortex available on various branches of the phylogenetic tree varies greatly.

Cerebral Cortex: Mostly "Softwired"

The cerebral cortex is the thin outer layer of the mammalian brain. In humans it has a convoluted or folded surface structure, consisting of gyri and sulci, that maximizes its surface area within the fixed volume of the cranium. While it has precursors in the form of the pallium, early in the phylogenetic tree, its differentiation and development vary widely among taxonomic orders. It emerges most fully developed in mammals and shows much more variation than brain stem, diencephalon, and subcortical structures, though, interestingly, its thickness varies only within a narrow range both within and across species. For example, it varies from 1 mm to 4.5 mm in humans, from 0.8 mm to 1.6 mm in dogs, and is approximately 2 mm in whales (DeFelipe, 2011). The cerebral cortex can be differentiated into the older allocortex, with three or four layers, and the newer neocortex, with six laminated layers. The hippocampus is considered part of the older allocortex. Some neuroscientists argue that the six-layered neocortex has no clear homolog (counterpart or even precursor) in the brains of reptiles and birds and is a new addition to the mammalian brain (Briscoe & Ragsdale, 2018). The neocortex can be subdivided in many different ways. Four common subdivisions are in terms of the two hemispheres (left hemisphere and right hemisphere), four (reflected) lobes (frontal lobes, parietal lobes, temporal lobes, and occipital lobes), 52 cytoarchitecturally distinct numbered regions, known as "Brodmann areas" after pioneering German neurologist Korbinian Brodmann, and three types of cortices: primary, secondary, and association.

The primary cortex consists of the sensory and motor areas. The primary sensory areas receive sensory inputs via the thalamus and do the initial modality-specific processing (like luminance, spatial frequency, orientation, and motion in vision). The primary motor areas execute voluntary movements. As we have already seen in the case of vision, the primary cortex is hardwired but differs from subcortical structures in terms of delayed maturation timelines and the need for sensory environmental input for parameter settings (figure 10.7 and chapter 14). Lesions to the primary cortex can lead to specific sensory (and motor) deficits.

After processing via the primary cortex, sensory information passes on to secondary cortices for more abstract, higher-level processing. The secondary cortex does require considerable environmental input as part of the maturation process.

Large parts of the cerebral cortex are not committed to any specific sensory modality or motor operations and constitute a genuine, uncommitted association cortex. The association cortex is "softwired." Softwired systems are considered to be *experience dependent* (Greenough et al., 1987) in that they are not precommitted and require shaping by environmental interaction. Thus, unsurprisingly, softwired systems mature postnatally. In the case of the human prefrontal cortex, these maturation processes can extend into early adulthood. This means that while some rough parameters of the wiring plan are genetically encoded, most of the sculpting or shaping occurs through extensive environmental interaction. It is also the case that the association cortex is not limited to modality-specific inputs and outputs. The same specific regions are recruited for many different functions. It should be obvious why such systems are not suitable for controlling specific sensory, motor, autonomic, or instinctive functions. Softwired cortex constitutes a more "general-purpose" neural resource. As with hardwired systems, there is considerable plasticity prior to maturation. After maturation, lesions will still result in deficits, but they are more nebulous than those found in hardwired systems. We will revisit developmental trajectories of different brain systems in greater detail in chapter 14.

The most dramatic examples of softwiring can be seen in the human prefrontal cortex. Large ablations to the prefrontal cortex often seem to have so little effect on behavior and intellectual abilities (such as IQ and memory) that in the 1930s and 1940s many prominent neurologists and psychologists came to believe that the prefrontal cortex did not do anything of much importance (Hebb, 1939). This mistaken conclusion led to several decades of medically encouraged lobotomies, whereby large portions of the prefrontal cortex were intentionally destroyed in patients as a "treatment" for various psychiatric disorders (Freeman & Watts, 1942).

Softwiring of the association cortex is critical for its particular role in human cognition. It allows us to learn and revise beliefs throughout our lifetime. Without these properties, reason would not be possible. However, this reorganizational ability is not unlimited. We now know that lesions to the prefrontal cortex lead to significant deficits in generalized cognitive functions such as hypothesis generation, detecting inconsistency, and dealing with indeterminacy, all high-level cognitive abilities necessary for real-world functioning (Goel, 2019). These deficits can be hard to detect because of their generality but are very real and have real-world consequences. In chapter 14, I will argue that after maturation even a healthy association cortex is not very receptive to large-scale belief revision requiring architectural reorganization, because of a lack of neural resources.

I have now outlined a story of stepwise brain evolution where certain early emerging structures (brain stem, diencephalon, and subcortical structures) are conserved and propagated (with different levels of articulation), while others are modified and added at later points (cerebral cortex) in the phylogenetic tree. These different brain structures are associated with neural systems with different developmental trajectories (experience expectant and experience dependent), resulting in different functional properties (hardwiring and softwiring). Furthermore, the tethering of the later evolved systems to the earlier evolved systems is apparent in the anatomy and neurophysiology.

This evolutionary story of brain development allows for a reasonable mapping of brain structures to the different types of behaviors or kinds of minds that we began with, at least with the first three: autonomic, instinctive, and associative. But we still don't have a good sense of how reasoning emerges from these brain systems. If we associate reasoning with the emergence of the cerebral cortex, which is common among mammals, we cannot explain why we have propositional attitudes and they do not. So, the question now is, what is it about our brain that allows us to reason? We are not sure of the answer, apart from agreement that it is not God's grace. There must be something about the structure and organization of human brains that can account for it. But what? Over the past century, neuroscientists have puzzled over this question and proposed a number of answers.

Accounting for the Reasoning Brain

Most proposals to explain human reasoning abilities are variations of the claim that size matters: bigger is better. The most straightforward formulation is that animals with bigger brains are smarter. Humans have among the biggest brains, so we are the smartest. There is something to this. Chimpanzees (brain size 400 g) have bigger brains than chipmunks and chinchillas and are considered smarter by most measures. Similarly, we are smarter than chimpanzees and have even bigger brains. The development of the hominid brain has seen its size increase from 550–650 g in *Homo habilis* to approximately 1,450 g in *Homo sapiens*. So, is it as simple as bigger is better? Not quite. There are some very obvious and glaring counterexamples. Bottlenose dolphins have brains the same size as ours. African elephants have much larger brains (4,200 g), and sperm whales have enormous brains (9,000 g), but these species are not composing symphonies or mapping the stars. Cow and horse brains are larger than chimpanzee brains and four to five times the size of macaque monkey brains, yet no one would argue that cows and horses are more intelligent than chimpanzees and macaque monkeys.

Among humans, male brains are on average 200 g larger than female brains. Poet Lord Byron is said to have had a brain that weighed 2,238 g, while equally gifted novelist Anatole France had a brain with a paltry mass of 1,100 g (DeFelipe, 2011). Einstein had an average to small brain at 1,230 g. So the comparison of brain sizes within *Homo sapiens* indicates sex differences and considerable individual variation. Since elephants and whales are much larger than humans, and cows and horses are larger than chimpanzees and macaque monkeys, and human males are slightly larger than human females, perhaps we need to take body size into consideration and calculate a brain to body ratio.

Humans have a brain mass to body mass ratio of 1:40 (i.e., brain mass 2% of body mass), much higher than the brain mass to body mass ratio of elephants (1:560) and sperm whales (1:4,000) (Roth & Dicke, 2005). Among mammals roughly our size, we do have the most impressive brain to body size ratio. The ratio can also account for absolute brain size differences between human males and females. This makes more sense for the bigger brain hypothesis. But counterexamples persist; mice have a brain mass to body mass ratio of 1:40, the same as humans, and shrews have a body mass ratio of 1:10 (i.e., brain mass 10% of their body mass) (Roth & Dicke, 2005)! If we move beyond mammals, some insect species have a brain mass 15% of their body mass. Again, these counterexamples complicate the interpretation.

A third idea is to fine-tune the measure of brain to body mass ratio by introducing the construct of encephalization (Jerison, 1976). Instead of simply calculating the brain to body mass ratio, encephalization attempts to calculate and compare "excess" brain capacity beyond that required to innervate and control the organism's body. The larger the animal's body, the more neural resources will be required to innervate it. Anything beyond this requirement is deemed "surplus" neural capacity and should correlate with intelligence. The encephalization quotient (EQ) provides a measure of relative brain size, defined as the ratio between actual brain size and expected brain size based on a standard species of the same taxon.[5] A cat was chosen as a typical mammal and assigned an EQ value of 1. Humans have the highest EQ value on this measure, ranging from 7 to 8. The interpretation is that the human brain is 7 to 8 times larger than one would expect it to be, given our body size. A bottlenose dolphin has an EQ of 4–5, a chimpanzee has an EQ of 2.5, an elephant has an EQ of 1.3, and a rabbit has an EQ of 0.4. These measures seem consistent with the brain size hypothesis. However, there are still counterexamples. Chimpanzees and gorillas are considered to be more intelligent than New World capuchin monkeys, but the latter have the higher EQ.

What seems to be happening in each case is that the brain size (whether absolute, relative, or converted into an encephalization quotient) to cognitive capacity relationship seems to break down when species with similar brain sizes are compared across taxonomic orders. We believe the reason that increased brain size underlies increased cognitive abilities is that we assume that cognitive ability is a function of information processing capacity, that information processing capacity is (at least) a partial function of the number of neurons and the density of synaptic connectivity of the neurons, and that larger brains will have proportionately more neurons (with density of synaptic connectivity remaining reasonably constant across species). Given the various brain size and cognitive capacity relationships discussed, some of these assumptions must be incorrect.

Suzana Herculano-Houzel of Universidade Federal do Rio de Janeiro has developed a new methodology for more accurately counting brain cells that is helping to clarify the relationship between brain size and cognitive capacity (Herculano-Houzel and Lent, 2005). Her most basic finding is that even though the building blocks of all brains are similar, there are important differences in cellular scaling rules across taxonomic orders.

By counting neurons across different orders of mammals, initially rodents and primates, she discovered that the cellular scaling rules were different across them. An increase in the number of neurons in a rodent brain results in an exponential increase in brain size. Rodent brains increase in size through an increase in the number of neural cells, but there is also an increase in neuronal size and an even greater increase in nonneuronal (glial) cells, resulting in a decrease in neural density. A tenfold increase in the number of neurons in a rodent brain will result in a 35-fold increase in brain size. A primate brain has very different scaling rules. It scales up much more linearly. A tenfold increase in the number of neurons in a primate brain results in only an 11-fold increase in brain size. This means that the rate of increase in the number of cells matches the rate of increase in volume. Furthermore, there is no increase in the ratio of glial cells to neuronal cells in primate brains (as in the rodent brain), resulting in compact, economical brains with much higher neuronal density than in larger rodent brains. A rodent brain with the same number of neurons as a human brain would weigh 35 kg (Herculano-Houzel, 2009)! So the human brain is not a scaled-up rodent brain.

Instead, the human brain is a scaled-up primate brain (Herculano-Houzel, 2009). If a chimpanzee brain is scaled up to the size of a human brain, it will have a similar number of neurons and glial cells. If a human brain is scaled down to the size of a chimpanzee brain, it will have a similar number of neurons and glial cells. Human brains are primate brains with a large increase in the number of neurons. A chimpanzee brain has 28 billion neurons (Collins et al., 2016) and a gorilla brain has 33 billion neurons, compared to 86 billion for a human brain (Herculano-Houzel, 2012), so perhaps intelligence is a function of the absolute number of neurons. This might be a reasonable conclusion, but counterexamples are not far away. Elephant brains have 257 billion neurons (Herculano-Houzel et al., 2014), while sperm whale brains have even more neurons! So this leaves us in the same place as looking at brain mass. The number of neurons largely correlates with intelligence, but there are counterexamples that need to be accommodated.

Maybe "bigger is better" needs to be applied to the specific parts of the brain we have been discussing, such as the brain stem and cerebellum, diencephalon, subcortical structures, and cerebral cortex. It has been noted that these structures developed and differentiated at different points in the phylogenetic tree and vary in terms of the degree of hardwiring (i.e., predetermined functioning). Perhaps they scale up differently as well (Herculano-Houzel et al., 2015). Of the 86 billion neurons in the human brain, 16 billion (19%) are in the cerebral cortex, less than 1 billion (1%) in the subcortical regions, and 69 billion (80%) in the cerebellum. This distribution is consistent with that found not only in other primates but across most mammals. Across mammalian species, the cerebral cortex constitutes 50%–80% of brain mass but only 15%–25% of brain neurons, meaning that the expanded human cerebral cortex does not have *relatively* more neurons than other primate brains, even rodent brains. The cerebellum, which constitutes only 10%–20% of brain mass, contains 70%–85% of brain neurons. African elephants (and perhaps whales) remain outliers to these rules, with only 2% of their 257 billion neurons located in the cerebral cortex and 97% confined to the cerebellum. The scaling rules seem to preserve the relative number of neurons in each structure, across not only primates but most mammals, with the noted exception of African elephants and whales.

Perhaps we can set aside brain stem, cerebellum, diencephalon, and subcortical structures and focus on the cortex, or parts thereof. For decades, we have taught students that intelligence and reasoning are a function of frontal lobes and that our frontal lobes are much larger than any other animal's. This idea seems to have originated with Korbinian Brodmann. He estimated that the frontal lobes occupied 3% of the cerebral cortex in the rat, 11% in the macaque monkey, 17% in the chimpanzee, and 29% in the human (Herculano-Houzel, 2020).

The question of relative size of frontal lobes is not as straightforward as it appears. There are several complicating issues, including delineating the frontal cortex across species, choosing to measure mass, volume, or surface area, and determining the unit of comparison for "disproportional" expansion (should we make comparisons to all the cerebral cortex or some specific area?). Modern imaging and measurement techniques have generated mixed results. One group that used MRI techniques to measure frontal lobe volume in macaque monkeys, gibbons, orangutans, gorillas, chimpanzees, and humans found no staggering differences among primates (Semendeferi, Damasio, Frank, & Van Hoesen, 1997; Semendeferi, Lu, Schenker, & Damasio, 2002). The frontal lobes constitute 37% of the human cerebral cortex, compared to chimpanzees at 36%, orangutans at 35%, and macaque monkeys down at 28%. The researchers then further subdivided the frontal lobes into smaller regions, which are referred to as the dorsal, mesial, and orbital cortices. In humans, 59% of the frontal lobes made up the dorsal region, 26% the mesial region, and 15% the orbital region. The numbers for the other primates were almost identical. Other studies, using different measurement techniques, do suggest statistically significant enlargement of human frontal lobes (more specifically the prefrontal cortex), but compared to the great apes, the difference is small to modest (Donahue, Glasser, Preuss, Rilling, & Van Essen, 2018; Passingham & Smaers, 2014).

What about the number of cells in the frontal cortex? The human prefrontal cortex contains 8% of the neurons of the cerebral cortex, the same as in other primate species. However, the 8% translates into 1.3 billion neurons for humans compared to 590 million for chimpanzees (Collins et al., 2016) and 230 million for baboons (Herculano-Houzel, 2020). So even though percentagewise our frontal lobes may not be bigger than those of other primates, they are much larger in the sheer number of neurons.

So, again we return to the concept of absolute size, specifically the number of neurons, to differentiate human brains from nonhuman brains. The human brain seems to be a typical primate brain, with twice as many neurons in the prefrontal cortex as in a chimpanzee brain and six times as many neurons in the prefrontal cortex as in a baboon brain. In this account, the difference is purely quantitative. Is it possible that such quantitative changes at the level of brain size and structure can result in such massive qualitative changes at the level of behavior?

The issue of size (be it overall structure or subcomponents, and be it measured in mass or number of neurons), while subject to some counterexamples, is not without *some* merit. We have reviewed some cross-species comparisons of brain size and "intelligence." There are even some weak to modest correlations between overall human brain size and measures of IQ reported in the literature (Willerman, Schultz, Neal Rutledge, & Bigler, 1991). This general correlation even holds between gray matter density in specific brain areas and general IQ measures (Haier, Jung, Yeo, Head, & Alkire, 2004). These studies remain controversial not only because they are purely correlational but also because of a lack of agreement surrounding what IQ actually measures and how it relates to real-world functioning. Even more interesting than whole brain size correlations with behaviors are the correlations between specific brain structures and specific abilities. One of the most studied examples in the literature is the hippocampus.

We've already mentioned a relationship between relative size of the hippocampus and spatial navigation expertise in some animal species. Eleanor Maguire and her colleagues (2000) at the Wellcome Department of Cognitive Neurology, University College London, carried out a structural MRI study to clarify the neural correlates of spatial navigation in humans. Previous animal and human studies had indicated that the posterior part of the hippocampus was preferentially involved in spatial navigation. Maguire et al. (2000) compared the hippocampus size of licensed London taxi drivers with that of non–taxi drivers. London taxi drivers must undergo several years of arduous training involving the learning of spatial routes in the city of London. Beyond the years of training, the researchers' sample of taxi drivers had been driving London cabs for a period ranging from 1.5 to 42 years (average of 14.3 years). The non–taxi drivers had no such experience. It was reported that the posterior hippocampus of the taxi drivers was significantly larger than that of the non–taxi drivers. Furthermore, within the taxi drivers, the size of the posterior hippocampus positively correlated with the number of years they had been driving taxis.

In a follow-up study, the hippocampus size of London taxi drivers was compared to that of London bus drivers (Maguire, Woollett, and Spiers,

2006). While both groups do an equivalent amount of driving, there is a major difference in terms of spatial navigation. Bus drivers repeatedly navigate the same route every day. They do not possess the prodigious spatial knowledge of taxi drivers. Maguire, Woollett, and Spiers (2006) reported that the taxi drivers had larger gray matter volume in the midposterior hippocampus compared to the bus drivers, and only the gray matter volume of taxi drivers covaried with the number of years of driving. The bus drivers did not show this relationship. This suggests that the size varied not as a function of driving, stress, or other occupation-related factors but rather because of spatial navigation expertise.

A very different line of evidence that size matters—that quantitative differences in size can result in qualitative differences in behavior—comes from the study of computational neural networks. An early pioneer was Frank Rosenblatt. His goal was to understand and model neural processes involved in perceptual learning. In the late 1950s and early 1960s, he developed a simple neural computational model that has come to be known as the Perceptron (see the appendix in chapter 5 for the basics of neural models). It could do simple tasks such as letter recognition, but with modest success. Yet this simple modeling technique has scaled up—to the surprise of many computer and cognitive scientists (Minsky & Papert, 2017)—to perform impressive tasks such as facial recognition, verbal speech recognition, beating the world Go champion, detecting fraud in credit card transactions, and driving cars on city streets. What allowed for this phenomenal scaling? Surprisingly, a number of modest modifications and a massive increase in the number of nodes and connections.

The major conceptual advances included replacement of step functions with sigmoidal activation functions, incorporation of nonlinear learning functions, employment of recurrent neural networks, and auto encoders. Switching to sigmoidal activation functions allows for greater learning sensitivity. The incorporation of nonlinear learning functions allows for nonlinear classification. The innovation of recurrent neural networks allows for the use of sequential information (for example, to predict the next word in a sentence, it is useful to know the word that preceded it). Autoencoders use hierarchically organized hidden layers to pass inputs from a large number of neurons to a smaller number of neurons, through multiple successive levels, generating a compressed, more abstract representation at each successive level. The training occurs by running the system "backward" and comparing the output generated from the abstract representations with the original input and modifying connections accordingly until the original representation can be reconstructed from the abstract representation. This provides for a sort of "unsupervised" learning, with the original input acting as the teacher. Each of these contributions is important in its own right and part of the story of how the humble Perceptron scaled to perform real-world tasks.

One of the most impressive of these systems was the Google network that pulled off the amazing feat of correctly categorizing human faces and cat faces from YouTube videos. It used a nine-layer network with one billion connections, trained on 10 million images, running on 1,000 machines with 16,000 cores for three days (Le, 2013)! So, perhaps sheer size, computing power, and volume of data do make a qualitative difference.

But brain size without structural and organizational changes cannot be the complete story. In addition to size, the issue of organization may be equally important. If neural networks are to model human mental states, they must exhibit the properties of compositionality, systematicity, productivity, and inferential coherence (Dauphin et al., 2017; Shen et al., 2019; Smolensky, 1988; Socher et al., 2011). As argued in chapter 5, association on its own is insufficient to capture most relations needed for human reasoning. Neural networks need to be organized and structured in certain ways to accomplish this. This is noted in figure 10.7 in terms of the "innate constraints" that structure some cortical networks.

There is also some emerging evidence of differences at the level of microcircuitry and biochemistry of neurons that may help to account for structural and organizational changes at the cortical level. In terms of biochemistry, there is some evidence that neurons in the human temporal cortex have membrane properties different from those of other mammals. This affects their abilities to transfer electrical charges from dendrites to cell body and to generate and propagate an action potential along the axon, which in turn may affect their computational properties (Eyal et al., 2016).

In terms of microcircuitry, there is some evidence for species-specific differences in synaptic organization (Defelipe, Alonso-Nanclares, & Arellano, 2002; DeFelipe, 2011). For example, while the human cortex has less neuronal density than the mouse cortex, it does have greater synaptic density (approximately 30,000 synapses per neuron versus 21,000 synapses per neuron). Synapses are points of communication between neurons via passage of neurotransmitters from a presynaptic member to a postsynaptic member at dendritic spines (see the appendix in chapter 5). Dendritic spines are small protrusions on the dendrites of neurons. They receive excitatory input from synapses of other neurons. There is considerable variation in the number of dendrites, their size, and branching as a function of cortical areas and across species. Dendritic spines in human neurons have much larger volume and longer necks than in mice. Pyramidal cells in the human prefrontal cortex have 70% more dendritic spines than found in the corresponding cells of macaque monkeys and 400% more than found in the corresponding cells of mice. These differences affect the electrical, biochemical, and biophysical properties of neurons.

Finally, there is also some recent evidence for the existence of a special type of neuron not yet found in nonhuman brains (Boldog et al., 2018). This neuron, called the rosehip neuron, has compact bushy dendrites with lots of branching points rather than having long dendrites. It is not clear what this neuron actually does, but it is an inhibitory neuron, perhaps involved in regulating the flow of information to other parts of the brain. It seems to make up about 10% of the human neocortex. Again, while it is not certain, it is not unreasonable to believe that these physiological differences will translate into differences that will affect the overall computational and behavioral properties of the system.

Our current best account for the reasoning brain invokes a larger number of (hierarchically organized) neurons in the neocortex than in any other animal; different membrane properties and greater synaptic density of neurons compared to other mammals; and some unique types of neurons. This view is perfectly consistent with the evolutionary story and allows us to complete mapping our four different kinds of minds to four different kinds of brains.

A Tethered Brain for a Tethered Mind

We now have in place an account of the tethered brain that underlies the tethered mind. The overall system is diagrammed in figure 10.7. The left-hand column of the diagram identifies the phylogenetically distinct brain structures introduced in the chapter: brain stem, diencephalon, subcortical structures, and cerebral cortex (primary, secondary, and association). These brain systems have different functional and computational properties because of hardwiring (brain stem, diencephalon, most subcortical structures, and primary cortex) and softwiring (hippocampus and association cortex). This is indicated in the second column of the figure. Hardwired systems are largely experience-expectant systems that come genetically pre-committed (but may require some environmental parameter settings during a critical period), have very specific inputs and outputs, mature very early, and lose plasticity very quickly after maturity. Softwired systems are experience dependent in that they are largely uncommitted and require shaping or training by environmental interaction. They mature late and

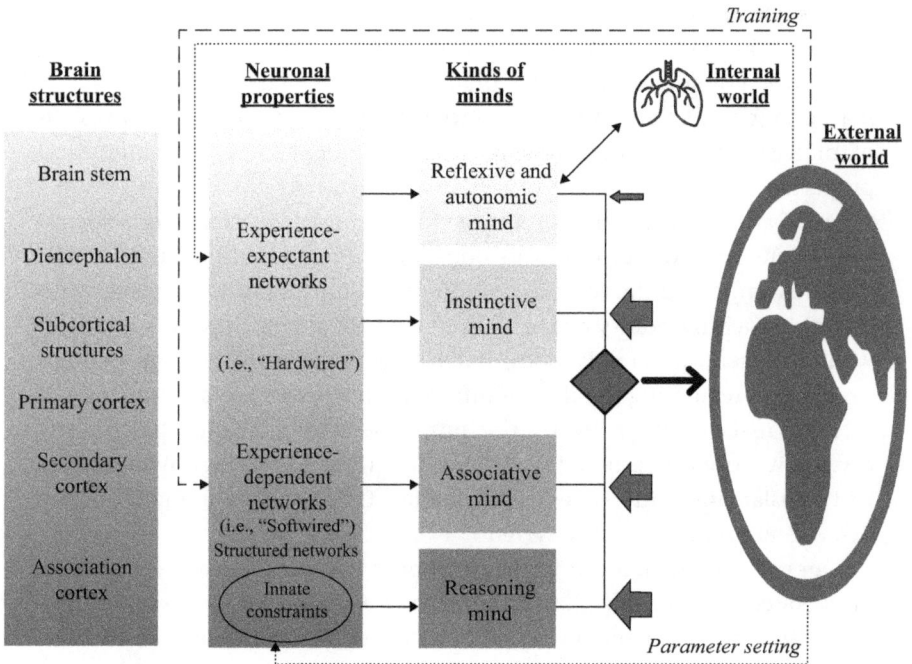

Figure 10.7
Brain systems and neural network properties associated with the four behaviors of interest. The qualitative differences in the four behaviors or minds under consideration are underwritten by the appearance and variability of brain structures in the phylogenetic tree and their corresponding levels of experience-expectant and experience-dependent maturation schedules. Innate structural constraints apply to certain parts of the cortex, so they exhibit the properties of compositionality, systematicity, productivity, and inferential coherence required by the reasoning mind. The tethering of the various behaviors and the underlying neural systems allows for a single blended or integrated response.

are thought to allow for lifelong learning and belief revision (though I will suggest some limits on this plasticity in chapter 14).

This still leaves us with the dilemma that the association cortex is widely available among mammals, while reasoning is not. To account for this, it is proposed that neural networks underlying human reasoning need to be further constrained and structured so that they exhibit the properties of compositionality, systematicity, productivity, and inferential coherence necessary to represent propositions and encode various coherence relations. It is reasonable to assume that some of the special features of the human neocortex provide innate constraints that along with some environmental parameter settings facilitate this structuring. This is also depicted in the second column of figure 10.7.

The third column of figure 10.7 shows the mapping from brain systems to behaviors. A critical feature of this mapping of phylogenetically distinct brain systems onto phylogenetically distinct behaviors is that the systems do not float on top of one another. Newer systems are tethered to older systems. One example of this tethering is illustrated in the neural wiring across various levels in the motor control example in figure 10.4. At the brain level, the interconnections are readily apparent and must be acknowledged. However, at the behavioral level, the tethering is usually ignored. One reason for this may be that while there are multiple systems in play, the organism is generating one blended response, also illustrated in figure 10.7. A blended behavioral response from these distinct systems requires a common vocabulary and some sort of control structure. The common vocabulary is addressed in chapter 11 and the control structure in chapter 12.

*   *   *

There are qualitative differences in the types of behaviors exhibited by brains, and these differences are underwritten by differences in neuroanatomy and neurophysiology. Brains specializing in autonomic functions and instincts can largely get away with hardwired systems. Brains that allow for associative learning need memory, which requires uncommitted, softwired neural hardware such as found in the hippocampus and cortex. Brains that can reason require even more malleability and flexibility for the acquisition, updating, and revision of beliefs. This function is served by the neocortex. However, most primate brains have a large, well-developed neocortex but lack the ability to reason. There is currently no consensus on what differentiates the human reasoning brain from a nonhuman primate brain. However, the evidence suggests that the combination of a very large number of neurons along with some special types of neurons and unique structural properties, and innate constraints, allows for the organization and structuring of some neocortical networks such that they exhibit the generativity, productivity, compositionality, and inferential coherence necessary for processing propositional contents. All of these brain systems are hierarchically organized such that phylogenetically newer systems are tethered to phylogenetically older ones.

Taking this neuroanatomy seriously has far-reaching consequences for our understanding of human behavior. It provides an underlying biological basis for the qualitative differences in autonomic, instinctive, associative, and reasoning behaviors. The biology suggests that the cognitive characterization of rationality as unhindered by more earthly concerns is not rooted in reality. The neuroanatomy paints a picture of hierarchically organized systems but with a clear tethering of newer systems to older ones. It is these interconnections provided by the tethering that allow a low blood sugar level, cheater punishment instincts, and my learned aversion to advertising to enter the world of reason. But what is the proximal mechanism for this? In particular, what is the nature of the "information" that is flowing between these various systems? This is not a trivial question. There are qualitative differences in what and how information is processed at each level. How can the various systems possibly interact? What is the common currency? I think the answer to this question has to be framed *not* in terms of information processing but in terms of *feelings*. It is feelings of pleasure and pain that allow phylogenetically older systems to intrude into the reified space of reason, and vice versa. We are now ready to examine how this might occur.

11 Feelings: Chocolate, Lust, and Coherence

Nature has placed mankind under the governance of two sovereign Masters, pain and pleasure.
—Jeremy Bentham

The postulate that affective processes have an objective existence and that they intervene between stimulus and response has great utility. . . . Experimental findings practically demand such a hypothesis.
—Paul Thomas Young

Almost every feeling of physical pleasure or pain felt by your forebrain has climbed its way there through the brain stem.
—Kent Berridge

At this juncture in our story, we are approaching a dilemma. The model of mind that we have been developing to explain organismic behavior postulates multiple qualitatively distinct systems based on different principles and mechanisms. The autonomic mind is based on homeostasis and reflex arcs. The instinctive mind is explained as consisting of fixed action patterns triggered by (metaphorical) action-specific energy reservoirs. The associative mind is about learning through reinforcement. The reasoning mind is about coherence relations between propositional attitudes. Despite multiple systems, behavior consists of a single blended response. This implies a global integration function that takes input from each system and determines the blended response. The conundrum is to explain how there can be integration across the levels in the absence of a common language. The solution to this problem requires a bold, speculative conjecture: what is common to each system are *feelings*. Feelings provide the common currency allowing for communication across levels and for calculation of the overall response to a given situation.

This is a pivotal chapter in our story. I want to make the controversial case for the central role of feelings in the behavior of all mammals, perhaps even all vertebrates. I will characterize feelings, differentiate them from emotions, and argue that feelings are generated in old brain stem, diencephalon, and subcortical systems that have been widely conserved across large parts of the phylogenetic tree. This means that they are available to each of our four minds. I will propose that feelings are the solution evolution has come up with to solve the two critical problems of *selecting* behavior and *initiating* behavior. Feelings serve these critical roles in the internal operations of each of the four different kinds of minds we have been discussing and provide the common currency for the integration of a single behavioral response.

Who's Afraid of Feelings?

Feelings span the range of sensations associated with the kink in my neck, the warmth of sunshine on my face, the pain in my knee, my heart racing after sprinting, hunger pangs, bowel distention, the taste of chocolate cake, sexual arousal, the desire to see a loved one, fear of lions, pangs of jealousy, anger and remorse, and my very sense of being. Feelings are undoubtedly the most vexing and shunned topic in psychology and neuroscience.

The problem with feelings is that they are *feelings*. They have a first-person ontology, meaning they are inherently subjective. I'm certain of the feelings associated with the kink in my neck, the taste of chocolate cake, and my fear of heights through my direct first-person experience and only through my direct first-person experience. If this is the case, how can feelings be studied objectively? Modern empirical psychology began in the 1870s as an exploration of conscious feelings. Wilhelm Wundt, the founder of the first empirical psychology lab, is attributed to have said that "when we study a living system from the outside, we are doing physiology; when we study it from the inside, we are doing psychology." Since the only method for directly accessing feelings is through introspection, and introspection does not generally permit repetition and verification of experiments, the hallmarks of the scientific method, this initiative was short-lived and was soon overwhelmed by behavioral psychology.[1]

Since then, many serious, accomplished psychologists and neuroscientists have argued that there is no independent, objective evidence for the reality and causal efficacy of feelings. They note that what can be observed and measured are behaviors and their underlying physiology; let us stick to these measures in our theorizing (LeDoux, 2012; Skinner, 1953). In fact, we have learned a great deal about living organisms through studies of behaviors and physiology, without any appeal to feelings. If feelings do exist, perhaps they are epiphenomenal. That is, they may be like the sound generated by hammering a nail into a piece of wood. It is a natural consequence of metal striking metal but has no part to play in the causal story of driving the nail into the wood and thus has no part in scientific explanation.

Indeed, physics has made enormous progress by redefining intuitive concepts to eliminate feelings. Consider the notion of heat. Heat was initially defined as the feeling of the sunshine on your face or the feeling of putting your hand in a pot of boiling water. This did not turn out to be particularly useful for understanding the world. As physics progressed, it redefined heat as the transfer of mean kinetic energy of the object's component particles. This redefinition carved off the subjective feeling component and provided an objective, measurable concept that actually deepened our understanding of the world. Skeptical psychologists and neuroscientists have taken a similar approach to feelings. They may be real, but they are not causally relevant in explaining the world and therefore not a subject matter in and of themselves. What needs to be studied is the accompanying behavior. The danger with this position is that if feelings turn out to be essential to the subject matter of psychology and neuroscience, carving them away also means carving away our subject matter (Searle, 1992). To confront this issue is to confront the problem of consciousness. The philosophical problem of consciousness is beyond my paygrade. Nonetheless, feelings are central to my story. By feelings I'm referring to the pain sensation of an electric shock resulting in withdrawal and avoidance behavior (both in myself and in the rat) and the pleasurable taste of chocolate cake resulting in consumption behavior (again, both in myself and in the rat).

Fortunately, a number of psychologists and neuroscientists are gravitating away from this skepticism and are willing to acknowledge feelings both in their informal conversation and formal theory development (Berridge & Kringelbach, 2013, 2015; Bindra, 1974; Craig, 2009; Damasio & Carvalho, 2013; Leknes & Tracey, 2008; Panksepp, 2011; Young, 1959). The main reasons for this radical shift are a combination of behavioral and neurophysiological data and some basic tenets of the theory of evolution. These scientists are beginning to accept that one can explain more data more coherently by *positing* feelings rather than ignoring them. However, sharp internal divisions remain as to the source and nature of feelings (Barrett, 2006; LeDoux, 2012; Lindquist, Wager, Kober, Bliss-Moreau, & Barrett, 2012; Panksepp, 2007). I agree that feelings are not just phenomenologically real but also causally efficacious in behavior. In fact, they may be the primal essential feature of all mammalian, perhaps even all vertebrate, life. The role that they play may be that of initiating and guiding behavior.

Characterizing Feelings

Feelings are sensations. Pioneering neuroscientist Charles Sherrington (1952) grouped sensations into the following five categories: teloreceptive (vision and hearing), proprioceptive (limb position), exteroceptive (touch, temperature, and pain), chemoreceptive (smell and taste), and interoceptive (visceral). More recently, some neuroscientists have been suggesting reorganization of the interoceptive category to include all aspects of the physiological condition of the body, not just the viscera (Craig, 2002). For our purposes, it may be adequate to group feelings into two broad categories: interoceptive, referring to all internally generated homeostatic and visceral bodily sensations,[2] and exteroceptive, referring to the sensations emanating from the impact of the external environment on the five senses. It may also be useful to recognize, as a third category, the "feeling of effort," or what John Searle (1983) called "intention in action," associated with volitional action.

Feelings are typically characterized along two orthogonal dimensions: valence and arousal. Valence constitutes a scale along a positive/negative or pleasant/unpleasant dimension. For many, eating chocolate cake is at the positive end of the scale and eating bitter melon is at the negative end of the scale. The second dimension is arousal. It ranges from high to low intensity. Sexual orgasm is at the high end of the intensity scale and the positive end of the valence scale. Experiencing an electric shock is also at the high end of the intensity scale but at the negative end of the valence scale. Feeling depressed implies both low arousal and negative valence. Feeling calm and content implies low arousal and positive valence. Some researchers argue that each and every feeling, whether it be the kink in my neck or the feeling of disappointment in the peer reviews of my last manuscript, can be captured along the two dimensions of valence and arousal, but others disagree (Colibazzi et al., 2010; Panksepp, 2007; Panksepp & Biven, 2012; Russell, 2003).[3] Either way, these two dimensions are a useful organizing tool. Feelings also have a third dimension, temporal duration, which is often ignored. Feelings have a beginning and an end. They can wax and wane. The pleasurable taste of chocolate cake commences with the first bite (or perhaps even first aroma or sight) and subsides postconsumption.

From Feelings to Emotions

In addition to these basic feelings associated with internal bodily regulation and exterior sensory inputs, there is another set of feelings or affective states that all humans are familiar with: *emotions*.[4] Our emotional states encompass feelings, but feelings are not emotions. Examples of common human emotions are fear, rage, disgust, hope, and jealousy. Like interoceptive and exteroceptive feelings, emotions have valence, physiological arousal, and duration components, but unlike interoceptive and exteroceptive feelings, they also have intentional objects, action tendencies, physiological expressions, and cognitive antecedents associated with them (Elster, 1998).

To say that emotions have intentional objects is to say that emotional states are representational states. That is, they are directed at, refer to, or represent objects, individuals, and states of affairs in the world.[5] Hunger in itself is not a directed state; I can feel hungry without wanting to eat anything in particular. My craving for chocolate cake, though, is a directed state. Recall the discussion about the nature of contents of intentional states from chapter 6. I made a distinction between the directedness of the intentional states of a cat tracking a mouse and human intentional states that have sophisticated propositional contents. We need to maintain this distinction here to allow for emotional states in animals. My directedness toward an immediately present object, such as the slice of chocolate cake on my plate, may be similar to the type of directedness that a hungry dog has to that same slice of cake, but I am also capable of more sophisticated forms of directedness that require propositional content:

> I pushed my fork through the top layer of creamy frosting, then all three layers of the cake. Keeping my eyes down, I put the fork to my mouth. He'd used good chocolate, I knew, and after a moment, I picked up a note of coffee, which only intensified the flavor of the chocolate. The frosting was decadent and smooth, but not cloying. In fact, the entire bite struck the precise balance of sass and sweet. (Stuart, 2017)

Both types of mental states directed at the chocolate cake qualify as referential or directed mental states, but they are qualitatively different. The former type of directedness should be widely available on the evolutionary tree. The latter type requires mental states with propositional content, presumably only available to humans, and allows emotions to enter the reasoning mind.

Physiological expressions such as bodily posture, pitch of voice, blushing, smiling, baring of teeth, laughing, frowning, weeping, and crying are usually associated with emotions. There are various degrees of conscious control over these expressions. No one can blush on command, but some people can cry on cue. Actors can learn to imitate many of these facial expressions and body postures. It has been argued that specific facial expressions are associated with specific emotions across all human cultures and even in nonhuman animals, providing evidence for a basic set of emotions (Berridge & Kringelbach, 2015; Ekman, 1993).

Emotions are also associated with action tendencies. Action tendencies are not unlike the "fixed action patterns" discussed in the context of instinctual behaviors (chapter 4). Nico Frijda, quoted in Elster (1998, p. 51), described them as "states of readiness to execute a given kind of action. . . . Action tendencies have the character of urges or impulses." Such "urges and impulses" also appeared in Lorenz's instinct model as pent-up energy reservoirs. For example, fear may lead to fight or flight, lust may lead to actions to possess the object of sexual desire, shame may lead one to hide or disappear, guilt may lead to atonement or confession, envy and malice may lead one to destroy, love may lead to approaching and touching the other person, and anger may lead to hurting the person who has hurt you. But don't fixed action tendencies violate the "gap" that plays a critical role in our conception of rationality (chapter 6)? This need not be the case if no *specific* stimuli are necessary and sufficient to trigger a *specific* emotion and if no *specific* action patterns are associated with each emotion. This is illustrated in the example below from *King Lear*.

Many human emotional states are triggered by other intentional states, typically beliefs and desires. Why did Shakespeare's King Lear become angry with his youngest, favorite daughter, Cordelia? When asked to profess her love for him alongside her two sisters, Cordelia has no words to compete with the insincere flattery offered by Goneril and Regan and, when pressed, replies as follows:

> Good my lord,
> You have begot me, bred me, loved me: I
> Return those duties back as are right fit,
> Obey you, love you, and most honour you.
> Why have my sisters husbands, if they say
> They love you all? Haply, when I shall wed,
> That lord whose hand must take my plight shall carry
> Half my love with him, half my care and duty:
> Sure, I shall never marry like my sisters,
> To love my father all.

This is not enough for Lear. He desires more obsequious displays of her love. Failing to receive them, he comes to believe that she does not really love him, and being particularly wounded because she is his favorite, he flies into an angry rage and disowns her and divides his kingdom among her sisters:

> Let it be so; thy truth, then, be thy dower: . . .
> Here I disclaim all my paternal care,
> Propinquity and property of blood,
> . . . thou my sometime daughter.

This illustrates not only the triggering of the human emotion (anger) by beliefs and desires but also the triggering of accompanying action tendencies. Notice that there is nothing that *compels* Lear to be angered by any *specific* antecedent belief and desire. However, once the anger is triggered, the action tendency unfolds. It is not, however, stereotyped as in the case of instincts. There are numerous actions, ranging from a verbal expression of disappointment to execution, for Lear to express his anger. The rationality gap is intact. The duration component of emotions is also illustrated when Lear belatedly regrets his anger and actions:

> O Lear, Lear, Lear!
> Beat at this gate that let thy folly in
> And thy dear judgement out!

This characterization of emotions in terms of intentional or directed states, associated with action tendencies, physiological expressions, and the fact that they are usually triggered by other intentional states (in humans), differentiates them from sensations associated with interoceptive and exteroceptive systems and volitional motor actions. The status of emotional states is a highly contentious and debated issue in the literature. Some researchers believe that emotions are high-level cognitive constructs computed or inferred from core interoceptive biofeedback signals by neocortical structures (Barrett, 2006; Barrett, Quigley, & Hamilton, 2016; LeDoux & Brown, 2017; Lindquist et al., 2012; Ortony, Clore, & Collins, 1988; Seth, 2013). As such, they are available only to humans and perhaps some other primates. Other researchers believe that some primal emotions are internally generated in specific deep brain stem, diencephalon, and subcortical regions of mammalian brains (perhaps even vertebrate brains generally) and thus are a common heritage of large parts of the evolutionary tree (Berridge & Kringelbach, 2015; Damasio & Carvalho, 2013; Ekman, 1993; Kringelbach & Berridge, 2009; Panksepp, 2007; Panksepp, Lane, Solms, & Smith, 2017; Toronchuk & Ellis, 2013). I believe that the bulk of the evidence supports the latter position, and I also adopt it. However, there is also a potential for confusion here that needs to be preempted. When talking about emotions in nonhuman animals, we are referring to emotions *without* propositional contents. I will shortly introduce some vocabulary to distinguish full-blown human emotions, which have propositional content and participate in the reasoning mind, from nonhuman emotions. We will also see that the latter have a characterization similar to instincts.

Origins of Feelings

You are sure that you have feelings. Based on the observation that I'm physically very much like you, and belong to the same species, you are probably prepared to accept that I have feelings. But what is the evidence that nonhuman animals also experience feelings? How can you know whether your dog feels pain or is capable of loving you?

The first-person ontology of feelings makes this a difficult question to answer, but, in reality, we can answer the question of feelings in nonhuman animals in the same way we answer it in fellow humans: through behavioral and anatomical/physiological observations. We have already noted some of the pitfalls of relying purely on behavioral observations (chapter 7). However, combining behavioral with anatomical/physiological observations and noting the basic engineering principle that structure is not unrelated to function allows us to make some headway. In biology, many functional homologies can be mapped onto structural homologies. Similarities in motor function or visual function across species are underwritten by similar neural architecture. If we can identify the source or generators of feelings in the brain, we can then see how widely these structures are available on the evolutionary tree.

Where feelings first appeared on the phylogenetic tree is relevant to understanding their potential functions. If feelings are generated in the neocortex, then only humans, and perhaps some other primates with well-developed neocortices, will have access to them. In that case, their role may be confined largely to cognitive systems. However, if feelings are generated in the brain stem, diencephalon, and subcortical structures—which appeared very early and have been conserved, certainly in mammals and perhaps in all vertebrates—feelings may be more widely available and have much more basic functions affecting many survival systems. There is evidence from electrical stimulation studies, decorticate (removal of cortex) studies, chemical stimulation studies, and conditioning studies to suggest that feelings are indeed generated in deep brain stem, diencephalon, and subcortical structures, though they can be represented in higher-level cortical structures for various purposes and combined into sophisticated, complex emotions, available only to humans (Berridge & Kringelbach, 2013; Damasio & Carvalho, 2013; Kringelbach & Berridge, 2009; Panksepp, 2007, 2011; Panksepp & Biven, 2012; Pfaff, Martin, & Faber, 2012; Venkatraman, Edlow, & Immordino-Yang, 2017).

The study of feeling systems in the brain began with a classic study from the 1950s where electrodes were placed in the septal region in the brains of rats (and other animals) and the animals were placed in a Skinner box with a lever that, when pressed, generated an electrical stimulation in the electrode (figure 11.1a). The rats soon learned to press the lever and then use it extensively to obtain continuous stimulation (Olds & Milner, 1954).[6] Not only would animals work for the stimulation, the stimulation could be used as a substitute for food as a reward in classical conditioning and operant conditioning experiments (Ross et al., 1965; White & Milner, 1992). As electrical stimulation serves no biological homeostatic function (unlike food), animals are expending effort to receive it presumably because they find it pleasurable or rewarding in itself (Olds & Milner, 1954). Deep electrode stimulation of the same septal brain region in humans (figure 11.1b) was found to be associated with sexual arousal in two patients (Heath, 1972). More recent studies note that these electrodes were placed very close to the nucleus accumbens and suggest that it, not the septal region, is the source of the rewarding arousal. There is also some current reconsideration as to whether the studies have activated pleasure centers or motivation centers (Berridge & Kringelbach, 2015). We will return to this important distinction.

Electric brain stimulation in the septal region elicits these rewarding feelings not only in intact rats but also in decorticate rats that have had their neocortex surgically removed (Panksepp, 2007). These animals still work to receive electrical stimulation, and they continue to engage in many pleasurable behaviors, such as play (Panksepp, Normansell, Cox, & Siviy, 1994) and sexual lordosis (Carter, Witt, Kolb, & Whishaw, 1982). Another strong source of evidence that the generators for these feelings are in the brain stem, diencephalon, and subcortical regions comes from clinical cases of children born without a neocortex. These children are still capable of conscious experience and emotional reactions (Merker, 2007; Shewmon, Holmes, & Byrne, 1999).

Rats are known to emit frequency-modulated 50 kHz calls during positively valenced appetitive behaviors such as sex and play. Electrical brain stimulation in deep brain structures such as the lateral preoptic area, lateral hypothalamus, and ventral tegmental area elicits the same 50 kHz vocalizations, suggesting that the animals find the stimulation equally pleasing

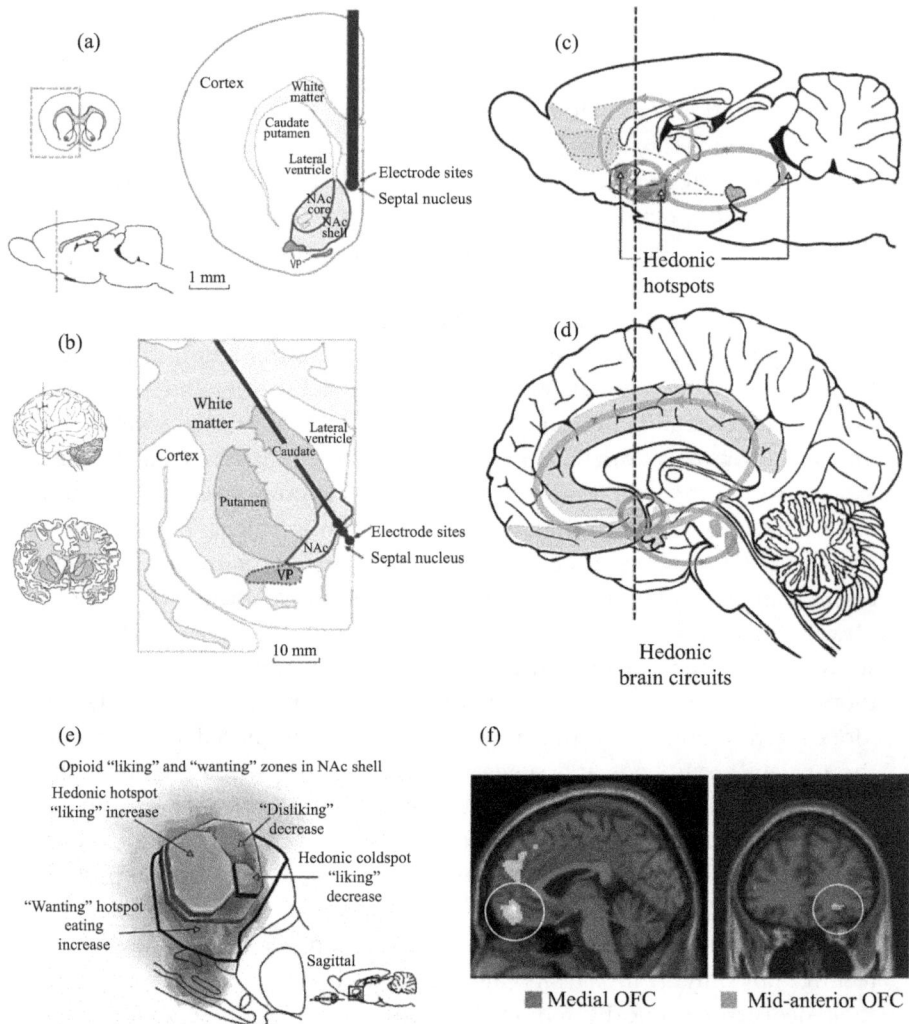

Figure 11.1

(a) The location of the hedonic "hotspot" discovered by Olds and Milner (1954) in the septal nucleus of rats has been reconstructed and found to be very close to the nucleus accumbens. (b) Placement of electrodes in or near the nucleus accumbens in one patient reported by Heath (1972) resulted in feelings of sexual arousal. (c), (d) Hedonic reward systems in both rodents and humans involve similar interlinked brain stem, diencephalon, and subcortical networks, with some cortical representation, particularly in humans. The systems involved include the periaqueductal gray (PAG), the ventral tegmental area (VTA), ventral pallidum, nucleus accumbens, amygdala, hypothalamus, insular cortex, cingulate cortex, and orbital frontal cortex. (e) Shows the identification of distinct "liking"/"disliking" and "wanting" hotspots in the nucleus accumbens of a rat brain. (f) Reward regions are represented in the human orbital frontal cortex. Figure reproduced (with some reorganization) from Kringelbach and Berridge (2010) with permission of the authors.

(Burgdorf, Wood, Kroes, Moskal, & Panksepp, 2007; Burgdorf et al., 2008, Burgdorf, Panksepp, & Moskal, 2011).

Opiate drugs such as morphine and heroin mediate sensory pleasure and positive social bonding systems in the brain (Panksepp, 1981; Panksepp & Biven, 2012). Like humans, animals will seek out and work for these drugs (Ikemoto, 2010), presumably for similar reasons: consuming them is pleasurable. When these drugs are infused directly into animal brains, the animals prefer morphine infusions into primitive brain stem regions such as the periaqueductal gray and the ventral tegmental area over infusions into other regions, even though these other regions also have abundant opiate receptors. This preference suggests that the preferred brain stem and subcortical regions for infusion may be the main source of the pleasurable affect generation associated with the drug (Olmstead & Franklin, 1997).

Finally, it is well known that animals develop preferences for places where they have had positive experiences such as food and sex (conditioned place preference) and avoid places where they have had negative experiences such as an electric shock or the odor of a predator (conditioned place aversion). They develop the same place preferences and aversions, respectively, to artificial electrical and chemical stimulation of the relevant brain systems, suggesting that they find the brain stimulations equally as pleasing as food and sex or as displeasing as an electric shock (Olmstead & Franklin, 1997; Panksepp & Biven, 2012; Pfaus et al., 2012).

Since these pioneering studies, neuroscientists have made considerable progress in identifying and mapping out reward (and aversion) systems in both nonhuman and human brains. There is considerable consensus that hedonic systems in human and nonhuman animals involve overlapping hierarchically organized neural nets in the periaqueductal gray (PAG), ventral tegmental area (VTA), ventral pallidum, nucleus accumbens, amygdala, hypothalamus, insular cortex, cingulate cortex, and orbital frontal cortex (figure 11.1c, d). Many different types of rewarding stimuli (food, sex, addictive drugs, even art and music) activate this same common system (Berridge & Kringelbach, 2015).

As noted in chapter 10, brain stem, diencephalon, and subcortical structures are highly conserved across mammals and even across vertebrates. The data reviewed in chapter 10 and the present chapter suggest that they serve similar essential functions across species. If one of these functions is the generation of feelings in humans, it is reasonable to hypothesize that homologous regions serve homologous functions in large parts of the phylogenetic tree.

This is a hypothesis. Do not bet the house on it. But it is a reasonably robust hypothesis; I would bet my car on the grounds that we can make sense of more behavioral and neurophysiological data with this hypothesis than without it. It is also important to note that these considerations do not speak to the question of whether your dog can actually love you. They only indicate that dogs have an affective life. The particulars of that life will undoubtedly be shaped by the evolutionary niche of each species.

But why should these feeling circuits be conserved? Nature is not prodigal. What critical role do feelings play in enhancing survival and reproduction (i.e., fitness)? Why should brains, which are very expensive to maintain—comprising 2% of body weight but consuming 25% of energy (in humans)—conserve structures and processes required to generate feelings? Why do we need feelings? The answer may lie in the fact that survival and propagation of organisms depends on the selection and initiation of appropriate behaviors or actions in response to environmental (internal and external) change. Feelings may be the solution that evolution has converged on to detect certain changes and select and initiate actions. This hypothesis is particularly robust if feelings are generated in phylogenetically old brain stem, diencephalon, and subcortical structures, as the data indicate.

Function of Feelings: Motivate and Guide

I propose that feelings evolved to allow organisms to detect changes in their environments and to select and initiate appropriate actions.[7] Consider the oral sensory system of taste discrimination as an illustration (figure 11.2). It begins with a chemical reaction activating taste buds, proceeds via sensory neurons to brain stem nuclei, to the thalamus, and then to the insular cortex (Matsumoto, 2013). The pathway is very similar for rodents and primates, except it bypasses the parabrachial nucleus in primates and humans. It allows for differentiation between the sweet buttery taste of chocolate cake and the salty taste of crackers and the sour taste of lemons.

Not only can I differentiate between different tastes, I also have different preferences for them. I like certain tastes more than others. I will go out and purchase and consume chocolate cake more frequently than lemons. The taste differentiation system alone cannot explain this. A notion of reward or pleasant affect is needed to account for preference. I prefer chocolate cake to lemons because it tastes better. We can *feel* this preference in ourselves and we can see it behaviorally in others. Similar facial liking and disgust reactions to sweet and bitter tastes can be elicited from children on

Ascending gustatory neural pathway

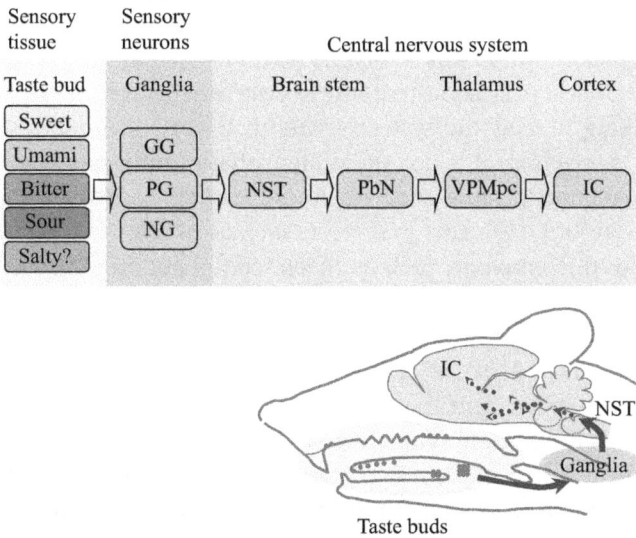

Figure 11.2

The rat's system for taste discrimination. Chemical reactions in the oral cavity between food and the taste buds are recognized as distinct tastes in the insular cortex after processing in the brain stem and thalamus. GG = geniculate ganglia; IC = insular cortex; NG = nodose ganglia; NST = nucleus of the solitary tract; PbN = parabrachial nucleus; PG = petrosal ganglia; VPMpc = ventral posterior medial nucleus of the thalamus. Figure reproduced with permission from Matsumoto (2013).

the first postnatal day and are homologous across humans, primates, and even rodents (Berridge & Kringelbach, 2015).

Neuroscientists Kent Berridge of the University of Michigan and Morton Kringelbach of Oxford University, along with many colleagues, have spent decades studying the neural basis of the brain's reward system. They propose that it can be subdivided into two distinct (but interrelated) components, the "wanting" system and the "liking" system. They characterize *wanting* as the incentive salience or motivational magnet component of reward. *Liking*, by contrast, is a hedonic reaction (i.e., feels pleasurable) and is detectable both behaviorally and in neural signals generated by subcortical brain structures (Berridge & Kringelbach, 2015).

At the neural level, taste sensations are generated by a small set of discrete "hotspots" (liking) and "coldspots" (disgust) located in the nucleus accumbens (figure 11.1e), ventral pallidum, parabrachial nucleus in the brain stem, and perhaps also in the orbital frontal cortex and insular cortex, at least in humans (figure 11.1f). Activation of hotspots for liking in the nucleus accumbens amplifies pleasurable reactions, while activation of the coldspots dampens pleasure and initiates disgust reactions. Lesions in the ventral pallidum result in loss of hedonic response to taste (Berridge & Kringelbach, 2015).

The liking or hedonic system seems to have two functions. The first function is selection. It is not an accident that omnivores such as humans innately respond positively to sweetness and fat. Sweetness signals fast-releasing carbohydrate energy sources, such as in ripe fruit, and motivates consumption. Herbivores such as sheep, cattle, and rabbits will eat more forage grasses cut later in the day, when sugar content is highest, than cut earlier in the morning. Carnivores, by contrast, are indifferent to sweet tastes. The taste and texture of fat signals high-density energy sources. Similarly, it is not an accident that humans generally find bitter tastes unpleasant and objectionable. Bitterness signals the presence of noxious toxins and poisons. Taste is the interface (and guardian) between the external environment of potential foods and our internal bodily environment. It maximizes an organism's chances of survival (Prescott, 2012).

The second function of the liking or hedonic system is modulation of the duration and intensity of an activity. As we will see here and in chapter 12, how hard and long mice and men work at an activity is a direct function of how pleasurable they find the reward (Yeomans, 1996; Young, 1959).

The other component of the reward system is the wanting system. I may find chocolate cake pleasant if it is placed in my mouth, but why should I get up and make the effort to acquire it and place it in my mouth? Motivating me to do so is the job of the wanting or incentive-salience system. It is the motivational magnet for action. The brain systems for wanting are more broadly distributed and involve opioid- and dopamine-sensitive sites in the brain stem including the ventral tegmental area, with mesolimbic projections to ventral pallidum, nucleus accumbens, and amygdala and extend into the orbital frontal cortex, anterior cingulate cortex, and insular cortex, particularly in humans (see figure 11.1c, d). But the evidence suggests that the actual *generators* of these feelings reside in the brain stem, diencephalon, and subcortical structures rather than in the neocortex (Berridge, 2009; Berridge & Kringelbach, 2015). To serve its motivational function, the wanting system must also be able to trigger certain action tendencies (discussed shortly).

In normal cases, liking and wanting act together and constitute the reward system of the brain, which compels the organism to act, and act appropriately. But they seem to have distinct neural bases, and in certain pathological conditions they can become dissociated. Rewarding (liking plus wanting) experiences naturally lead to Pavlovian and operant learning (as in conditioned place preference and conditioned place aversion reported earlier) (Berridge, 2009; Berridge, Ho, Richard, & DiFeliceantonio, 2010; Berridge & Kringelbach, 2015).

There is also a third related system—the general *arousal* system—which has been studied by Donald Pfaff and his colleagues (Pfaff et al., 2012, p. 468), and characterized as "the most powerful and essential force in the nervous system for activating behavior." A more highly aroused animal is more responsive to sensory stimuli, displays greater voluntary motor activity, and is more emotionally reactive. One can imagine the general arousal system modulating the intensity of the liking and wanting systems and serving as the neural basis of the "energy reservoir" in Lorenz's model of instincts (figure 4.1). It has its origins in medullary brain stem structures so it is widely available across branches of the phylogenetic tree.

Having discussed feelings, differentiated them from emotions, identified their origins in phylogenetically old, widely conserved brain stem, diencephalon, and subcortical structures, and proposed that their function is to guide and motivate behavior, I now want to illustrate how feelings are instantiated and operate within each of the four types of minds that account for human behavior.

Feelings and the Autonomic Mind

The autonomic mind is usually considered to be beyond volitional control. This is largely true. While I generally cannot consciously modulate autonomic processes, I can make behavioral and environmental changes that will affect autonomic systems. For example, it was noted earlier that biochemical reactions are critical for maintaining homeostasis in the digestive system (chapter 3). When blood glucose levels drop below a set point, the pancreas will secrete glucagon into the bloodstream, signaling the liver to start converting the stored glycogen into glucose and releasing it into the bloodstream. There is no conscious awareness (feelings) associated with this process. (The same is true of reflexes and many other autonomic processes.) It just happens. I cannot will it or unwill it.[8] If all is going well, I am not, and do not need to be, aware of it.

However, at a certain point, the stored energy reserves will be insufficient to maintain energy requirements so intervention will be required. A meal will need to be ingested. Without this intervention, the system will eventually break down. This intervention will require engagement of one or more of the systems that evolved to interact with the external environment. For nonhuman animals, these will be the instinctive and associative systems. For humans, it will be these plus the reasoning system. But how do I know *when to eat, what to eat*, and why should I make the *effort* to do so? Why should I bother, especially if it requires effort and may expose me to predation?

The solution evolution has converged on—perhaps with mammalian brains, perhaps earlier—is the utilization of *feelings* of reward (wanting and liking) and aversion (disgust) as intervening variables between stimulus and response. In the case of energy management, a homeostatic system signals it is time to eat by generating hunger pangs. This is a restless, unpleasant, agitating feeling, which the organism wants to get rid of. Its function is to make us care about initiating or stopping an action. It does so by being directly connected to certain behaviors controlled by instinctive, associative, or reasoning systems. These feelings are drivers, motivators, and inhibitors. They activate certain action tendencies, which result in the organism eventually undertaking the actions to procure and ingest food. Feelings will also guide (via taste) food selection. The control and operation of this energy management system is considered in some detail in chapter 12.

Feelings and the Instinctive Mind

Instinctive behavior is one important way in which organisms satisfy biologically critical goals through interaction with the external environment. These goals include acquisition of food, water, a sexual partner, a nest, or a home territory, securing well-being of offspring, predator avoidance, avoiding injurious levels of hot and cold, and even cheater detection and punishment, among many others. Some of these goals will be species specific, others are available across species.

It is one thing for the autonomic system to signal hunger or sexual arousal but another thing for the organism to get up and actually acquire food or sex. The ethologists struggled with an explanation of why an animal would do this. The initial models of robot-like chain reflex arcs were one solution to the problem, but they did not have the degrees of freedom necessary to accommodate intention actions and vacuum activities, where behavior remains incomplete or is initiated in the absence of the stimuli, respectively (chapter 4). As already noted, Wallace Craig's insight that "an element of appetite, or aversion, or both" (i.e., feelings) between a stimulus and the instinctive behavior was instrumental in the development of the Lorenz and Tinbergen model of instincts (figure 4.1). We can now redescribe this model in neuroscientific terms.

In Berridge and Kringelbach's vocabulary, we would describe Lorenz's action-specific energy reservoir as a specific appetitive (wanting) or aversive (avoiding) state (figure 4.1). An appetite has a positive valence (e.g., sexual arousal), an aversion a negative valence (e.g., hunger pangs). The volume of pent-up energy (i.e., "built-up pressure") in the reservoir corresponds to the level of arousal. Pfaff's general arousal system would serve this function (Pfaff et al., 2012). Separate arousal systems may also be associated with each specific instinctive system, or there may be a generalized arousal system that when coupled with specific feeling systems leads to specific motivated behavior (Garey et al., 2003). The presence of the appropriate stimuli initiates the wanting system (releases the action-specific energy), resulting in execution of the fixed action pattern. The consummatory response (associated with increase of liking or decrease of disliking) is modulated by the level of arousal (volume of pent-up energy) and the properties of the stimuli. The greater the level of arousal, the greater the urge to act (the need to relieve the pressure) by engaging in the consummatory behavior. An animal will actively seek environments in which the behavior can be discharged. The execution provides increased pleasure (or relieves the agitation) and returns the animal to a state of equilibrium.

Not only does this redescription of instincts convert a metaphorical model into a plausible biological model while preserving the critical insights, it also highlights that the model of instincts overlaps with the model of emotions described earlier in that both involve directedness, valence and arousal, physiological expressions, such as relaxed facial muscles and licking of the lips in response to sweet taste (Berridge & Kringelbach, 2015), and triggering of action tendencies. (It differs from human emotions in not involving propositional attitudes.) Neuroscientist Jaak Panksepp has made this connection very explicit by proposing a model where specific appetitive wanting systems (Lorenz's action-specific energy reservoirs) correspond to discrete "primordial emotion" systems (Panksepp, 2011). To minimize confusion between Panksepp's primordial emotion systems in rats and human emotions, I will use the hyphenated form: primordial-emotion. The activation of a primordial-emotion state by the presence of appropriate environmental stimuli, in conjunction with sufficient levels of arousal, will release the fixed action behavioral patterns, resulting in the consummation of the action pattern and the associated feelings of pleasure and relief.

Panksepp identified the following seven systems in rats and proposed that they constitute a basic blueprint applicable to all mammals, including humans: SEEKING, RAGE, FEAR, LUST, CARE, PANIC, and PLAY (Panksepp, 2011). The terms are capitalized, following Panksepp's convention, to indicate that they are being used to label *specific brain systems associated with specific neural networks and neurochemicals* rather than human emotions that might be similarly labeled. Judith Toronchuk and George Ellis (2013) subsequently amended these basic primordial-emotion systems to comprise nine systems by adding DISGUST and POWER systems and relabeling the PANIC system as the NEED/ATTACHMENT system. Table 11.1 lists these systems, grouped into categories of basic functioning, survival, reproduction, social bonding and interaction, and group conflict regulation. The functions of each system are highlighted along with how they interact with other systems to achieve various behaviors. The neuroanatomy and neurochemistry underlying each system are also specified.

What is important for our purposes is not whether there are seven systems, nine systems, or 90 systems. The interesting point is that these systems delineated along neurophysiological lines correspond to the types of things identified as instincts by ethologists and evolutionary psychologists. There is no harm in referring to these states in rats as primordial-emotions so long as my earlier admonition is heeded: any reference to any type of emotion in nonhuman animals is a reference to a directed mental state but one that lacks propositional content.[9] Let's examine one such primordial-emotion or instinct.

The LUST System

I will use the LUST system to illustrate the workings of instinctive systems and the central role of feelings.[10] Since there are behavioral, neuronal, and hormonal gender differences in the operation of the LUST system, the discussion will be confined to males. In males, the appetitive phase commences with wanting and leads to a feeling of sexual arousal, easily recognizable in ourselves.[11] Nonimpotent, postpubescent human males experience specific powerful and pleasurable feelings associated with sexual arousal as a prelude to copulatory behavior. The consummatory phase (copulation) is dominated by liking, building up to orgasm and ejaculation, followed by satiety or restoration of equilibrium. Figure 11.3 graphs these three phases in human males in relation to reported pleasure, along with the associated brain area activations.

Not only is the pleasurable arousal a "prelude" to courtship and copulation behavior, it imparts a degree of urgent compulsion to the initiation of the behavior. The level of desire will differ among individuals as a function of the level of arousal and the quality of the stimuli. The slightest thought, visual perception, even a picture or a dream of nubile females can result in sexual arousal and erection in postpubescent males. Men reportedly think about sex on average 19 times per day (Fisher, Moore, & Pittenger, 2012) and experience sexual desire 37 times per week (Regan & Atkins, 2006).[12] Sometimes it seems that men's whole world revolves around seeking out and discharging these feelings. Once aroused, the only relief comes through ejaculation, which can occur through sexual intercourse or, in the absence of a partner, through masturbation. Sometimes no cost seems too high to incur for this experience. Men are willing to not only expend considerable resources but also bypass social and legal prohibitions at great personal risk to experience and consummate these feelings, as indicated by the John Edwards example in chapter 1 and the following excerpt from *Lolita* (Nabokov, 1991, p. 285):

> I recall certain moments, let us call them icebergs in paradise, when after having had my fill of her—after fabulous, insane exertions that left me limp and azure-barred—I would gather her in my arms with, at last, a mute moan of human tenderness (her skin glistening in the neon light coming from the paved court through the slits in the blind, her soot-black lashes matted, her grave gray eyes more vacant than ever—for all the world a little patient still in the confusion of a drug after a major operation)—and the tenderness would deepen to shame and despair, and I would lull and rock my lone light Lolita in my marble arms, and moan in her warm hair, and caress her at random and mutely ask her blessing, and at the peak of this human agonized selfless tenderness (with my soul actually hanging around her naked body and ready to repent), all at once, ironically, horribly, lust would swell again—and "oh, no," Lolita would say with a sigh to heaven, and the next moment the tenderness and the azure—all would be shattered.

Great novelists, playwrights, and poets make the most perceptive psychologists, but in line with scientific methodology, we must ask to see the hard evidence for the objective existence of such feelings and their causal role in behavior. The same conclusion has been reached less eloquently in behavioral and neuroscience animal research.

Behaviorally, in lab animals the most common and reliable measure of sexual arousal in males is penile erection (tumescence of tissue) (Sachs, 2007). Levels of arousal are measured in various ways, including the time it takes the male to begin mounting an estrus female, the time between intromissions, the time from first intromission to ejaculation, and the time to resume copulation after ejaculation (Clark, 2013). Males in many species display erections in response to remote sexual stimuli: visual erotica for humans, inaccessible estrus females for rhesus monkeys, estrus odors for rats. Humans and rhesus monkeys will masturbate after arousal if a copulatory partner is unavailable (Slimp, Hart, & Goy, 1978). Male rats will mount

Table 11.1
Panksepp's primordial emotions (or instincts)

Evolutionary needs met	Primordial-emotion systems (instincts)	Putative neurochemicals	Putative key components of neural networks	Works with:	Functions
Individual Needs					
Basic functioning	E1: SEEKING system (hedonic appraisal, "liking" component)	endorphins (+), GABA (+,–), enkephalins, DA (?), endocannabinoids (+)	nucleus accumbens, ventral pallidum, brain stem nuclei	E2–9	situation evaluation, hedonic appraisal, learning
	E1: SEEKING system (incentive motivation "wanting" component)	DA (+), glutamate, Ach, CCK (+,–), neurotensin, endorphins	nucleus accumbens, ventral pallidum, lateral hypothalamus, and VTA to PAG		incentive salience
Basic survival	E2: DISGUST system (repulsion, avoidance)	serotonin (+), substance P (+)? endocannabinoids (–)	anterior insular cortex, putamen, lower brain stem (area postrema, NTS)		avoiding harmful foods, substances, environments
	E3: RAGE system	substance P (+), Ach (+), glutamate (+), vasopressin (+)	medial amygdala, BNST, medial perifornical hypothalamus, dorsal PAG	E4,E9	defense: protection of organism, resources, and conspecifics, limiting restraint on movement
	E4: FEAR system	glutamate (+), DBI, CRH (+), CCK (+), α-MSH, NPY	lateral and central amygdala, medial and anterior hypothalamus to dorsal PAG and pontine nuclei	E3, E9	defense: flight, limiting of tissue damage
Social Needs					
Reproduction	E5: LUST system (sexual desire)	steroids (+), vasopressin (+), LH-RH (+), DA (+)	basal forebrain, amygdala, BNST, anterior cingulate, medial preoptic, and VMH to ventral PAG	E6, E7	ensuring procreation, enhancement of bonding

	Sexual satisfaction	opioids (+), oxytocin (+)	septum, medial preoptic (VMH in ♂?), VTA to PAG		
Group cohesion: bonding and development	E6: NEED/ATTACH-MENT (separation distress)	opioids (–,+), oxytocin (–,+), prolactin (–,+), CRH	anterior cingulate, BNST, POA, VTA to PAG	E5, E7	protection of vulnerable individuals; creates bonding through need for others
	E7: CARE system	oxytocin (+), prolactin (+), DA, opioids (±), glutamate (+)	anterior cingulate, BNST, preoptic hypothalamus to VTA and PAG	E5, E6	caring for others, particularly offspring
	E8: PLAY system	opioids (+,–), DA (+), Ach (?)	dorsomedial diencephalon (thalamic nuclei) to ventral PAG	E6, E7	bonding with conspecifics, development of basic adaptive and social skills, creativity
Group function: regulating conflict	E9: POWER/dominance system (rank, status, submission)	serotonin (±), DA(±), testosterone (±), vasopressin (±), CCK, CRH (±)	medial prefrontal cortex, ventral pallidum and other basal ganglia, hypothalamic nuclei to PAG	E3, E4, E5	limiting aggression in social groups: allocating resources, especially sexual ones

Note: The nonspecific effects of serotonin and norepinephrine are omitted, as are higher cortical areas. *Key*: Ach=acetylcholine; BNST=bed nucleus of the stria terminalis; CCK=cholecystokinin; CRH=corticotrophin releasing hormone; DA=dopamine; DBI=diazepam binding inhibitor; GABA=gamma-aminobutyric acid; LH-RH=luteinizing hormone, releasing hormone; MSH=melanocyte stimulating hormone; NPY=neuropeptide Y; NTS=nucleus tractus solitarius; PAG=periaqueductal gray; POA=preoptic areas; VMH=ventromedial hypothalamus; VTA=ventral tegmental area.

Source: Reproduced with slight modifications from Toronchuk and Ellis (2013).

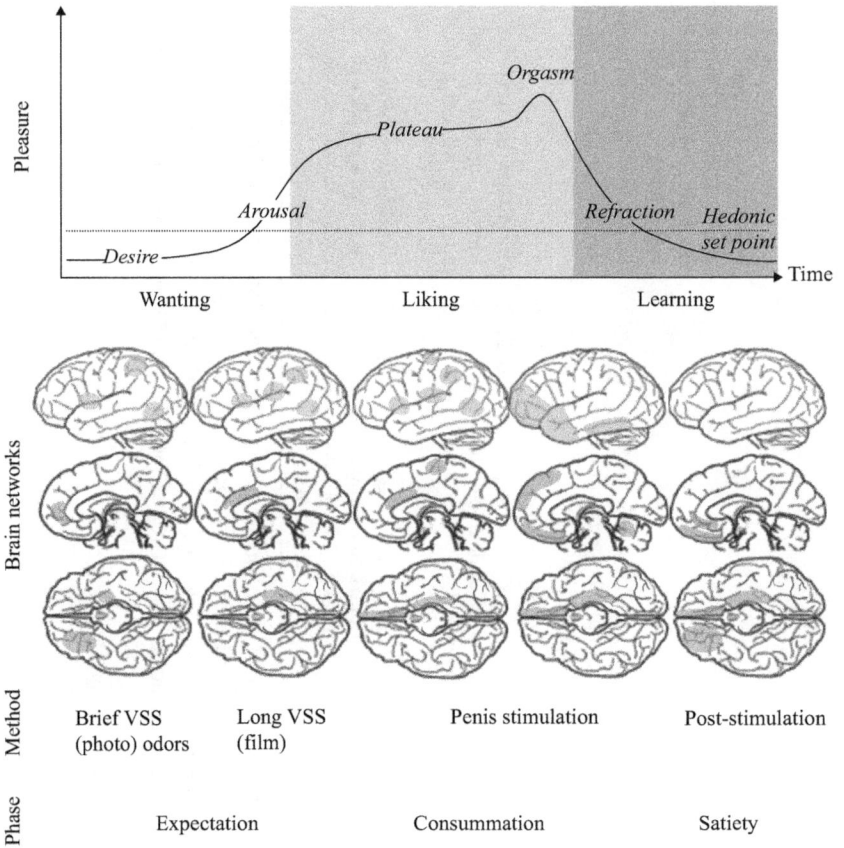

Pleasure

Orgasm

Plateau

Arousal

Refraction

Hedonic
set point

Desire

Time

Wanting   Liking   Learning

Brain networks

Method

Brief VSS   Long VSS   Penis stimulation   Post-stimulation
(photo) odors   (film)

Phase

Expectation   Consummation   Satiety

Figure 11.3
The top graph tracks the pleasure modulation associated with the appetitive (wanting), consummatory (liking), and satiety (equilibrium) phases of the LUST system
in humans. The brain images show the activation of brain regions as a function of
three phases of pleasure. Both subcortical and cortical regions are involved. Given
the ubiquity of the LUST system in large segments of the phylogenetic tree, the generators of the feelings and behaviors will be in the brain stem, diencephalon, and
subcortical regions, while their representations also involve cortical regions. Reproduced with permission from Georgiadis and Kringelbach (2012).

a female much sooner after noncontact exposure to estrus females than rats not preexposed to the odor of estrus females (Sachs, 2007). The compulsive desire for consummation after arousal can be measured in experiments such as the following. If a partition with a hole is placed between a male rat and an estrus female, the male will become agitated and sniff, chew, and nose poke the partition in continuous efforts to access the female. Its blood testosterone levels will increase. It will linger 12 times longer in the vicinity of the hole when an estrus female is present than in the control condition (Amstislavskaya & Popova, 2004).

There are similarities in hormonal and neural systems and pathways in rats and humans, and sex differences in both. Similar drugs delivered to similar deep brain sites will stimulate sexual desire in both rats and humans (Pfaus, 2009). In human males, the epicenter of sexual arousal seems to be located in the medial anterior hypothalamus, with some variability across species. In rats, this region is the preoptic area.

In adolescent males, sexual maturity commences with the production of testosterone by the testicles. The production of testosterone activates a number of neuropeptides, including vasopressin, which in animal models initiates sexual arousal and courtship. Males produce twice as much vasopressin as females. Testosterone also activates a gaseous nitric oxide transmitter in the brain, which is thought to enhance sexual arousal and aggressiveness. Key areas of the anterior hypothalamus contain such testosterone receptors. Testosterone produces a greater effect in males than in females because male brains have larger areas of testosterone receptors in the anterior hypothalamus than female brains. Thus, unsurprisingly, male rats will work to have testosterone injected into their preoptic area (POA) (Georgiadis & Kringelbach, 2012; Panksepp & Biven, 2012; Pfaus et al., 2012).

In animal models, lesions to the testicles and to the key anterior hypothalamus regions produce similar effects: both weaken sexual urges and abilities. Interestingly, there is an inverted effect of sexual maturity. Preoptic area lesions in sexually naive male animals weaken sexual arousal more than lesions in sexually experienced animals. The latter will continue to work to access estrus females, but their consummatory behavior will be sluggish. One explanation for this is that while sexual urges and responses are being generated in deep brain regions, in the experienced animals they have also become encoded in higher-level cortical regions and can continue in spite of lesions to the subcortical generators.[13]

While we cannot directly access the feelings of sexual arousal in nonhuman animals (or indeed other humans), the behavioral, hormonal, and neuroanatomical homologies reviewed here suggest that feelings are pivotal in driving behavior in both human and nonhuman animals. It is certainly a plausible working hypothesis.

I chose LUST as a paradigmatic example of how hedonic reward systems initiate action because it is something most people have experienced; it has powerful positive valence and arousal components that motivate, even compel, subsequent behavior (mounting and intromission in male rats and lordosis in female rats); and we understand a great deal about the anatomical pathways and neurotransmitter systems involved in each of the anticipatory motivational (wanting), rewarding (liking), and satiety phases of the behavior (Pfaff, 2009). Not only is LUST relevant to explaining the Edwards example from chapter 1, it provides a readily comprehensible blueprint for *all* instinctive behaviors. In particular, it highlights the fact that all instinctive behaviors will be initiated by the reward system and driven by specific interoceptive and exteroceptive feelings.

Similar appetitive feelings drive the compulsion in teenagers to groom, the compulsion in parents to alleviate a child's suffering, the compulsion to punish a cheater, the compulsive attraction to alpha male cues, and so on. The feelings will vary in valence and arousal, may have specific flavors associated with them, and will trigger different action tendencies. The consummatory behavior will be accompanied by feelings of liking and return the human or nonhuman animal to equilibrium.

It is also worth mentioning Panksepp's SEEKING system here. The SEEKING system is the general-purpose wanting, seeking, exploring, searching, motivating, and interest system. It is associated with positive, even euphoric, hedonic value. It is this system that energizes me to get up and pursue everything from chocolate cake to PhDs. The pleasure associated with the SEEKING system is not the sensory pleasure of consummating the act (e.g., eating the chocolate cake) but rather the excitement of pursuit and anticipation (Hamburg, 1971). It interacts with and facilitates the operation of all other primordial-emotion (instinctive) systems (table 11.1) and drives behavior at all levels. It seems very similar to Berridge and Kringelbach's wanting system and will need to contain components of Pfaff's general arousal system. Panksepp suggests that the "generalized reward" system discovered by Olds and Milner was actually the SEEKING system. Anatomically, the SEEKING system involves connections running from the ventral tegmental area to the medial forebrain bundle and lateral hypothalamus, nucleus accumbens, and medial prefrontal cortex. The main neurotransmitter involved is dopamine (Panksepp, 2011; Panksepp & Biven, 2012). The SEEKING system will reemerge at the cognitive level as *desire*.

Feelings and the Associative Mind

How are associative behaviors learned and initiated? Associations are formed through positive and negative reinforcement and punishment. The reader will recall that the behaviorists defined reinforcement and punishment without appealing to feelings. Positive reinforcement occurs when some stimulus event (e.g., picking up and comforting a baby) increases the probability of repetition of the behavior that precedes it (baby crying). Negative reinforcement occurs when the termination of a stimulus event (e.g., baby crying) increases the probability of the repetition of the behavior that follows the stimulus event (e.g., picking up and comforting the baby). A punishment event occurs when the presentation of a stimulus following a behavior results in a decreased probability of that behavior reoccurring. This formulation defines reinforcement and punishment as probability occurrences. It tells us nothing about the reinforcing event itself.

This vacuous construal persisted despite the fact that the reinforcers utilized were pleasing or discomforting biologically important stimuli, such as food, water, sexual partner, harmful levels of heat or cold, odor of a predator, or distress call of an offspring. The natural way to describe an animal's reaction to such stimuli is in terms of appetitive and aversive motivation. Positive reinforcers produce approach behaviors (appetitive motivation) associated with "pleasant hedonic impact," and negative reinforcers and punishers produce withdrawal behavior associated with aversive, unpleasant feelings. Yet subjective experience of the reward or punishment was not allowed to play any role in the theoretical accounts of the behaviorists (Bozarth, 1994).

Despite the failure to acknowledge the rewarding properties of reinforcement, behavioral psychologists ironically continued to measure the amount of reinforcement in grams or count them in pellets, in the number of fluid drops, or as the concentration of sugar in solution. The rats continued to consume the reinforcement! Such vocabulary only makes sense if there is an underlying understanding that the reinforcer is a food object that the animal finds rewarding (Young, 1959). The same is true for a punishing reinforcer.

Several behaviorist psychologists did notice the shortcomings in this approach to reinforcement and suggested that the purpose of reinforcement is "not response strengthening, but the creation of a motivational state that influences a wide variety of subsequent behavior of the animal" (Bindra, 1974, p. 200). Two psychologists who attempted to rectify this shortcoming were Paul Thomas Young, a student of Edward Titchener (himself a student of Wilhelm Wundt), and Dalbir Bindra. The former provided behavioral evidence for the role of feelings and reinforcement (Young, 1959), and the latter developed a brain-based theoretical model of how such a system might work (Bindra, 1969, 1974), not unlike the model developed by Tinbergen and Lorenz already discussed. We will focus on the behavioral data.

Paul Young and his colleagues carried out carefully controlled behavioral experiments in rats to distinguish between food palatability (preference or liking) and appetite (quantity eaten) and sensory intensity and hedonic intensity (e.g., the distinction between detecting the concentration of sugar or salt in a water solution versus preference for the solution), and then used these distinctions to carry out further studies to demonstrate that the hedonic value or valence (pleasant or unpleasant) and arousal or intensity of the stimuli (solution) affect performance (Young & Falk, 1956; Young & Greene, 1953). These experiments provided objective behavioral measures for the causal efficacy of what in humans are referred to as subjective feelings.

In subsequent experiments, rats were offered a choice between a 1% salt solution and sugar solutions of different fixed concentrations ranging from 2% (very weakly sweet) to 54% (very sweet). All rats developed a preference for the sugar solution. However, the speed at which they learned to discriminate between the sugar and salt solutions was a function of the concentration of the sugar solution. Rats in the 54% sugar solution condition required only 17 trials to discriminate between the salt and the sugar solutions, rats in the 18% sugar solution condition required 38 trials, rats in the 6% sugar solution condition required 66 trials, and rats in the 2% sugar solution condition required 122 trials to discriminate (Young & Asdourian, 1957).

In another experiment, rats were trained to run down a runway to a circular platform around the circumference of which were placed five evenly spaced cups. During pretraining, all animals received one drop of 10% sugar solution in each of the five cups. In the first phase of the experiment, one cup was baited with the sugar solution (the others were empty) and the animals learned to run to the baited cup. In the second phase, the animals were divided into four groups, and animals in each group learned to run to a baited cup containing either a 20%, 10%, 5%, or 0% sugar solution. The rate of learning was a function of sugar concentration in the baited cups. The 20% sugar solution resulted in better performance than the 10% solution, which resulted in better performance than the 5% solution, which in turn resulted in better performance than the 0% solution. In fact, the 0% sugar solution resulted in an extinction of the behavior learned in the original training period (Dufort & Kimble, 1956). Both these experiments demonstrate that preference discrimination learning is a function of not only practice (i.e., number of trials run) but also concentration of sugar solution (i.e., preference or palatability of the stimuli).

It is not only the valence or palatability of the reinforcer that is relevant to behavior; the intensity and duration also matter. In another experiment, food pellets were used to train rats to run back and forth from testers containing sucrose, wheat, and casein solutions. Independent tests determined that the rats preferred sucrose to wheat and wheat to casein solutions. Two interesting results were reported: learning was a function of the amount of practice in running the pattern, not the palatability of the incentive solutions. However, the rats ran *faster* for the preferred incentive solution (Young, 1947).

In a follow-up experiment, the reinforcing stimuli controlled for affective intensity (via concentration of sugar solutions), affective duration (by varying the number of seconds in contact with solutions), and frequency of affective arousal (by varying the frequency of access). These factors modulated the rate of running for the food incentive. That is, the intensity, duration, and frequency of affective arousal determined how hard the animals tried (Young & Shuford, 1954). Similar conclusions should apply to other reinforcers, such as sexual behavior, play, and exploration.

As in instinctive behaviors, feelings (appetite for the pleasant feelings of food and sex and aversion to toxins and electric shock) are critical in forming and triggering associative behaviors. The discovery that artificial stimulation of the medial forebrain bundle and lateral hypothalamus regions in rats has an effect on learning similar to that of positive reinforcement suggests that the SEEKING system may be of particular importance in motivating and driving reinforcement behavior (Burgdorf, Knutson, & Panksepp, 2000; Panksepp & Biven, 2012). All these data suggest that, as in the autonomic and instinctive minds, feelings are also the internal currency of the associative mind. What about the reasoning mind?

Feelings and the Reasoning Mind

Rationality is about recruiting reason in the service of a goal or desire. The machinery of reason consists of propositional attitudes and the coherence relation (chapter 6). Is there a role for feelings in the reasoning mind? Can we *feel* the coherency?

Cognitive scientists are happy to commit to representational or intentional mental states but have been as adamant as the behaviorists in avoiding any commitment to affects or feelings. There are no chapters on feelings in cognitive science textbooks. The rationale is that, in the cognitive account, it is the representational content of our mental states that is causally efficacious. Feelings, if they exist, are superfluous.

The most obvious entry point for feelings into the reasoning mind is via emotions. As already noted, emotions are a subset of propositional attitudes such as hope, jealousy, love, fear, shame, surprise, pride, regret, happiness, anger, disgust, contempt, sadness, guilt, and resentment. To say that they are emotional states is to say not only that they are referential or directed, but that they also have valence, arousal, duration, physiological expressions, cognitive antecedents, and action tendencies associated with them.

Emotions, broadly construed, are not unique to the reasoning mind. We have already encountered precursors and noted they are cognized versions of Panksepp's primordial-emotions or instincts. The cognization occurs through the appearance of propositional contents. Once propositional contents are involved, we are largely confined to the hominina or homo branch of the phylogenetic tree. The finer-grained distinctions made possible by propositional contents allow for multiplication of emotions in humans through cognitive construction.

The cognitive strategy with respect to emotions has been to either ignore them or strip off valence, arousal, and duration components and redefine them in terms of complex series of beliefs and desires. For example, my *hope* that P can be redefined as a *desire* for P and a *belief* that the probability of P is very low. My *surprise* that P can be redefined as a *belief* (up to now) that P is not the case and the *belief* (now) that P actually is the case. The issue of affect is not usually raised in the context of beliefs and desires. I am, of course, challenging this approach.

Desires should be another obvious entry point for feelings into the reasoning system. My desire to complete this book has valence and arousal components associated with it. It *feels* like something. It motivates and drives me daily. It is a source of considerable pleasure and satisfaction and occasional frustration. This seems obvious, but cognitive science, restricted to mere information processing, does not have the theoretical machinery to deal with it. In fact, in their information processing theory of problem solving, Allen Newell and Herbert Simon (1972) situated desires or goals *outside* the organism, as part of the task environment! We can remedy this shortcoming by recruiting Panksepp's SEEKING system. This is exactly what desires are. Human desires differ from nonhuman animal desires in that they have sophisticated propositional content that allows them to be directed at an unlimited number of states of affairs in the world, but ultimately, they are a sophisticated cognized variant of the SEEKING system.

What about beliefs? Does my belief that "all apples are fruit" have a feeling associated with it? There is a tradition within philosophy, and certainly within cognitive science, that assumes the answer is "no" (Horgan & Tienson, 2002; Kriegel, 2003; Searle, 1992), yet there may be reasons to reconsider this answer. In 1980, UC Berkeley philosopher (and my mentor) John R. Searle offered the world an argument that came to be known as the Chinese room argument. It was intended to show that a rule-based computational system could not be considered intelligent. There was more to intelligence than following rules. The argument became infamous. It is probably fair to say that no other argument or claim so exercised and consumed the cognitive and artificial intelligence communities during the decade as the Chinese room argument. At the heart of Searle's argument was the following thought experiment (Searle, 1980).

Imagine that you are locked in a room. (It is important that it is you.) You do not know any Chinese, written or spoken. Chinese characters are just meaningless squiggles for you. Inside the room is a box full of cards with Chinese characters written on them. There is also a book of instructions inside the room that gives you rules (in English, which you understand) on how to correlate one set of Chinese symbols with another set. There is a slot through which you can receive and pass out cards. You are now handed a card from outside the room containing a set of Chinese symbols. You consult your rulebook and find in it the symbols appearing on the card. The rulebook tells you the symbols that are correlated with the input symbols. You find the card containing the correlated symbols in your box of symbols and hand it back to the person outside the room. (With practice, you may become so proficient that you have memorized all the rules in the book, so you may not even need to consult the rulebook to generate the correct answer.) To the person outside the room, who speaks Chinese, the card he is handing you contains a question, and the card that you hand back contains the answer to the question. Given that the answers are correct or sensible, that person may well draw the inference that you understand Chinese. Your behavior is certainly consistent with understanding Chinese.

But do you really understand Chinese (or even the symbols written on the cards that you are manipulating)? This is the key question of the thought experiment. Searle concludes that when he places himself in the room he does not understand the "questions" he is being asked and the "answers" that he is providing. He does not understand Chinese. If he (Searle) does not understand, then a rule-based computer program will not understand either.

My concern is not with claims regarding computers and artificial intelligence techniques. It is with why Searle concludes that he does not understand

Chinese. An essential part of the thought experiment that is often overlooked by critics is that Searle himself is in the room (and asks you to put yourself in the room). Searle concludes that he does not understand Chinese because from his first-person perspective it does not *feel* like he understands. If you put yourself in the room, from your first-person perspective, you will most likely draw a similar conclusion. I certainly do. All the vociferous objections to the argument, and Searle's conclusion, ignored first-person experience and relied on behavioral or functional accounts of semantics to argue that Searle did indeed understand Chinese.[14] This thought experiment opens up the possibility that perhaps feelings permeate all propositional attitudes, not just desires and emotional states.[15]

The final component of the reasoning mind is the coherence relation. Can we *feel* coherence? I think so. In introducing the coherence relation, I noted that, ultimately, it is an intuited relation. An argument or a set of beliefs is coherent if it *feels* right. Coherence feels right (positive valence); incoherence or inconsistency feels wrong (negative valence). For example, if Mary is taller than George and George is taller than Michael, it feels right to say that Mary is taller than Michael. It feels wrong to say that Michael is taller than Mary. As with this example, the bathtub analogy from chapter 1, and all self-evident postulates, we cannot prove the rightness of the answer. It is a feeling that presumably all humans with normal cognitive capacity will share. Even in more complex situations, where we appeal to formal, normative rules, these rules are accepted because we can break them down into simple components and test them for the feeling of rightness against our intuitions. As logician Clarence Irving Lewis reportedly noted, when a point of logic is in question, the only thing we can do is appeal to intuition.

Why should this be? Why should there be a feeling associated with coherence relations? As in the case of taste, the feeling is fitness enhancing. Coherency feels good because representations that are internally consistent and veridical will enhance survival. Incoherency feels unpleasant because it can be harmful.

We are creatures whose behavior is a function of our beliefs about the world, rather than the world itself. If I have the belief that there is a tiger under my desk, then I am asserting a certain state of affairs is the case in the world (viz. that there is a tiger under my desk). The source of this belief can be direct perception or an inference based upon perception and/or other beliefs. Irrespective of source, to be useful in the facilitation of my survival and thriving, beliefs need to be veridical. In the case where beliefs are formed by direct perception, there is considerable sensory machinery devoted to getting this largely right, most of the time. In the case of inference, consistency will facilitate veridicality (assuming veridicality of perceptions and existing beliefs).

The importance of veridicality is largely self-evident.[16] Where the beliefs are based on inferences from perceptions and/or other beliefs, the consistency of these inferences becomes critical for guaranteeing veridicality and appropriate actions. For example, if my inferences lead to the belief "tigers are extremely dangerous" and also the belief "tigers are not extremely dangerous," what is it that I believe? More importantly, what do I do when confronted by a tiger: approach or run away? Two different actions are mandated; one will lead to survival, the other to death. For creatures with propositional attitudes, the ability to distinguish between coherency and incoherency is as important as distinguishing between sweetness and bitterness, and for similar reasons. Furthermore, we do it with the same common currency: feelings.

While cognitive scientists studying reasoning have thus far ignored the crucial role of feelings, they do seem to play a central role in Festinger's very influential theory of cognitive dissonance (Festinger, 1957; Harmon-Jones & Mills, 2019). The key idea here is that the presence of discordant or inconsistent beliefs and desires results in cognitive dissonance or discomfort that can sometimes be minimized by changing beliefs. For instance, I believe that I'm overweight and I believe that eating an extra slice of chocolate cake is detrimental to my health. I desire to maintain good health. I also desire an extra slice of chocolate cake. The combination of these beliefs and desires results in cognitive discord or dissonance. I can of course simply choose not to eat the slice of chocolate cake. This follows rationally from my beliefs and my desire to maintain good health. In this case, the theory has nothing to say. Alternatively, I can eat the cake and reduce the cognitive dissonance by telling myself that this cake is made with artificial sweetener and half the regular amount of butter, so it has far fewer calories and will have minimal negative health consequences. This allows me to eat the cake.

There are two possible interpretations of the role of cognitive dissipation of the dissonance in the situation where I eat the cake. First, the introduction of the new beliefs that the cake is made with artificial sweetener and has fewer calories allows me to reason away (reformulate) my previous belief that eating the cake is detrimental to my health, and this allows me to rationally pursue my desire to eat the cake. Second, it may be a cognitive rationalization occurring after the fact. It may make me feel better but provides no explanation for my eating of the cake. In the first interpretation, the behavior is still driven by the reasoning mind. In the second

interpretation, the behavior is left unexplained. So while cognitive dissonance theory may recognize the critical role of feelings in reasoning, the
actual machinery and control structures underlying behavior are very different from those envisioned by tethered rationality (chapter 12).

Finally, unlike in the case of autonomic, instinctive, and associative systems, we know next to nothing about the brain systems involved in generating the feelings associated with propositional attitudes and coherence
relations. There is some evidence for the involvement of the right lateral
prefrontal cortex in the detection of inconsistency or incoherency (Goel et
al., 2000; Goel & Dolan, 2003; Stollstorff ,Vartanian, & Goel, 2012; Tsujii,
Masuda, Akiyama, & Watanabe, 2010; Tsujii, Sakatani, Masuda, Akiyama, &
Watanabe, 2011). Making progress along these lines will not be easy because
unlike the work on reward systems, primordial-emotions, and interoceptive
and exteroceptive affects, animal models cannot be utilized in the study of
reason, and the animal-testing techniques cannot be adopted for human
participants, for obvious ethical reasons. But because propositional attitudes
and coherence relations belong to the reasoning mind, we would expect the
corresponding feelings to be constructed in higher cortical and neocortical
systems, with input from older subcortical systems (Lieberman & Eisenberger,
2009).

*   *   *

Feelings permeate all levels of behavior—autonomic,[17] instinctive, associative, and rational. They are not only integral to the operation of each type
of mind but also provide a common currency for the *interaction* of the different levels. Accepting such an account paints a very different picture of
human choices and decisions than postulated by standard theories. Tethered rationality views behavioral responses to be a blend of the responses
generated by the different systems available to organisms. This requires
some sort of common currency that can be used for global integration. The
suggestion in this chapter is that feelings provide this common currency
and allow for global integration of responses of each behavioral system. But
we still do not know how the system is controlled. Who is in charge of the
tethered mind? This question is addressed in chapter 12.

12 Control Structures: Who Is in Charge of the Tethered Mind?

But I see another law in my members, warring against the law of my mind, and bringing me into captivity to the law of sin which is in my members.
—St. Paul, Romans 7:23

If, in short, there is a community of computers [instincts] living in my head, there had also better be somebody who is in charge; and, by God, it had better be me.
—Jerry Fodor

I did not direct my life. I didn't design it. I never made decisions. Things always came up and made them for me. That is what life is.
—B. F. Skinner

One final factor to consider in the development of tethered rationality is the control structure; who is in charge of the tethered mind? That is, when multiple responses are queued by the different systems, how is behavior determined? Different models of human behavior implicitly or explicitly come with different control structures. Several different models have been referenced en route. In this chapter, I explicitly examine the control structures of these various models so we can compare and contrast them with the control structure of tethered rationality.

The control structure for tethered rationality calls for not a selection of response but rather a blending of responses. This blending will be illustrated with three examples. The first example will be the decisions of the parole judges before and after lunch that we encountered in chapter 3. The second will be a literary example from Jane Austen that highlights the temporal component of feelings. The third example will be a detailed consideration of my propensity to overeat despite the fact that I'm overweight. This is followed by a brief consideration of the predictive power and falsifiability of the model. I conclude the chapter by returning to and completing the explanations of some of the other example behaviors that were introduced earlier.

Western-Christian Model Endorsed by St. Paul (and God)

In chapters 1 and 2, we met the Western-Christian model of human behavior articulated by St. Paul (Romans 7:15–25) and, presumably, endorsed by God (figure 12.1a). In this account, God has endowed us with both reason and "animal passions." The "animal passions" in our terminology consist of the autonomic, instinctive, and perhaps associative minds. God has also provided a set of laws for us to live by. Reason is the machinery that allows us to act in accordance with the laws. Our animal passions are independent of reason (and thus not sensitive to the laws) and also have access to actions. The reasoning mind is endowed with the ability to detect conflicts between actions triggered by animal passions and those mandated by the laws, use willpower to inhibit and override the former in favor of the latter, and develop avoidance strategies so we are not tempted in the first place. It is a free choice, but God provides the necessary tools and makes no secret which is the preferred choice. If we fail to make the correct choice, it is only because we are rejecting God's laws (and thus God) or we are not trying hard enough. "Trying hard enough" is the notion of effort. It incorporates both willpower and judicious utilization of reason to minimize the effect of the animal passions.

If I am presented with a slice of chocolate cake, my reasoning mind is fully in charge of whether I choose to consume it or not. My reasoning mind is aware that I'm overweight and suffer from the related consequences (and knows that God does not approve of gluttony). In the absence of a death wish, reason dictates that I set the cake aside. If I fail to do so, my endowment of reason—being a gift from God—cannot be questioned (except in clinical cases), so the only explanation must be that I'm not trying hard enough. At the expense of repetition, it is important to appreciate that "not trying hard enough" is not just about willpower in the face of temptation but also about utilizing reason to avoid situations of temptation (e.g., not keeping chocolate cake at home), or exposing yourself to the temptation but in a "safe" or controlled context (as in Odysseus's siren song strategy).

Many socially encoded norms are about utilizing reason to avoid temptation of the animal passions. For instance, many religions require separation of men and women during prayer gatherings so one is not tempted

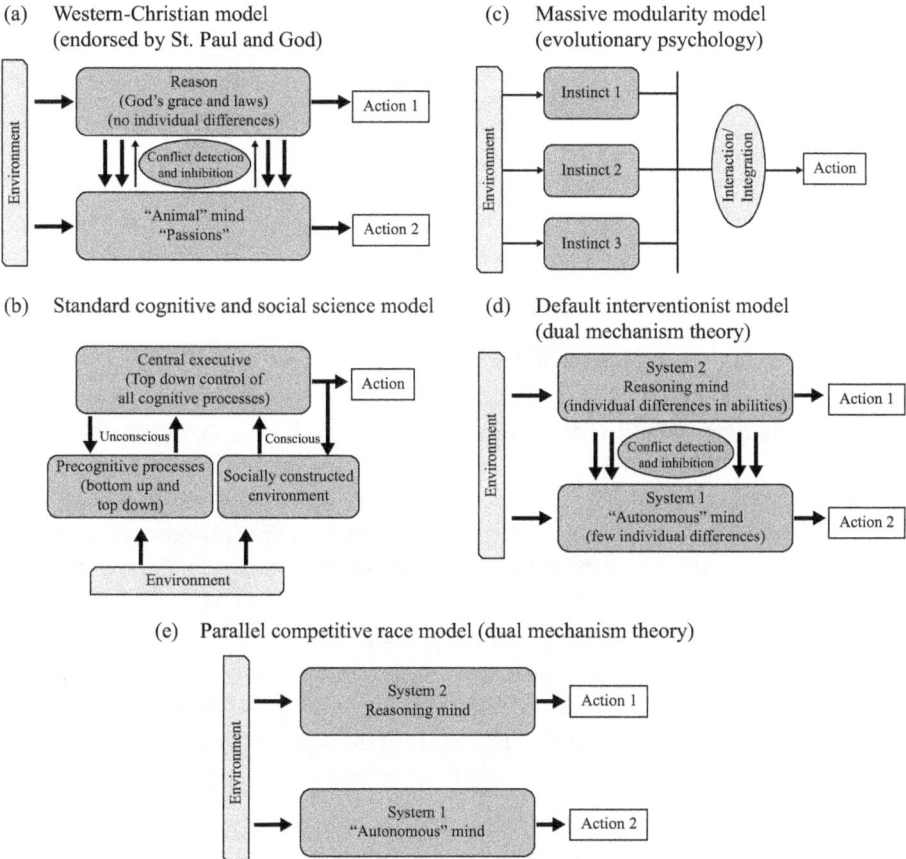

(a) Western-Christian model (endorsed by St. Paul and God)

(b) Standard cognitive and social science model

(c) Massive modularity model (evolutionary psychology)

(d) Default interventionist model (dual mechanism theory)

(e) Parallel competitive race model (dual mechanism theory)

Figure 12.1
Various models of control structure found in the literature. (a) This is the control structure endorsed by St. Paul and God and is still the basis of much Western societal, religious, and legal norms. (b) This is the generic control structure of the standard cognitive and social science model. Reason is the only route to action. (c) The massive modularity model has no reasoning component. The action is determined by a set of interacting instincts. (d) The flow of control for the default interventionist model advocated by many adherents of dual mechanism theories. (e) The parallel competitive model advocated by other adherents of dual mechanism theory accounts.

or distracted by one's attractive neighbor when one should be focusing on God. There are societal dress codes to attenuate (or accentuate) sexual attraction. The model is silent on whether there are individual differences at the level of the animal passions, but in terms of following the laws it may not matter, because God has given each of us sufficient reason to override them. We just have to make the effort and use the tools of reason to avoid situations of temptation. If social, cultural, and legal norms are substituted for God's laws, it is fair to say that this model still forms the fabric of Western social and legal norms.

Standard Cognitive / Social Science Reasoning Model

In its American instantiation—as reflected not only in large parts of academia, but perhaps more importantly, the Walt Disney Corporation—the Western-Christian model emerges in two related forms. What is common across the two variants is that there are no meaningful differences at the level of "animal passions"—where this is broadened to mean that *there are no meaningful individual differences in terms of biological endowment*. In the popular American conception, all individual differences occur at the level of reason and effort (both construed nonbiologically). We can use reason and effort to be anything and everything we want to be, as proclaimed in the beginning narration of the Disney film *Zootopia* (Howard and Moore, 2016):[1]

> Fear. Treachery. Bloodlust. Thousands of years ago, these were the forces that ruled our world. A world where prey were scared of predators. And predators had an uncontrollable *biological* urge to maim and maul and. . . . But over time we evolved and moved beyond our primitive, savage ways. Now predator and prey live in harmony. . . . [We formed] the great city of Zootopia, where our ancestors first joined together in peace and declared that "Anyone can be anything!" (emphasis added)

Why can't I ride a bicycle like Lance Armstrong or be as rich as Jeff Bezos? Determination and effort. I don't try as hard as they do. If I practiced as much and tried as hard, I, too, could win the Tour de France, be a billionaire or the president of the United States. As Thomas Edison is reported to have said, "Genius is 1% inspiration and 99% perspiration." This may at times be comforting to believe, but alas, it is not true (Davids & Baker, 2007; Moran & Pitsiladis, 2017; Sandel, 2020; Tucker & Collins, 2012). Believing it can also have deleterious effects. Working very hard at something—that your biological endowment does not support—without success can lead to frustration, self-incrimination, and despair. It also makes for bad science.

The standard cognitive and social science reasoning models are largely consistent with this worldview but focus on the reasoning machinery rather than effort: given the right beliefs and basic reasoning machinery, any behavior can follow. Not only are autonomic, instinctive, and associative systems not recognized, reason itself is specified independently of biology—so no biological constraints need apply (figure 12.1b).[2] In this account, environmental information comes into the reasoning mind already in propositional format or via perceptual processes that have unconscious top-down and bottom-up components and results in the formation of beliefs and desires (propositional attitudes). Once these are formed, the reasoning machinery is applied to them, resulting in decisions, choices, and actions. The reasoning mind is the central executive and the only basis for initiating action. The notion of effort (while always in the background) does not explicitly come up in this model because there is no such construct in cognitive science theories. Slight variations of this model appear in many places in the cognitive and social science literature (Newell & Simon, 1972).

Massive Modularity Model

The massive modularity model highlights the bottom half of the Western-Christian model (figure 12.1c). There is no reason; all behavior is a function of instincts. Environmental inputs trigger various instincts to varying degrees, and some will have a greater impact on my behavior in certain situations than others. There is some sort of integration of the various instinctual responses resulting in an action. Again, the issue of effort does not arise, as there is no mechanism to deal with it.

To illustrate the standard cognitive and social science and massive modularity models and explore their consequences, let's apply them to a common but sensitive topic. There was a recent newspaper story of a clothing policy instigated on an Arctic research vessel on a prolonged voyage, where the crew contained both men and women (Oakes & Last, 2020). Halfway through the voyage, the crew was informed that "no leggings, no very tight-fitting clothing—nothing too revealing—no crop tops, no hot pants [and] no very short shorts" could be worn. The policy applied to both sexes, but presumably the rationale was that covering of bodily forms would minimize instances of staring, sexual overtures, and harassment of women by men. An individual who believes that human behavior is driven strictly or largely by instinctual systems might justify this policy by citing the LUST system discussed in chapter 11. An individual who believes that human behavior is strictly reason-based will not accept this rationale (because the cognitive social science model does not recognize instincts) and in fact may be offended by the policy and even conclude that it implies "that women should be responsible for managing the behavior of men." Indeed, one sociologist was quoted as having remarked that "the assumption that . . . if a woman wears something fitted, then . . . she's inviting harassment— that's just so, so gross" (Oakes & Last, 2020). The view that follows from tethered rationality is that advocates of both these positions are making the same mistake: embracing half a model. The massive modularity models are embracing only instincts, while the standard social and cognitive science models are embracing only reason. Both are ignoring the tethered nature of the human mind and the blended response that is human behavior. All models of behavior have consequences. We return to the implications of tethered rationality in chapter 15.

Dual Mechanism Models

For completeness, I will also discuss the control structures of dual mechanism models introduced in chapter 7, even though they do not address the behaviors of interest. The default interventionist model (figure 12.1d) is actually the Western-Christian model applied to the cognitive mind, segregating it into "conscious" and "autonomous" processes. The "autonomous mind" is sometimes meant to incorporate the "animal mind," but as we saw in chapter 7, all the empirical work and proposed machinery is confined to the cognitive mind. Dual mechanism theory has little to say about the types of processes encompassed by the autonomic, instinctive, and associative systems. In the default interventionist model, the stimuli are received and processed by both systems. System 1 is the default, "autonomous" route. If System 2 detects a conflict, it has the power to inhibit System 1 and generate a response of its own. In the absence of intervention by System 2, System 1 will initiate the response. As in St. Paul and God's model, responses by System 2 are preferred, since System 1 responses may not respect the laws of rationality.

Any individual differences in behavior (decisions or actions) are attributed to individual differences at the level of System 2, the reasoning mind. These differences consist of cognitive factors, such as memory and IQ measures, and top-down control factors, such as conflict detection, inhibition, and rerouting information. These factors account for individual differences in reasoning abilities. Few or no differences are recognized at the level of System 1, the "autonomous" mind (Evans & Stanovich, 2013). The claim that there are no individual differences in lower-level systems is perhaps influenced by the American-Western-Christian model and is simply untrue. The energy regulation example that I will give shortly provides a decisive counterexample.

Another model endorsed by dual mechanism theories is the parallel competitive race model depicted in figure 12.1e. Here, both the System 2 reasoning mind and System 1 "autonomous" mind are activated in parallel by environmental stimuli. Whichever arrives at a conclusion first wins the race and determines the response of the organism (Sloman, 1996; Trippas et al., 2017). Sources of individual differences have not been discussed with respect to this model, but one can imagine individual differences in the sensitivity and efficacy of the two systems. Neither of the two dual mechanism theory models have the machinery required to deal with the concept of effort.

Control Structure for Tethered Rationality

Unlike the academic models mentioned earlier, but consistent with St. Paul's model, tethered rationality assumes that human behavior is a function of multiple systems, including autonomic systems, instinctive systems, associative systems, and reasoning systems. Previous chapters have discussed the properties and internal functioning of each of these systems and how each generates behavior. Evidence confirming that human behavior is indeed a blended response of these different systems has been presented throughout with anecdotal stories and more formally with data from economic decision-making studies in chapter 9. The key insights of this model are that despite the four different systems working along different principles, they all share the common currency of feelings, and this common currency allows the organism to generate a single response through an integration or blending function. How does this work, and what is the control structure for tethered rationality? Who is in charge of the tethered mind?

Different aspects of the environment differentially trigger different systems. All or a few systems may be triggered, depending on the environmental cues. Each system that is triggered generates a response in the currency of feelings, with particular valence, arousal, and duration values. In some cases, the same action may be triggered by multiple systems, while in other cases different, even contradictory, actions may be triggered. Either way, the overall response is determined by a blending function that factors in the output of the individual systems, guided by the principle of maximizing positive feelings and minimizing negative feelings (figure 12.2). (This is a very old idea most closely associated with Jeremy Bentham ([1789] 1823), who even developed a "felicific calculus" to undertake the calculation.

However, his concerns were normative while mine are descriptive.) There is no central executive in charge. All systems contribute. Depending on the situation, some will be more active than others. It may be possible to construct a notion of effort in the model by associating it with levels of arousal of the different systems (Pfaff et al., 2012). This construct would not be the same as that associated with the "I" in the Western-Christian model, but it may be the best we can do without abandoning a mechanistic explanation. Based on empirical evidence, we might even differentially weight the outputs of various systems as an architectural feature of the model. This model is depicted in figure 12.2. I will now offer three examples—the differences in decisions of parole judges before and after lunch, a literary example from Jane Austen incorporating a temporal component, and my perennial struggle with weight management—to flesh out the model.

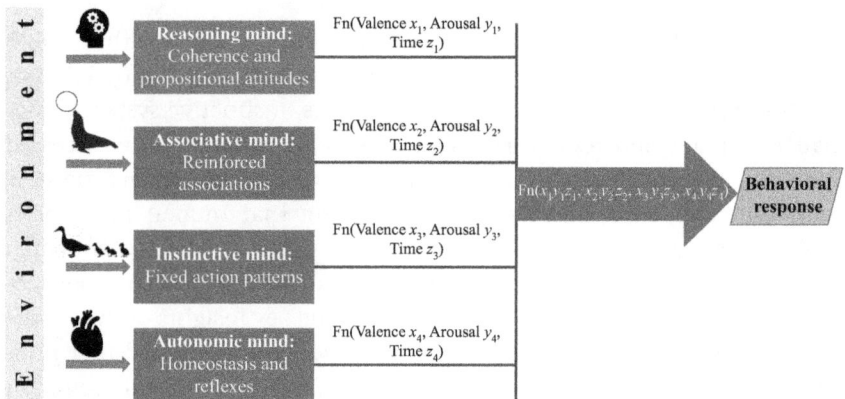

Environment

Reasoning mind: Coherence and propositional attitudes	Fn(Valence x_1, Arousal y_1, Time z_1)	
Associative mind: Reinforced associations	Fn(Valence x_2, Arousal y_2, Time z_2)	
Instinctive mind: Fixed action patterns	Fn(Valence x_3, Arousal y_3, Time z_3)	
Autonomic mind: Homeostasis and reflexes	Fn(Valence x_4, Arousal y_4, Time z_4)	

Fn($x_1, y_1, z_1, x_2, y_2, z_2, x_3, y_3, z_3, x_4, y_4, z_4$) → **Behavioral response**

Figure 12.2
Control structure for tethered rationality. Human behavior is a function of the responses of the autonomic system, the instinctive system, the associative system, and the reasoning system to any given situation. While the mechanism underlying each system is different, all utilize the common currency of feelings. The response generated by each system is also in the currency of feelings, with valence, arousal, and duration components. The system is set up to maximize pleasure and minimize pain or displeasure. The selected behavior will usually be a blended response based on the output of all systems. There is no central executive in charge. The reasoning system has an input into the response, but so do the other systems. Individual differences in behavior are explained not just in terms of individual differences at the level of reasoning but also individual differences at the level of the autonomic, instinctive, and associative systems. A notion of variable individual effort may be captured in this model by variability in levels of arousal associated with different systems.

Parole Judges

The reader will recall the data from the study of parole judges, where the judges made significantly more favorable parole decisions just after lunch than just prior to lunch (Danziger et al., 2011). If the judges are considering an application with some merit, the reasoning system may generate a rating ranging from neutral to positive. This will be associated with a positive valence and some amount of arousal. If, however, the judge adjudicating the case also happens to be hungry, the autonomic system will be signaling negative valence and high arousal associated with hunger. The two issues (the merit of the case under consideration and the judge's hunger) are logically unrelated, but neurologically it is one system. Therefore, one can imagine the negative feelings associated with the hunger overpowering, or at least dampening, the neutral to positive feelings associated with the evaluation of the case, resulting in more negative parole decisions. On the other hand, if the judge has had a good meal prior to the evaluation of the case, the positive feelings of fullness and contentment may well push a neutral to negative feeling associated with the evaluation of a case over the top, resulting in more positive parole decisions.

The Inheritance

Consider an example from Jane Austen's novel *Sense and Sensibility* that involves several systems and plays out over an extended period of time, illustrating the importance of the temporal component of feelings. John Dashwood promises his father on his deathbed to share his large inheritance with his half-sisters (Austen, 2013, p. 40):

> When he gave his promise to his father, he meditated within himself to increase the fortunes of his sisters by the present of a thousand pounds a-piece. . . . "Yes, he would give them three thousand pounds: it would be liberal and handsome! It would be enough to make them completely easy. Three thousand pounds! he could spare so considerable a sum with little inconvenience."—He thought of it all day long, and for many days successively, and he did not repent.

The promise that John Dashwood made to his dying father was itself a cognitive act, inert on its own, its implementation to be guided by the resolve and action tendencies of filial love and duty. Filial love and duty are the reciprocal of parental care and duty. They need to be present to enable the parent-child bond that is critical to the survival of offspring until such time as they can fend for themselves. There are profound positive feelings of attachment and commitment associated with filial love, especially at a deathbed. John's wife, Mrs. Dashwood, has no such attachment to the father-in-law or his wishes and is primarily concerned with *self-maximization* of resources for herself and *her* son. She is unhappy with the promise (p. 45):

John Dashwood: It was my father's last request to me. . . . The promise, therefore, was given, and must be performed. Something must be done for them.

Mrs. Dashwood: Well, then, LET something be done for them; but THAT something need not be three thousand pounds. . . .

At the early stage, the positive feelings associated with the promise dominate. But with time, as the intensity of John Dashwood's feelings supporting the resolve and action tendencies naturally diminish, his wife's counterarguments vis-à-vis the promise, and arguments supporting the action tendencies associated with self-maximization, begin to sound more persuasive (p. 45):

John Dashwood: Perhaps, then, it would be better for all parties, if the sum were diminished one half.—Five hundred pounds would be a prodigious increase to their fortunes!

Later . . .

John Dashwood: I believe you are right, . . . A present of fifty pounds, now and then, will prevent their ever being distressed for money, and will, I think, be amply discharging my promise to my father.

Later still . . .

Mrs. Dashwood: To be sure it will. Indeed, to say the truth, I am convinced within myself that your father had no idea of your giving them any money at all. The assistance he thought of, I dare say, was only such as might be reasonably expected of you; for instance, such as looking out for a comfortable small house for them, helping them to move their things, and sending them presents of fish and game, and so forth, whenever they are in season.

John Dashwood: I believe you are perfectly right. My father certainly could mean nothing more by his request to me than what you say. I clearly understand it now, and I will strictly fulfil my engagement by such acts of assistance and kindness to them as you have described.

The rational mind is now beginning to reinterpret and re-remember beliefs so as to minimize the unpleasant feelings (i.e., cognitive dissonance) associated with any lingering conflict between the promise and current actions, until finally (p. 45):

Mrs. Dashwood: Your father thought only of THEM. And I must say this: that you owe no particular gratitude to him, nor attention to his wishes; for we very well know that if he could, he would have left almost everything in the world to THEM.

> This argument was irresistible. It gave to his intentions whatever of decision was wanting before; and he finally resolved, that it would be absolutely unnecessary, if not highly indecorous, to do more for the widow and children of his father, than such kind of neighborly acts as his own wife pointed out.

Finally, the reasoning mind is not only able to minimize the conflict (and unpleasant feelings) associated with the breaking of the promise but also rationalize and construct a world (a set of beliefs) that actually enhances the pleasure associated with substituting self-maximization for the promise.

In this scenario, at least two instincts come into play, self-maximization and filial love and duty (kinship bonds). The two push in opposite directions. The latter dominates at the deathbed of the father (greater positive valence and greater arousal) and results in the cognitive act of a promise to share the estate with his half-sisters. The intensity of the positive feelings associated with filial love and duty (perhaps generated by Panksepp's NEED and CARE systems) is essential to sustain and carry out the promise and to initially discard his wife's counterarguments and self-serving rationalizations. But with time, the intensity of feelings associated with filial love and duty naturally fade. As the intensity diminishes, so does the support for sustaining the promise. Self-maximization (perhaps involving Panksepp's SEEKING and POWER systems) persists throughout life. It is not triggered by specific events and does not usually diminish with time, though it may wax and wane with age and health. In addition, Mrs. Dashwood's rationalizations serve to enhance the positive feelings associated with self-maximization and diminish any of Mr. Dashwood's negative feelings associated with breaking the promise to his father. The promise is broken and the rational mind is utilized to reinterpret events such that breaking of the promise is consistent with maximizing positive feelings and minimizing negative feelings. This type of interplay within and across levels is the norm in human decision-making. Behavior is a natural consequence of integrating feelings associated with each system's output in a manner that maximizes pleasure and minimizes displeasure for the organism. There is no central executive guiding behavior or changing "its" mind.

Both these examples are little more than "just so" stories. Is it possible to move beyond purely descriptive stories and provide a more quantitative account that actually illustrates at the level of mechanisms how the whole system might actually work in a particular case? We can do this in certain cases where we have sufficient knowledge of the underlying biochemistry and neurophysiology involved. For this detailed example, I return to my propensity to indulge in slices of pizza and chocolate cake even though I'm overweight and know they are not good for my health. This telling will require some patience on the part of the reader as the necessary biochemistry and neuroscience of energy or weight management are introduced. The biological details and nuances are important because they determine the particulars of the story.

Detailed Example: Why Do I Overeat?

Energy management is perhaps the most fundamental process for any organism. Life-forms from plants, to plankton, to primates must engage in it. It is a fixed problem with solutions invariant across generations. As such, in the (nonhuman) animal kingdom, it is an excellent candidate for implementation in autonomic, instinctive, and associative systems. The autonomic systems control the conversion of food to energy, plus or minus fat deposits, and signal the initiation and termination of food acquisition. The initiation signals result in the organism actively seeking, selecting, pursuing, procuring, and ingesting food. There are various species-specific instinctual (and some learned) mechanisms available for this purpose. Quantity of food consumption is modulated not simply by the hunger signals but also by the palatability of the food—how good it tastes. Once a sufficient quantity of food has been ingested, feelings of satiety and fullness (generated by the autonomic system) signal termination of food acquisition and consumption. In humans, the reasoning system also comes into play. The overall human energy management system is diagrammed in figure 12.3 and discussed here. The reader should take special note of the following four points:

(1) all four behavior-initiating systems that we have been discussing are intimately involved;

(2) they communicate with each other via feelings;

(3) the "wanting" and "liking" reward systems drive the process by triggering instinctive behaviors, conditioned cues, and cognitive/reasoning systems; and

(4) there is no central executive in charge, meaning that in humans reason has a say in the process but does not dictate it.

Determining correct weight Organisms consume food for immediate energy and store excess energy as fat deposits for future use, but they also

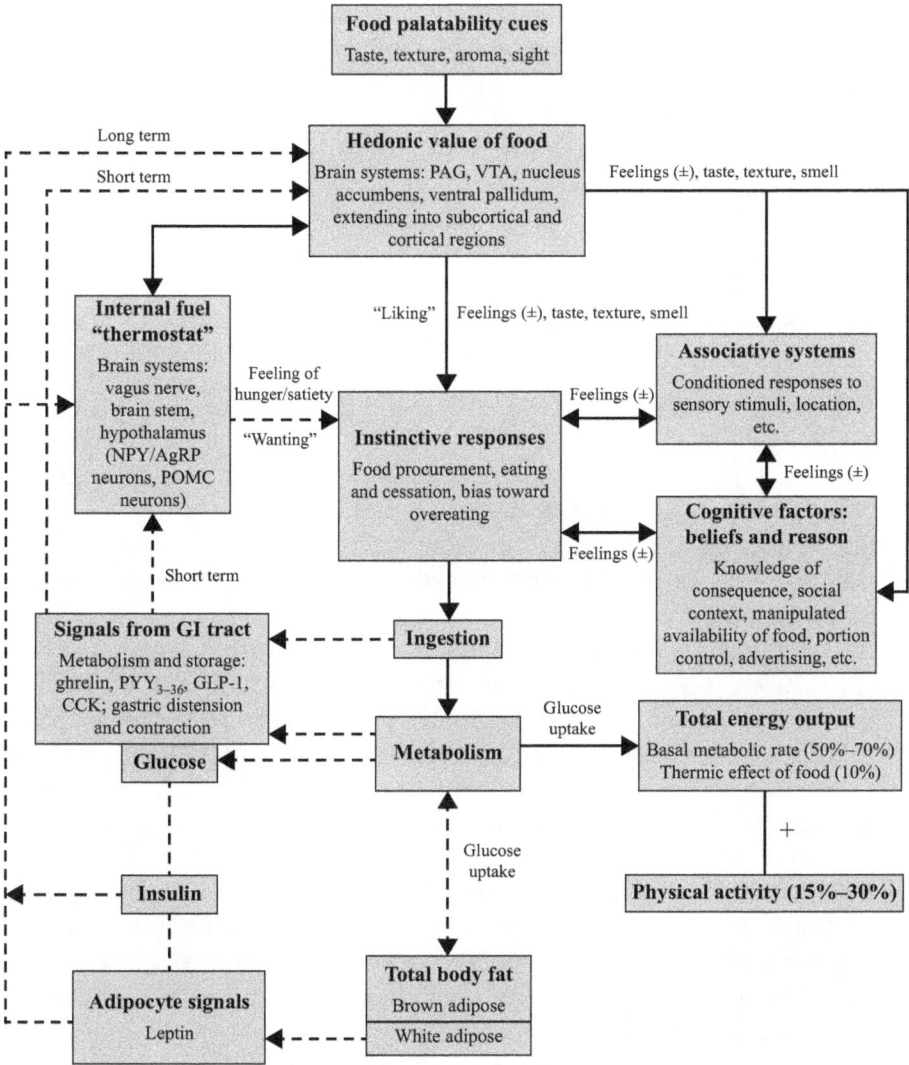

Food palatability cues
Taste, texture, aroma, sight

Hedonic value of food
Brain systems: PAG, VTA, nucleus accumbens, ventral pallidum, extending into subcortical and cortical regions

Long term

Short term

Feelings (±), taste, texture, smell

Internal fuel "thermostat"
Brain systems: vagus nerve, brain stem, hypothalamus (NPY/AgRP neurons, POMC neurons)

"Liking" Feelings (±), taste, texture, smell

Feeling of hunger/satiety

"Wanting"

Instinctive responses
Food procurement, eating and cessation, bias toward overeating

Feelings (±)

Associative systems
Conditioned responses to sensory stimuli, location, etc.

Feelings (±)

Cognitive factors: beliefs and reason
Knowledge of consequence, social context, manipulated availability of food, portion control, advertising, etc.

Feelings (±)

Short term

Signals from GI tract
Metabolism and storage: ghrelin, PYY$_{3-36}$, GLP-1, CCK; gastric distension and contraction

Ingestion

Glucose

Metabolism

Glucose uptake

Total energy output
Basal metabolic rate (50%–70%)
Thermic effect of food (10%)

+

Physical activity (15%–30%)

Glucose uptake

Insulin

Total body fat
Brown adipose
White adipose

Adipocyte signals
Leptin

Figure 12.3
Tethered rationality control structure for human energy management. Energy management is reasonably well understood, though details continue to be refined and updated. It is in part a homeostatic system sensitive to nutrient signals, gastrointestinal signals, adiposity signals, and signals from other organs. The dashed lines indicate homeostatic processes. CCK = cholecystokinin; GI = gastrointestinal; GLP-1 = glucagon-like peptide-1; NPY/AgRP = neurons expressing neuropeptide Y/agouti-related protein; PAG = periaqueductal gray; POMC = neurons expressing pro-opiomelanocortin; PYY$_{3-36}$ = peptide YY$_{3-36}$; VTA = ventral tegmental area.

have a set weight that is vigorously defended by homeostatic systems (Keesey & Powley, 1975, 2008). For many animals, the weight is constant throughout the year and indeed much of their lifetime. It may vary only by a few percentage points over many years. In other animals, such as those that hibernate (e.g., Siberian hamsters), it will vary seasonally. If this set weight is artificially manipulated by starving or overfeeding the animal, it will revert back to its seasonally adjusted normal once the artificial perturbation is removed (Morgan, Ross, Mercer, & Barrett, 2003). The set weight is maintained by multiple autonomic processes interacting with instinctive, associative, and (in humans) reasoning systems. The connections between the autonomic system and these other systems run through the reward systems.

Autonomic homeostatic systems: Short-term weight regulation A number of details of the workings of this system are known (Harrold, Dovey, Blundell, & Halford, 2012; Näslund & Hellström, 2007; Roh, Song, & Kim, 2016; Sam, Troke, Tan, & Bewick, 2012; Yeo & Heisler, 2012). The process begins with the metabolism of ingested food (figure 12.3). I use "metabolic process" as a general term to refer to all chemical reactions that break down food into energy forms that can be utilized by cellular processes; synthesize proteins, carbohydrates, lipids, and nucleic acids; transport nutrients between cells; and eliminate nitrogenous wastes. The consumption and metabolism of food results in physical changes in terms of contraction and distention of the stomach and the release of certain hormones from the gastrointestinal (GI) tract, including cholecystokinin (CCK), glucagon-like peptide-1 (GLP-1), peptide YY_{3-36} (PYY_{3-36}), and ghrelin. Nutrients such as glucose, fatty acids, and amino acids signal nutrient availability to the hypothalamus. All these components play a role in modulating food intake.

Hunger is signaled by ghrelin, a hormone synthesized largely in the stomach and circulated throughout the body. Ghrelin has orexigenic (appetite stimulant) properties and modulates meal initiation via the nucleus tractus solitarius in the brain stem and the arcuate nucleus (ARC) in the hypothalamus (Näslund & Hellström, 2007). Levels of ghrelin peak prior to meals and decrease during and after meals, only to increase for the next meal. Artificial administration of ghrelin increases appetite and food intake in both rodents and humans (Tschöp, Smiley, & Heiman, 2000; Wren et al., 2001).

Satiety is signaled by several GI tract peptide hormones, including CCK, GLP-1, and PYY_{3-36}. CCK inhibits food intake. Intravenous infusion of CCK significantly reduces food intake in humans (and other animals) compared to a saline solution (Ballinger, McLoughlin, Medbak, & Clark, 1995; Gibbs,

Young, & Smith, 1973). Decreasing plasma levels of CCK correlate with increased hunger, while increased levels correlate with increased fullness (Näslund & Hellström, 2007). CCK acts on the GI tract through the vagus nerve and the nucleus tractus solitarius in the brain stem and on appetite suppression via the hypothalamus (Harrold et al., 2012).

GLP-1 is another peptide produced in the GI tract that inhibits hunger. It is part of the system that controls the movement of food through the gut, along with acid secretion, and regulates blood glucose levels (Näslund & Hellström, 2007). Delivery of GLP-1 directly into cerebral spinal fluid (via intracerebroventricular injections) of fasted mice significantly reduces feeding behavior. It exerts its influence on GI tract functions through the vagus nerve and inhibits feeding via the hypothalamus (Turton et al., 1996; Wettergren, Wøjdemann, & Holst, 1998).

A third inhibitor of appetite is the peptide YY_{3-36}. Injections of PYY_{3-36} into rats reduce food intake and result in weight loss. In humans, infusion of PYY_{3-36} significantly decreases appetite and food intake (Batterham et al., 2002). It regulates appetite via the ARC of the hypothalamus. The physical distention and contraction of the stomach as a function of food intake and outtake also signals hunger and satiety to the brain stem (nucleus tractus solitarius) via the vagus nerve (Näslund & Hellström, 2007). All these are components of the homeostatic system responsible for regulating the body's immediate short-term energy needs.

Autonomic homeostatic systems: Long-term weight regulation Defending a set weight over years also requires sensitivity to long-term modulation of food intake and energy expenditures. Long-term energy management involves the body's fat reserves, which consist of white and brown adipose tissue. Adipose tissue constitutes 20% of total tissue mass in ideal weight adult males (body mass index, BMI = 22) and up to 50% of total tissue mass in obese (BMI = 30) adult males. Brown adipose is primarily concerned with thermogenesis. White adipose is largely responsible for storage of excess energy.

White adipose is not just a fat reserve but also a major part of the endocrine system. It produces the hormone leptin, which, along with insulin, plays a role in regulating appetite, largely through hypothalamic neuroendocrine pathways. It inhibits appetite-stimulating peptides and stimulates appetite-reducing peptides. Leptin and insulin are critical for long-term eating behavior and energy management (Heisler & Lam, 2017; Porte, Baskin, & Schwartz, 2002; Trayhurn & Bing, 2006). There is a direct correlation between BMI and circulating levels of leptin (Considine et al., 1996). Fasting and feeding lead to rapid decreases and increases, respectively, in levels of circulating leptin (Hardie, Rayner, Holmes, & Trayhurn, 1996). Genetically modified mice with a mutated *obese* gene and consequent leptin deficiency dramatically reduced food intake when injected with recombinant leptin (Mercer, Moar, Rayner, Trayhurn, & Hoggard, 1997). Obese children with a mutation in the leptin gene resulting in a nonfunctional hormone dramatically reduced food intake and lost body weight and fat subsequent to treatment with recombinant leptin (Farooqi et al., 1999; Montague et al., 1997). On the other side of the equation, leptin also affects energy expenditure, perhaps through thermogenesis (Trayhurn & Bing, 2006).

These various signals converge within the ARC of the hypothalamus (figure 12.3). Two neural populations within the ARC are the control center for the system of energy management. One group is called NYP/AgRP neurons, because they express neuropeptide Y/agouti-related protein. The second group is called POMC neurons, because they express pro-opiomelanocortin. Both respond to the signals from circulating glucose, leptin, insulin, ghrelin, PYY_{3-36}, CCK, and GLP-1 (Sam et al., 2012; Yeo & Heisler, 2012). They also inhibit each other and send projections to other hypothalamic areas, including the paraventricular nucleus (PVN), and on to higher brain centers, including reward centers. The NYP/AgRP neurons are orexigenic, meaning they stimulate appetite. Ablation of this region in adult mice causes reduced feeding and rapid starvation (Luquet, Perez, Hnasko, & Palmiter, 2005). The POMC neurons are anorexigenic, meaning they inhibit appetite (Sam et al., 2012). This is all part of the homeostatic system that signals hunger and satiety, but to actually start and stop feeding, it needs to access instinctive, associative, and reasoning systems. This access runs through the reward systems.

Reward systems: Liking and wanting   Homeostatic systems are necessary but not sufficient to account for energy management. They can signal hunger and satiety, but the reward system is critical for selecting and initiating behaviors. As noted in chapter 11, the reward system itself can be subdivided into two distinct (but interrelated) components—wanting and liking. In very broad terms, we could say that wanting initiates feeding behavior, while liking guides food selection and affects the quantity consumed.

Given that the wanting system is the initiator of behavior, it should be unsurprising that there are connections from the hypothalamus feeding system to the mesial limbic dopamine-based reward system (Douglass et al., 2017). Additionally, ghrelin and leptin receptors are also expressed in the ventral tegmental area (VTA), where they modulate dopamine activity. Ghrelin directly administered into the VTA binds to neural receptors in this region and triggers increased dopamine activity. Ghrelin administration in mice triggers a proportional increase in feeding and causes food-restricted rats to work harder for available food pellets. Blocking ghrelin receptors in VTA causes mice to display attenuated feeding behavior (Abizaid et al., 2006; King, Isaacs, O'Farrell, & Abizaid, 2011). Direct administration of leptin into the VTA leads to decreased food intake in mice, while deficiency leads to increased food intake (Hommel et al., 2006). These and other results suggest a role for ghrelin and leptin not only in signaling hunger but also in modulating the reward system.

The liking system is concerned with hedonic value activated by the palatability of food (taste, texture, aroma, visual cues) and greatly affects eating behaviors. Its critical role often is not fully appreciated. Palatability does not just reflect underlying nutritional need. Palatability ratings of food are positively correlated with the amount consumed, even when nutritional value and satiety levels are controlled for (Yeomans, 1996). Subjectively rated levels of hunger increase with the palatability of the food being consumed. Even the sight of a preferred food can increase hunger. Modulating palatability of foods results in corresponding changes in the eating rate and duration of meals (Hill, Magson, & Blundell, 1984; Yeomans, Blundell, & Leshem, 2004; Yeomans & Gray, 1996). Loss of taste and smell that affects food palatability leads to poor appetite (Schiffman & Graham, 2000).

Berridge has further characterized food reward systems as "go" systems. This means that once they are activated, satiety signals can diminish them but not fully stop them (Berridge et al., 2010). For example, rats satiated by milk or sugar solution dripping directly into their mouths (up to 10% of their body weight within half an hour) exhibited reduced "liking" reactions to sweetness immediately afterward but did not exhibit "disliking" reactions (Berridge, 1991). Similarly, satiating humans on chocolate mousse can reduce the "liking" value to near zero but not into negative "disliking" territory (Lemmens et al., 2009).

One potential neural pathway for oral sensory reward mechanisms is endogenous opioid peptides. Blocking this pathway with opioid receptor antagonists reduces food palatability and appetite in both human and non-human animal studies (Holtzman, 1979; Yeomans et al., 2004; Yeomans & Gray, 1996). Furthermore, as noted earlier, ghrelin initiates feeding response via AgRP neurons. Interestingly, work with animal models shows that if the AgRP system is disabled, palatable food on its own is sufficient to initiate feeding (Denis et al., 2015)! The extent to which the homeostatic system and the hedonic system interact is a point of some debate, but the fact that they both play a critical role in modulating food intake is widely accepted

(Berridge, 2009; Berthoud, Münzberg, & Morrison, 2017; Epstein, Truesdale, Wojcik, Paluch, & Raynor, 2003; Münzberg, Qualls-Creekmore, Yu, Morrison, & Berthoud, 2016; Yeomans et al., 2004).

Reward-driven instinctual systems   When the need for energy is signaled by the autonomic system, instinctive systems are the first to be activated. They are universally available. The drive is provided by the negative feelings of hunger pangs and the positive consummatory feelings associated with feeding. The appetitive phase is initiated and dominated by hunger pangs that activate the SEEKING system, among others, including specific hunting or foraging systems. The activities involved are highly species specific, ranging from drinking blood, to filtering plankton, to selecting and consuming vegetation, to hunting prey, to consuming carrion. Once food is acquired, the consummation phase is dominated by the hedonic pleasure (oral sensory properties) of the food.

Reward-driven associative systems   Rewarding experiences naturally lead to Pavlovian and operant learning (Berridge, 2009; Berridge et al., 2010). The most obvious manifestation of this is where visual images of food (rather than the food itself) trigger wanting behaviors (Pelchat, Johnson, Chan, Valdez, & Ragland, 2004; Spence, Okajima, Cheok, Petit, & Michel, 2016). Other common examples are where one food may trigger desire for another if they are usually paired (e.g., wine and cheese), or consumption patterns may become associated with different times and environments (e.g., home versus restaurant versus cruise ship) or activities (e.g., eating popcorn while watching a movie). The whole behaviorist paradigm that we discussed in chapter 5 was an exploration of such cue-triggered associative behaviors.

Reward-driven reasoning systems   Humans also have access to the reasoning mind in controlling eating behavior. Societal beliefs such as "fat is a sign of wealth" or "thin is beautiful," or knowledge of the increased morbidity associated with obesity, may factor into eating decisions by activating related desires (perhaps with associative components). In response to such desires, we are capable of devising strategies to modulate food intake, such as in the pizza story from chapter 1, or the "out of sight, out of mind" strategy: do not keep palatable calorie-dense foods at home within easy reach. There are many others. But these strategies can also be used in reverse by advertisers to intentionally stimulate hedonic brain systems with visual and olfactory cues to increase food consumption (Spence et al., 2016). The effectiveness of these reason-based desires and accompanying strategies will depend on the relative hedonic reward values associated with pursuing the desire to lose weight versus those associated with the pleasure of consuming chocolate cake. This is why decisions such as "I will eat less" are usually ineffective and are abandoned when confronted with pizza and chocolate cake, or even a commercial for popcorn and Coke before the movie starts; unless you are like Kate Moss and "nothing tastes as good as skinny feels" (O'Malley, 2018). Understanding the underlying biology and taking tethered rationality seriously helps to explain why.

Individual differences in energy management   During the past 25 to 50 years, there has been a dramatic increase in food availability and a dramatic reduction in energy expenditure in developed Western countries, resulting in an increase in body mass index in the population (Yeo & Heisler, 2012). This is just a nice way of saying that there has been a dramatic increase of fat people in rich Western countries. I am among them. But interestingly, it is not a uniform increase. There is no upward shift in the set weight of all members of the population. There is considerable individual variability. Some people have become obese, whereas others have not. How do we account for these individual differences? The same systems are in play in everyone. Why are some people able to regulate their food intake to correspond to their energy output, while others, like me, are not?

The Western-Christian model does recognize the temptation or reward component of food, but control is simply a matter of willpower and judicious use of reasoned strategies. In the standard cognitive and social sciences model, there is no explicit recognition of temptation, or indeed of effort. Choices are constrained only by beliefs and desires (many socially determined), and the principle of rationality, so individual differences in eating behavior have to be explained and modulated only in these terms. Beliefs and desires can be revised and updated at will (though see part V). Reasoning ability can be improved by education. With the revision in beliefs and desires and sound reasoning will come revision in eating behavior.

In the massive modularity account, humans evolved in an environment of feast and famine, so it is adaptive, or fitness enhancing, to overeat when food is available and store the excess as fat deposits for times of famine, as we saw in the case of bears and lions in chapter 4. We happen to find ourselves in an unprecedented environment of excessive, easily available, calorie dense, strategically designed, palatable foods. It is true that excessive fat deposits can have a negative effect. But instinctive systems evolve slowly and remain responsive to historical environments, so their performance may be counterproductive in suddenly occurring, radically different environments. This may well be true, but it fails to address the issue of individual differences. Many individuals are able to adjust their calorie intake, while some, like me, struggle. The dual mechanism theory models are also not particularly helpful in this context, for the reasons already noted.

Now that we have an understanding of how the energy management system actually works, we can see why the explanations offered by these models are at best incomplete and at worst incorrect. Numerous interacting systems are involved in energy management. We are all operating with the same systems, functioning under the same biological principles. So why the individual variations in outcome? The most plausible explanation for different outcomes is in terms of individual differences in one or several component systems resulting in different outcomes with "similar effort." Some obvious candidates for individual differences include the following:

1.   Differences in a number of genetic and/or epigenetic factors could simply establish a higher (or lower) set weight for an individual, resulting in increased (or decreased) food consumption (Ravussin et al., 2011).

2.   Differences in production of the relevant hormones (leptin, ghrelin, and others) will affect the system's ability to defend a given set weight (Farooqi et al., 1999; Montague et al., 1997).

3.   Genetic differences in oral sensory abilities are now well accepted. Approximately 25% of the population are considered "super tasters," an equal number "nontasters," and the others "average," based on their ability to taste the chemical 6-n-propylthiouracil (PROP). Super tasters have more visible taste papillae on their tongues, making them overly sensitive to many foods (Bartoshuk, 2000; Prescott, 2012). Sensory experiences affect food intake, even in the absence of need (Yeomans, 1996).

4.   Differences in the general arousal system, discussed in chapter 11, will result in different levels of "liking" and "wanting" associated with foods and will affect food intake, independent of homeostatic need. There is evidence of a genetic basis for differences in the general arousal system (Gray, Braver, & Raichle, 2002).

5.   Differences in interaction of "liking" and "wanting" systems will affect feeding behavior.

6.   Easy availability of unlimited energy-dense palatable foods will affect feeding behavior.

7.   Sensitivity to different conditioned cues (e.g., advertising) will affect feeding behavior.

8.   Individual decisions (e.g., goal to lose 10 pounds) and choices (e.g., bypass dessert today) made by the reasoning mind will affect eating and exercising behaviors.

9. Socioeconomic factors such as the availability of time, money, and resources will affect eating and exercise behaviors.

10. Finally, immersion in societal norms, beliefs, expectations, and consensus will affect eating and exercise behaviors through the reasoning mind.

Several important points emerge from this detailed example. Autonomic, instinctive, associative, and reasoning systems interact for human energy management. Eating behavior is initiated and driven by reward systems. There may be no explicit stop switch. We have conscious control over only one of these systems, the reasoning mind, but the reasoning mind is not the CEO. In fact, no specific system is fully in charge. There is an overall blended response based on valence, arousal, and duration of feelings generated by each component. There will be individual differences based on genetic, epigenetic, and hormonal factors, in both the homeostatic and reward systems. Individual differences in conditioning histories, beliefs and desires, and social environment are also factors in weight management.

Cognitive effort, construed as the level of arousal associated with desires (e.g., "I want to lose 20 pounds so I look and feel better"), will have some impact, but only to the extent that you feel it. Saying it is not enough. You must *feel it* for it to be causally efficacious. Even then, the desire to lose weight is just one of several affective inputs into the system. It will need to compete with feelings associated with the taste of chocolate cake and pizza. It will need to compete with the innate bias toward overeating built into the reward system. It will need to compete with the ubiquitous availability of calorie-rich palatable foods. It will need to compete with the conditioned cues triggered by advertising and social behaviors.

Finally, the list of individual differences in the long list of noncognitive factors, noted earlier, also makes it clear that the same "cognitive effort" in two different individuals will have very different impacts on their eating behavior and explains why some people are good at weight management, while others struggle. For example, if your systems are set up such that, like Kate Moss, skinny feels better to you than the taste of chocolate cake or pizza, you will be much more successful.

The three examples of parole judges, the inheritance story, and energy management we have considered provide an indication of the engagement of all four systems (autonomic, instinctive, associative, and reasoning) in human behavior. They highlight that the systems interact via the currency of feelings and there is no overriding executive control. All systems contribute to the decision or action, which is based on the principle of maximizing pleasure and minimizing displeasure. The third example, of energy management, is particularly germane. Because we understand a great deal of the underlying biochemistry and neurophysiology of energy management, we can see that the control structure proposed for tethered rationality is not merely speculative but consistent with what is actually implemented in the biology.

Evaluating the Tethered Rationality Model

If tethered rationality is to be a serious contender as a model for human behavior, it must be falsifiable and be able to generate specific predictions. There are a number of straightforward ways to falsify the model. For example, if my blended response hypothesis used to explain economic decision-making data in chapter 9 is incorrect, and the behavior can be better explained just in terms of reasoning or just in terms of instincts, then the model needs to be reconsidered. Several neurological findings could also lead to the dismissal of the model. For example, if there are no physiological distinctions in the mechanisms underlying autonomic, instinctive, associative, and reasoning systems and/or there is no anatomical and physiological evidence for the tethering of these different systems, the proposed model fails. If the neurophysiological and behavioral data do not require the postulation of feelings (i.e., feelings do not exist), then the model can be dismissed. If feelings do exist but they originate in neocortical systems instead of in brain stem systems, the model fails. If feelings do exist but they are not common to all four systems, the model also fails.

Can the tethered rationality model make specific predictions, or does it provide just post hoc explanations? It can certainly make predictions as accurate as the standard cognitive science model based on beliefs, desires, and coherence relations because, after all, this system is still part of tethered rationality. But my claim is, of course, that it can make much more accurate predictions. To get this additional accuracy, we need to plug in the three other systems. Furthermore, the model requires knowledge of the physiology and neurophysiology of these systems to generate predictions. This information can range from the very general to the very specific. For example, at a very general level, it can predict that if I lose my sense of smell, this will diminish my ability to taste and enjoy chocolate cake and pizza, resulting in an increased ability to resist them and subsequent reduction in my caloric intake. At a more specific level, if we are able to plug in detailed information about personal biochemistry and neurophysiology in each of the components of the energy management model in figure 12.3, we can make much more accurate predictions about eating behavior and consequences than by utilizing information just about environments and/ or beliefs and desires. In addition, the model can account for many individual behavioral differences by appealing to individual differences in biochemistry and neurophysiology (points 1–5 listed earlier).

The reader may object that for many situations we do not have the biological knowledge to utilize the model to its full potential. True, but this is not a shortcoming of the model. This is the whole point of tethering. The reasoning system is tethered to the underlying biology, so accurate predictions of human behavior need to take the biology into account. In some cases, as in the energy management example, we have made progress in understanding the relevant biochemistry and neurophysiology to plug into the model. In other cases, we have to wait for the science to catch up.

*   *   *

En route to constructing the tethered rationality model, we have actually already addressed a number of the issues with which the book began. Tethered rationality provides a plausible account of my continuing to indulge in pizza and chocolate cake against the advice of my doctor. The example analyzing the prelunch and postlunch decisions of the parole judges also addresses the anecdotal example from chapter 3 of snapping at my wife while hungry. We can also finish addressing John Edwards's fatal decision to have an affair during his run to be the 2008 presidential nominee of the Democratic Party. Here it is important to highlight that I, of course, do not have any privileged insight or information about John Edwards or his state of mind at the time. This is merely a speculative, plausible reconstruction of some systems that may have been involved.

To begin, a number of basic instinctual systems would be involved (using the Panksepp vocabulary): drive for POWER, ATTACHMENT to his wife (perhaps waning), ATTACHMENT to his mistress (perhaps increasing), and the LUST system. At the level of instincts, there need be no conflict between the POWER system and the LUST system. They are often complementary and connected. At the cognitive level, there would have been beliefs such as that American voters will not accept an extramarital affair at the best of times, that his wife is dying of cancer and receiving a great deal of public support, and that American voters will never forgive a candidate for having an extramarital affair under such circumstances. At the level of desires, the main driving force would be to secure the nomination of the Democratic Party and eventually win the presidency (a cognitive representation of the

POWER system). Given these beliefs and desires, the rational conclusion is obvious: do not get involved in an extramarital affair. The feelings associated with this coherent conclusion of the reasoning system, along with the feelings associated with Edwards's ATTACHMENT to his wife, were washed aside by the higher valence and arousal associated with the activation of the LUST and ATTACHMENT systems directed at his mistress.

How do I explain the exchange with my teenage daughter about prioritizing grooming over a proper breakfast from chapter 6? In the case of teenage daughters, at least the following three factors are involved. Teenage bodies are flooded with sex-differentiating hormones and accompanying attraction to the opposite sex, triggering various behaviors, including grooming. There is peer pressure to belong to the in-group (discussed in the next chapter). There is pushback against parental authority as part of striving for increased autonomy. Against this backdrop, I made the statement that "what is inside your head is more important than what is on top of your head" with the intention of encouraging my daughter to spend less time on her hair and makeup in the mornings, leaving time for eating a proper breakfast, rather than rushing out for the school bus hungry and not being able to focus in class. Notice that there is no basis for her to directly evaluate the truth or falsity of the statement. She can take my word for it; or not if she is pushing back against parental authority. Also, the instinctive behaviors initiated by the hormonal surge and in-group peer pressure are driving her to prioritize her appearance, hence her expressed disbelief that I could be stupid enough to believe such a statement. My gullibility in accepting the statement when I heard it as a teenager can be explained by postulating similar systems, with individual differences. In my case, different levels of hormones may have been involved, I did not have an in-group of friends, and I was not told by my parents but rather it was constantly repeated and reinforced at school. These differences may account for why I believed it and my daughter knew better.

In each of these cases, the behaviors are not rational—that is, coherently determined from an explicit set of beliefs and desires. The various nonreasoning systems that we have been discussing are getting the upper hand in generating the response in these cases. There's nothing irrational about this. This is how the system is set up. I have suggested that we adopt the term *arational* to refer to these choices.

There are remaining the examples of my American friend's aversion to universal healthcare, society's failure to mobilize to address global warming, and the seemingly "irrational" (but successful) White House strategy to counter the impeachment effort in the court of public opinion that need to be more fully analyzed. Partial answers to these issues were developed in chapters 8 and 9. Global warming was discussed in the guise of the collapse of the Canadian Maritimes fisheries because of the advantage provided by 20/20 hindsight. But these discussions remain incomplete. A fuller understanding of these choices requires consideration of the phenomenon of boldly denying the facts, of being unconvinced by the evidence. We saw this at work in the exchange between Stephen Schneider and the climate change skeptics from Australia in chapter 1. They asked questions. He provided science-based answers. In the end, they left unconvinced and continued to believe that climate change is a hoax. We now apply the tethered rationality model to explain these examples of arational behavior.

V What Color Is Your Bubble? Why Changing Minds Is Hard

You can't use logic to dissuade someone who didn't use logic to reach their viewpoint in the first place.

—Anonymous

Opinions don't affect facts. But facts should affect opinions.

—Ricky Gervais

Not only is the standard academic cognitive and social science model committed to the idea that all behavior is determined by beliefs and desires, but it is also assumed that beliefs and desires can be easily changed through learning and reason, leading to arbitrary behavioral change. Is this actually true?

Let us begin by reminding ourselves that beliefs are psychological states with propositional content that have a mind-to-world direction of fit (i.e., they can be true or false). Desires are psychological states with propositional content that have a world-to-mind direction of fit. Both are causally efficacious in behavior. Rationality requires that as new information comes in, it be vetted and justified. If it is veridical, it is integrated into our worldview (and referred to as "knowledge"); otherwise, it should be discarded to maintain mind-to-world correspondence. Integration may require confronting inconsistencies and revising existing belief networks. This is a basic function of the reasoning mind, and we have already discussed why it is evolutionarily adaptive. Yet despite being endowed with the tools of reason, we all harbor beliefs that are questionable—if not downright false—given available evidence, and are reluctant to change them. This is especially the case when large-scale belief revisions are required late in life. These phenomena are extremely difficult to explain convincingly with just the rational mind. Appealing to the biology provides some interesting answers.

The first answer is provided in chapter 13 by an appeal to tethered rationality. In challenging the assumption that all behavior is belief-based, the tethered mind provides a natural explanation of failure to revise certain beliefs. Beliefs and desires are only one source of input into our behaviors. The other systems may therefore prevent belief revision, or belief revision may not be sufficient to change behavior.

The other neglected constraint on belief revision is neural maturation. This phenomenon is independent of the tethering of reason and largely comes into play where worldviews need to be revised, late in life. In chapter 14 I propose that with the maturation of association cortex in adulthood there are insufficient neuronal resources remaining for large-scale architectural neural reorganization, making global belief revision challenging.

13   When Failures of Belief Revision Are Less than Motivated Reasoning or Sloppy Reasoning

The ideal subject of totalitarian rule is not the convinced Nazi or the dedicated Communist, but people for whom the distinction between fact and fiction, and the distinction between true and false, no longer exists.

—Hannah Arendt

Man-caused, catastrophic global warming, folks, is a sham, a scam, and a hoax. Don't fall for it. Science has your back on that one.

—Bryan Fischer (American Family Association)

The most outrageous lies are the ones about Covid 19. Everyone is lying. The CDC, Media, Democrats, our Doctors, not all but most, that we are told to trust. I think it's all about the election and keeping the economy from coming back, which is about the election. I'm sick of it.

—@chuckwoolery, July 12, 2020

We have now developed and fleshed out the model of tethered rationality in reasonable detail and used it to account for a number of otherwise puzzling behaviors, but there are several outstanding examples that have been considered but not fully explained, particularly climate change denial, the impeachment "debate," and my friend's aversion to universal healthcare. In each case, the unexplained phenomenon is the failure to revise beliefs in the face of overwhelming evidence (i.e., not *believing* the scientists or political opponents). It is not rational to harbor false or inconsistent beliefs, but it is pervasive. We sustain them by living in "bubbles" or "echo chambers." How is this to be explained? There are some explanations for this phenomenon within the context of the rational mind, largely involving motivated reasoning or sloppy reasoning, but these explanations are unable to address the failure of belief revision in the examples of interest. A more satisfying explanation is provided by tethered rationality along with the introduction of the in-group/out-group instinct.

Explaining Failure of Belief Revision as Motivated Reasoning or Sloppy Reasoning

Consider the following beliefs held by many individuals:

1. (a) Excessive stomach acid causes peptic ulcers. (b) The measles, mumps, and rubella (MMR) vaccine causes autism. (c) "Man-made global warming is the greatest hoax ever perpetrated on the American people." (d) Gender identity is a socially constructed choice, independent of biology.
2. "They [Democrats] are evil people. They want to open the borders and let in rapists, drug dealers and terrorists. They hate America."[1]
3. God created the earth in six days, approximately 6,000 years ago.

These are all legitimate beliefs in the sense that they advance a certain proposition that can be compared with states of affairs in the world and determined to be either true or false. But each of these three types of claims is interestingly different. In the first set of claims, the relevant evidence comes from science.[2] In the second claim, the evidence involves simply listening to the position of the Democratic Party. For the third claim, regarding the creation and nature of the world, evidence comes from both the scientific community and observing the world for the best explanation. The religious belief also differs from the others in that it is universal, with local variants. The preponderance of evidence suggests that these particular beliefs are all false. So why do so many people hold them?

We begin by considering the first two belief types and then return to the third. Most psychologists and political scientists who study the first two types of claims have proposed that these false beliefs are explained as a form of motivated reasoning (Epley & Gilovich, 2016; Kahan, 2016; Kraft, Lodge, & Taber, 2015; Kunda, 1990). Others suggest that cognitive laziness or sloppiness accounts for the phenomenon (Gampa, Wojcik, Motyl, Nosek, & Ditto, 2019; Pennycook & Rand, 2019). These models are diagrammed in figure 13.1. Within these two camps there also seems to be some support for dual mechanism models of reasoning that we encountered in chapter 7, with some researchers arguing that the bias toward false beliefs is driven by heuristic processes.[3] Based on our discussion in chapter 7, the coherency of such a distinction in this particular context is less than clear.[4] I will not revisit the issue. What is relevant for the present purposes is that all these explanations fall within the realm of the reasoning mind.

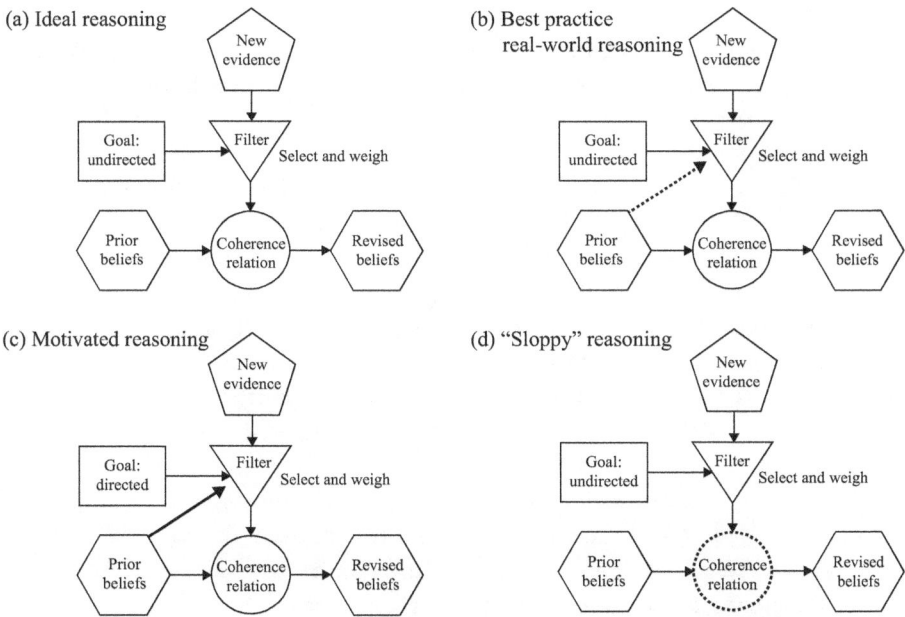

Figure 13.1
Process of belief revision under different models. As new evidence comes in, prior beliefs need to be updated and revised via the coherence relation. The models vary in terms of goal directedness or vested interest of the reasoner and how these interests and prior beliefs impact the filter. The purpose of the filter is to select and weigh the evidence. (a) This is reasoning in an ideal world. The reasoner is disinterested in the outcome, the filter lets through all relevant information, and the coherence relation revises beliefs accordingly. While no individual human can implement this model, scientific methodology over time converges toward it. (b) This model represents best practice real-world reasoning. The reasoner tries to maintain a disinterest in the outcome and minimize the impact of prior beliefs on the filter. This is indicated by the dashed line connecting prior beliefs to the filter. (c) The motivated reasoning model is characterized by goal directedness and a robust, active impact of prior beliefs on the filtering of evidence in the service of a directed goal. This can result in the shielding of belief systems from evidence. (d) In the sloppy reasoning model, the emphasis is on some sort of disruption in the determination of the coherence relation itself, indicated by the dashed circle, be it through lack of effort, intelligence, or succumbing to common reasoning fallacies, such as the confirmation bias, a natural tendency to look for confirming evidence and ignore conflicting evidence.

Motivated reasoning was introduced in chapter 1 as "reasoning with vested interests." Reasoning and argumentation in the courtroom provide a succinct illustration. The presiding judge enters the courtroom without any prior beliefs as to the guilt or innocence of the defendant and has no personal interest in the outcome of the case. She listens to the evidence, as impartially as possible, and afterward revises her beliefs to either guilty or not guilty. This reasoning should be captured by the ideal reasoning model (figure 13.1a) but is more realistically an instance of the best practice real-world reasoning model (figure 13.1b). The prosecutor, on the other hand, enters the courtroom believing that the defendant is guilty and has the very specific goal of convincing the judge of this. These prior beliefs and goal directedness will result in a calculated, judicious selection and weighted presentation of the evidence: "Your Honor, the defendant *lunged* across the counter to *threaten* the cashier." The defense lawyer, on the other hand, enters the courtroom believing that his client is innocent. His goal is to convince the judge of the same. His goal directedness and prior beliefs also result in a calculated selection and weighting of the evidence: "Your Honor, the defendant *leaned* across the counter to *speak* to the cashier." This is motivated reasoning, indicated in figure 13.1c.

The motivated reasoning explanation for false beliefs is that people reason like attorneys rather than like judges. They have certain beliefs and desires (the two being not unrelated) and filter, twist, and select the evidence to advance them. In the courtroom, there are certain constraints imposed by the legal system (enforced by the judge) on the level of filtering and quality of information allowed to be presented by the attorneys. The presence of these constraints and an impartial, disinterested judge usually allows for evidence-based decisions.[5] In the absence of these constraints, arriving at evidence-based decisions can be more challenging. Either way, it should be apparent that motivated reasoning is not flawed reasoning. It is rational. The reasoning mind is being utilized to achieve a particular goal. This is the purpose of the reasoning mind.

The sloppy reasoning model does not emphasize the impact of goal directedness and prior beliefs. Rather, it focuses on diminished analytical reasoning ability, attributed to various factors (figure 13.1d). There are a number of studies showing that logical reasoning (about neutral, nonpartisan content) is positively correlated with levels of education and measures of intelligence and memory scores (Stanovich & West, 2000). Consistent with this, a study involving participants' willingness to believe fake news headlines reported that individuals with higher analytic reasoning abilities were less likely to believe fake news headlines, even when political ideology was taken into account (Pennycook & Rand, 2019).[6] Is there a relationship between sloppy reasoning and motivated reasoning? Do sloppy reasoners engage in more or less motivated reasoning?

At least some studies suggest that sloppy reasoners are less able to engage in motivated reasoning. An experiment carried out by Daniel Kahan and his colleagues examined people's ability to draw correct causal inferences from empirical data in two conditions, a neutral condition involving the efficacy of a new cream in reducing skin rash and a numerically and logically identical but politically polarized question on the efficacy of gun control laws in curbing crime. The two conditions were identical in terms of the presented numbers and logical relationships. Participants were selected from both sides of the gun control issue and pretested for numerical abilities. Inferences from numerical data can be difficult to draw. Unsurprisingly, individuals with the higher numerical ability scores performed more accurately than individuals with lower numerical ability scores. However, the participants' accuracy decreased in the gun-control study, even though the numerical information and logical relations were identical to those in the skin rash study. Interestingly, the greatest deviation in the two conditions occurred among the participants with the *higher* numerical abilities. Presumably, they had greater cognitive capacity to filter and select the evidence to conform to their prior beliefs (Kahan, Peters, Dawson, & Slovic, 2017). While diminished reasoning abilities may increase the likelihood of accepting false beliefs, motivated reasoning requires high cognitive capacity to select data and construct a coherent case for maintaining existing beliefs. So, if motivated reasoning is the correct explanation for false beliefs, it implies higher, not lower, cognitive capacity in the reasoner. Let us now consider the preceding beliefs and see how much can be explained in terms of motivated reasoning and sloppy reasoning.

Belief 1a: Excess Stomach Acid Causes Peptic Ulcers
For decades the medical community was certain that peptic ulcers were caused by excess stomach acidity. There was widespread agreement among scientists and practitioners, and it was considered a closed issue. This conclusion was based on data, and reasoning captured by the best practice real-world reasoning model (figure 13.1b). In the 1980s, two Australian scientists, Barry Marshall and Robin Warren, resurrected an old discarded idea that ulcers were caused by bacteria, specifically, *Helicobacter pylori*. The bacteria theory of ulcers had been proposed multiple times over the previous 100 years but was repeatedly set aside due to lack of conclusive evidence. Scientists were unconvinced that bacteria could survive in the acidic environment

of the stomach. A particularly influential study in the 1950s had failed to find any bacteria in the stomach. So many in the medical community initially ignored, even ridiculed, Marshall and Warren and refused to believe their claims. The reputations and careers of individual scientists were at stake. The individuals who had developed, nursed, and defended the acid theory of ulcers were highly motivated to believe and advocate for it, as per the motivated reasoning model (figure 13.1c). Identities become associated with theoretical contributions and people resist falsification by questioning and rationalizing the evidence, just like attorneys. There is always *some* evidence for any claim and *some* evidence against any claim.

But the saving grace of science is that it doesn't care what individual scientists believe. Data are data, and in the end data always wins. Among any group of scientists there will be many different beliefs and goals. There will always be some individuals who are not hampered by strong prior beliefs and vested interests in any particular theory and will be guided by the data (or commitment to an alternative theory). This plurality of beliefs and interests overcomes individual shortcomings and allows science to approximate the ideal reasoning model (figure 13.1a). Once the evidence is overwhelming, scientists must (and do) accept it and change beliefs or are passed over. In 2005, Marshall and Warren shared the Nobel Prize in medicine for their conclusive demonstration that bacteria can survive in the stomach and cause ulcers. This changed the treatment of ulcers and improved the lives of millions of patients. Members of the lay public were not asked (and did not ask) to chime in with their opinions about the cause and treatment of ulcers before the medical community accepted that ulcers were caused by bacteria and could be treated with antibiotics. This is an example of the successful revision of a false belief in the face of motivated reasoning. This is how vetting, justification, and belief revision are supposed to work.

Belief 1b: The MMR Vaccine Causes Autism
The belief that the MMR vaccine causes autism seems similar to the claim that acid causes peptic ulcers in that it is a scientific medical question of fact to be settled by evidence. However, there is an important difference; the answer to the vaccine question has public-policy consequences, and this has transported the seemingly scientific question into the public domain, where a committed portion of the public feels not only justified but compelled to second-guess the scientists.

While there has always been some residual public resistance to vaccination (Hadwen, 1896), the current antivaccine movement started with the publication of a scientific article by Andrew Wakefield in the respected peer-reviewed journal *Lancet* (Wakefield et al., 1998). In the article, he and his coauthors claimed a correlational link between MMR vaccination and autism in children. In subsequent public appearances, he made very strong *causal* claims. Other scientists were skeptical, but data are data. However, the results of the study could not be replicated by any other lab, and it subsequently emerged that the paper may have been fraudulent. Wakefield was being paid by lawyers who were preparing a class-action lawsuit against the manufacturers of the MMR vaccine in the United Kingdom and may have "massaged" his data to fit the claim (Deer, 2011). The article was fully retracted by *Lancet* in 2010 when this information came to light. Wakefield was found guilty of serious professional misconduct and lost his medical license. Wakefield clearly engaged not only in motivated reasoning but knowingly reported unreliable, perhaps even false, results. For reasons already mentioned, such results can only survive briefly in the scientific methodology machinery. Once the evidence was confirmed to be suspect, any beliefs based on it were quickly revised. The question for us is that once the scientists accepted that these claims were false and revised their beliefs, why didn't a small but significant portion of the general population do the same?

The answer offered by the motivated reasoning model is that these individuals have prior diametrically opposed positions on vaccination policy. They are not looking to revise their beliefs in light of new evidence. They are actively using their prior beliefs and goal directedness to selectively sift through the evidence to support the case for their foregone conclusion: not vaccinating. If they do this and come up with some credible evidence, this would be a form of motivated reasoning, and as I have already stated, motivated reasoning is reasoning. The problem here is the persistence of the false belief despite the lack of any *credible* evidence.

Most of the general public does not have the knowledge and training in biology, medicine, biochemistry, pharmacology, experimental design, and statistical analysis to evaluate the data for themselves. They need to rely on the scientists. The scientific and medical establishments agree that there is no known credible link between the MMR vaccine and autism. Despite the lack of credible evidence in the peer-reviewed scientific literature, there is no shortage of "evidence" from unqualified sources (Scheibner, 1993). For example, the fact that a mother reports a diagnosis of autism in a child following immunization with the MMR vaccine might be *some* evidence to support a link between the MMR vaccine and autism, or it simply could be coincidence. The issue is always the overall quality and soundness of the evidence. While the notion of coherence is a basic intuitive mechanism that we all share, the role of education and training in drawing sound conclusions from complex scientific data cannot be overestimated, leaving open the possibility of some role for sloppy reasoning in individual cases. But, by and large, the motivated reasoning explanation seems thin. If there was some credible evidence linking the MMR vaccine to autism, then it would be appropriate. In the absence of such evidence, it is difficult to accept the adherence to the belief as rational, despite the motivation. So how can we explain the rise of the antivaccination movement and declining vaccination rates in the United Kingdom and United States (Pilkington, 2019)?

Belief 1c: "Man-Made Global Warming Is the Greatest Hoax Ever Perpetrated on the American People"

The global warming example was introduced in chapter 1, and some of the scientific evidence, along with the lay objections to it, was reviewed. Similar to the preceding two issues, the reality of global warming and its relationship to human activity is an empirical scientific question. We have already considered some of the basic questions and the answers offered by scientists. There is a consensus in the scientific community, as indicated by the joint statement of the Academy of Sciences of 17 countries, that human activity contributes to global warming (The National Academy of Sciences & The Royal Society, 2020). No academy of science of any country has contradicted this consensus. Science is self-correcting, so additional evidence could change these views, but in the meantime, the data are what they are.

As with vaccines, global warming has public-policy implications. In fact, it has overarching implications that require people to change their worldviews and lives. In chapter 9, global warming was also characterized as a classic tragedy of the commons problem. There is some evidence to suggest that such problems, while extremely difficult, can be overcome by punishing noncooperators and rewarding cooperators and making sure that people understand the long-term consequences of cooperating and not cooperating. These actions could change the payoff matrix so it is less of a dilemma, but the successful implementation of these strategies requires that participants accept the facts. We saw that the Canadian Maritimes fishermen refused to do so and suffered the consequences. Similarly, a substantial proportion of the population, particularly in the United States, refuses to accept the scientific evidence of climate change and revise their beliefs.

We certainly have a good sense of the strategy that is used to doubt the evidence from the exchange between Stephen Schneider and the good people of Australia (chapter 1): doubt the experts and then you can doubt the evidence. There is also experimental evidence to the same effect. In one study, members of the general public were shown pictures and credentials of fictitious scientists from three areas of expertise: global warming, nuclear waste, and gun control. The credentials indicated that the fictitious scientists graduated from Ivy League universities and became professors at different Ivy League universities. They were also members of the National Academy of Sciences. Each scientist had a position statement associated with them indicating whether they held a "high risk" or "low risk" position on the topic. The participants' assessment of the expertise of a scientist correlated with whether the scientist's position on a topic (high risk versus low risk) matched their own. They effectively discounted or increased the credibility of the scientists, and thus the evidence they were offering, based on whether that evidence was consistent with their prior worldview (Kahan, Jenkins-Smith, & Braman, 2011).

But the larger question is, why doubt the scientific consensus? In answering this question, it may be useful to differentiate between two groups of objectors. There is one group that is gaining immediate financial and/or political benefits from the denial. They may not actually believe the denial, but there are benefits in repeating it. So if you are Sen. James Inhofe from the state of Oklahoma, chairman of the Senate Committee on Environment and Public Works, and you receive millions of dollars in political contributions from the fossil fuel industry, you may find it reasonable to proclaim that "man-made global warming is the greatest hoax ever perpetrated on the American people" (Revkin, 2003). It is (locally) rational because it advances your personal interests. You can support your claim by cherry-picking data. If there are 100 scientific reports providing evidence for climate change and three reports that question it, you would ignore or cast aspersion on the 100 reports and seize on the three favorable reports to make your case.

You may also be a TV host or personality who is well paid to make shocking contrarian statements that attract audiences and sponsors. Again, you may not actually believe what you are saying, but there are people who will pay to hear you say it. You are an actor or entertainer. Your livelihood depends on continuously repeating that global warming is a hoax. It is locally rational for you to do so. It may even be more broadly rational if you judge your immediate financial and/or political gains to be of greater benefit to you than any cost that may be paid by your grandchildren several decades down the road. Motivated reasoning does capture the behavior in these cases.

But what about the 40% of Americans who believe that man-made global warming is a hoax? They derive no immediate, special personal benefit from harboring this belief. These individuals have a certain worldview and life based on burning fossil fuels. Many of them may derive their livelihood from businesses and industries related to fossil fuels. They may have financial or political investments in fossil fuel industries. Their goal is to continue enjoying an equal or better lifestyle. But the science is not trying to reduce their quality of life. It is saying that they will not be able to sustain their quality of life with the current technology. One could accept the scientific consensus and search for, invest in, and adapt alternative technologies to maintain lifestyles. This would be rational. It would also be rational to use motivated reasoning to question the scientific consensus by filtering, selecting, and marshaling the evidence that supports the contrarian view.

The problem is that, as with vaccination, there is very little credible evidence against the scientific consensus but an endless supply of more dubious "evidence" from which to choose. For example, as America experienced a particularly cold winter in 2019, the then president of the United States tweeted: "In the beautiful Midwest, windchill temperatures are reaching minus 60 degrees, the coldest ever recorded. In coming days, expected to get even colder. People can't last outside even for minutes. What the hell is going on with Global Warming? Please come back fast, we need you!" (@realDonaldTrump, January 28, 2019). Some fringe news sources ran headlines such as "Cold Sweat: Climate Alarmists Panic as America Freezes; Media Scrambles to Save [Global Warming] Narrative" (Nolte, 2019; Schlichter, 2019). Such responses give credence to the sloppy, even inept, reasoning model in individual cases. But again, given the overall paucity of credible evidence for the contrarian view,[7] one is hard pressed to call this reasoning of any kind. In the absence of *credible* reasons, how do we explain the loud, proud, patriotic, adamant insistence by these 40% of Americans that manmade global warming is a hoax?

Belief 1d: Gender Identity Is a Social Construct

The reader may object that the latter two examples of maintaining false beliefs in the face of contradictory scientific evidence largely cluster around individuals who would identify as conservatives (at least in 2020 America), but the issue being raised here is not confined to any particular ideology. It is a *Homo sapiens* phenomenon. To illustrate the perpetuation of a false belief from the liberal side of the aisle, let me revisit the gender identity example from chapter 4. The reader will recall that the question of interest was about the relationship between gender and sex. Gender refers to behavioral characteristics associated with being a male or female. Sex refers to one's endowment of reproductive organs. The issues are whether sex and gender are the same and whether gender is determined by biological factors or is a social construct or choice. Can one simply choose to be male or female?

The science was reviewed in chapter 4. It indicates that, despite XY or XX chromosomes, one can have a mismatch between external sexual organs and brain-based behavior (internal feeling of maleness or femaleness). The presence of the Y chromosome triggers two separate independent processes (one for masculinizing the body and the other for masculinizing the brain), which then have to unfold but may be disrupted by internal and external factors (e.g., genetic factors, prenatal exposure to hormones, medications, environmental chemicals, stress on the mother during pregnancy), resulting in gender ambivalence (Coolidge et al., 2002; Dessens et al., 1999; Zucker et al., 1996). Beyond this, there *may* be *some* scope for postnatal social and cultural shaping. The proximal mechanisms of how these latter factors interact with the biology have yet to be understood. In this sense, transgender is very real, but this is very far from the extreme, radical liberal position that gender is a socially constructed choice independent of biology. It is not. (Notably, the science is also inconsistent with the strict conservative and Vatican views (chapter 4) that gender is a black and white issue determined by chromosomes. The science allows considerable scope for gender ambivalence.)

But neither side is willing to accept the evidence and for the same reasons. They wish to enact very different social policies and have a vested or motivated interest in using (or misusing) the science to justify their predetermined positions. Again, can we explain this as motivated reasoning? These positions are certainly motivated, but as with the previous science-based examples, in the absence of some credible evidence, it is difficult to consider either position as "reasoned."

Belief 2: Democrats Want to Flood the Country with Rapists, Drug Dealers, and Terrorists

The political belief about (US) Democrats wanting to open the borders to rapists, drug dealers, and terrorists is equally interesting but in a different way.[8] It is currently repeated and reinforced by many (US) conservative news sites. Here there is no need to appeal to an expert source to accept or reject the statement. One just needs to listen to Democrats. No current US Democrat has said they want to flood the United States with rapists, drug dealers, and terrorists. This is not part of the current Democratic Party platform. The difference between the two ideological groups is in the particulars of immigration policies and quotas, but this is a far cry from the belief under consideration. How do we explain these false beliefs in terms of motivated reasoning? Politics is about power dynamics. Each group wants to win or dominate. Painting your opponent in the worst possible light, irrespective of the facts, may facilitate winning. Motivated or not, it is still not rational to harbor false beliefs.

Motivated Reasoning as Question Begging

Motivated reasoning is the dominant framework for explaining adherence to false beliefs in the face of counterevidence. It does work in certain cases, but in the more interesting cases it is largely unsatisfying. It may even be guilty of question begging. Suppose we hold the prior beliefs that MMR vaccination causes autism, man-made global warming is a hoax, gender identity is socially constructed, and US Democrats want to open the borders to rapists, drug dealers, and terrorists. In the face of counterevidence, why can't we use the reasoning machinery to change these beliefs? This is presumably prevented by the motivated goal or desire (i.e., policy preference). But why can't the reasoning machinery be recursively turned onto the desire (to hold that policy) itself to see whether *it* is rational?[9]

There is a common view that rationality is a tool or instrument for achieving some goal or desire. It is not for the tool to question the end. Hume ([1739] 1888, p. 416) famously put it thus:

> Where a passion is neither founded on false suppositions, nor chuses means insufficient for the end, the understanding can neither justify nor condemn it. 'Tis not contrary to reason to prefer the destruction of the world to the scratching of my finger. 'Tis not contrary to reason for me to chuse my total ruin, to prevent the least uneasiness of an *Indian* or person wholly unknown to me. 'Tis as little contrary to reason to prefer even my own acknowledg'd lesser good to my greater, and have a more ardent affection for the former than the latter.

Notice that Hume does qualify the unquestioned supremacy of desires, that they not be "founded on false suppositions." I think another qualification is necessary. Not all desires are equal. Some desires are a consequence of preexisting beliefs and desires. They are arrived at by reason. If they are so arrived at, surely they can be so modified. We've also seen that other desires arise from lower-level autonomic, instinctive, and associative learning systems in the brain stem, diencephalon, and subcortical brain systems. They will lack propositional content and as such not be susceptible to reason (coherence relations).

Let us revisit our prosecuting district attorney who, convinced of the guilt of a defendant, formulates the goal or desire to prosecute the individual. New compelling evidence arises indicating the individual is not guilty. This evidence forces the prosecutor to change his beliefs about the defendant's guilt, and this in turn changes his desire to prosecute the individual to a desire *not* to prosecute. In this classic paradigm of motivated reasoning, the machinery of reason can indeed be turned onto itself to modify false beliefs and any desires based on those beliefs. My desire to drop out of university as an undergraduate student to become a carpenter provides another example. My father provided evidence-based arguments highlighting career trajectory, income, and lifestyle differences between university graduates and high school graduates. Based on the information, I made a reason-based decision not to drop out but to switch programs and study architecture.

Now consider my desire to add sugar to my coffee, even though I'm diabetic. I'm capable of understanding the basic biology underlying diabetes, the need to limit carbohydrate intake, and I certainly don't have a death wish. Yet I do add sugar to my coffee because it tastes good. As noted in chapters 11 and 12, this desire is generated by primordial systems anchored in old brain neural circuitry and is not particularly amenable to reason. For another example, we can return to John Edwards and his desire to be with his mistress while his wife was battling cancer and he was seeking the 2008 Democratic Party presidential nomination. Edwards's desire for his mistress presumably involved the LUST system (chapter 12), characterized by an action-specific appetitive state with positive valence, pent-up arousal reservoirs, specific action tendencies triggered by specific stimuli, and the pleasurable feelings of the consummatory response. Individuals will seek environments in which the appetitive state can be discharged to derive the pleasure associated with the consummation. Again, these instinctual systems belong to primordial subcortical brain systems, though in humans they also have cortical representation and elaboration. We saw in chapter 12 that the potential of these systems to be modulated by reason exists but actually may never be fully realized. The tethered mind is set up to maximize pleasure and minimize pain. If the pleasure associated with consummating an appetitive state is greater than the pleasure associated with maintaining a coherence relation between evidence and beliefs, the former will dominate.

In short, what I'm suggesting is that some false beliefs are resistant to rational evidence because the motivating desire is not a cognitive desire. It is a desire propped up by instinctual (and other) low-level subcortical systems under discussion throughout much of this book. I believe that some

researchers are reaching for such an explanation when they invoke emotions, particularly anger (Mullen & Skitka, 2006), and "identity protection"
to buttress models of motivated reasoning (Kahan, 2016, p. 3): "The truth
independent goal of 'politically motivated reasoning' is identity protection:
the formation of beliefs that maintain a person's status and affinity in a
group united by shared values." The tethered rationality model provides
a much more natural way of encompassing and expressing these insights.

The model of tethered rationality is not restricted to coherence relations
among propositional contents. It has a larger repertoire of mechanisms,
consisting of the autonomic mind, the instinctive mind, and the associative
mind, in addition to the reasoning mind, to appeal to in explaining human
behavior. Chapters 3–6 introduced these different systems in some detail.
Chapter 9 discussed (the largely human) instinctual systems that modulate
economic decision-making, such as self-maximization, fairness, cheating,
and punishment. Chapter 11 introduced some of the basic instinctual systems that may be common to all mammals. Some of these systems are relevant to the examples introduced in this chapter. But there is a key system,
the *in-group/out-group*, that was mentioned during the discussion of the
impeachment debate but not properly introduced and discussed. I will first
introduce the in-group/out-group system, provide some evidence that it is
an instinct, and then argue that it allows tethered rationality to explain the
preceding examples, along with the White House impeachment defense.

Evidence for In-Group/Out-Group Bias Instinct

In-group/out-group formation is a pervasive human trait. We cannot help
but distinguish along these lines. The in-group is the group that we belong
to. It is always superior, noble, pure, righteous, and beloved of God. The
out-group is everyone else. Its members are inferior, contemptible, evil, and
not quite human (e.g., "no dogs or Indians allowed"[10]). The question of
interest is whether evolution has predisposed us to favor the in-group and
be wary of the other—where the other is anyone who is different (and thus
may do us harm)—or is in-group/out-group formation a social construct? If
it is a social construct, it belongs to the rational mind and involves propositional attitudes and coherence relations, and presumably the rational mind
can modify it at will. If, on the other hand, it is an adaptation, it belongs to
the instinctive mind and can be characterized in terms of specific valence-
laden appetitive states, pent-up arousal reservoirs, specific action tendencies, triggered by specific stimuli, and the release and satisfaction associated
with the consummatory response.

In chapter 4 I used the following questions to distinguish between social constructs and instincts: (1) Is the trait universally present in human societies or is it culture specific? (2) Is it available on other branches of the phylogenetic tree? (3) Does it emerge early in human infants, prior to any opportunity for extensive socialization? (4) Is it underwritten by implicit, automatic, low-level mechanisms? (5) Is it possible to trace specific neural circuitry devoted to it (and find homologous behavior and circuitry in other species)? Affirmative answers to all or most of these questions indicate instinctive systems.

With respect to the first question, the "us" and "other" formation is a human universal. It is found in every known human society (LeVine & Campbell, 1972). Precursors are clearly visible on other branches of the phylogenetic tree. Many species live in organized groups and favorably cooperate with other members of the group. There are also a number of cases of alliance or coalition formation for specific purposes among social carnivores. Lions, hyenas, and wolves hunt in cooperative groups. The spoils are shared based on the internal dominance hierarchy of the pride or pack. A small coalition of male lions (usually related but sometimes unrelated) will have an advantage in taking over a pride of females from a single male. Lionesses will group together in efforts to thwart infanticide from roaming males (Zabel, Glickman, Frank, Woodmansee, & Keppel, 1992). Among vervet monkeys, female kin-based groups are able to maximize access to dispersed food resources within the territory (Cheney & Seyfarth, 1987). Computer modeling provides evidence that group formation and favoritism increase cooperation, which in turn furthers personal fitness (Axelrod & Hamilton, 1981; Dugatkin, 1998).

In a few specific cases, these nonhuman coalitions also lead to open hostility and intergroup violence. Wolves mark their territorial boundaries with urine, feces, and scratch marks, and vigorously defend it with coalitionary violence against encroachment from other packs at the expense of injury and even death (Peters & Mech, 1975). Chimpanzees, our closest relatives on the phylogenetic tree, go even further. Male adults routinely patrol the boundaries of their territory. They not only defend it but on occasion make organized incursions into neighboring territories and, where opportunities arise, kill members of the other group and absorb their territory, along with any fertile females (Mitani, Watts, & Amsler, 2010). It is hypothesized that this is an adaptive strategy that allows successful killers to increase their fitness by increasing access to food, mates, and parenting resources (Wrangham, 1999). Coalitionary violence and territoriality are closely intertwined (Kelly, 2005; Wilson et al., 2014; Wrangham, 1999;

Wrangham & Glowacki, 2012).[11] Interestingly, coalitional out-group violence is not widespread, being largely confined to wolves, chimpanzees, and humans (Wrangham, 1999).[12]

Signs of in-group favoritism emerge very early in human infants. One-year-old babies display in-group preference, even based on trivial properties. In one experiment, 11-month-old infants were allowed to display a preference between two foods, graham crackers and green beans. They were then shown puppets eating either graham crackers or green beans. When given the opportunity to play with one of the puppets, they chose the puppet that preferred the same food that they did. The experiment was repeated with a preference for either orange or yellow mittens, with the same results. In-groups were established on the basis of similarity of preference to self and resulted in favorable treatment of the puppet with similar preferences (Mahajan & Wynn, 2012).[13]

This experiment provides some evidence for in-group favoritism, but this is only half the story. Not only do we favor in-group members, we also want to *harm* out-group members. In a follow-up experiment, a food preference for graham crackers or green beans was again solicited from 14-month-old infants, and then they watched a puppet show in which two puppets indicated their own food preference for graham crackers or green beans. The in-group puppet always displayed the same food preference as the infant. The out-group puppet always preferred the other type of food. Infants then viewed a second puppet show involving a third, neutral puppet. In yet a third puppet show, two new puppets either helped or harmed the original in-group and out-group puppets. When given a chance to play with the neutral puppet or the harmful or helpful puppets, infants did prefer to play with the puppet that helped their in-group puppet, but amazingly they preferred even more to play with the puppet that harmed the out-group puppet (Hamlin, Mahajan, Liberman, & Wynn, 2013)! These results emphasize that once groups are formed, it is not that we simply favor the in-group and are simply disinterested in the other group; we actively seek to harm the out-group:

> Believers, make war on the infidels who dwell around you. Deal firmly with them. Know that God is with the righteous. (Quran 9:123)

Nonreligious examples are provided by many national anthems, such as "God Save the Queen":

> O Lord our God arise,
> Scatter our enemies,
> And make them fall!

Confound their politics,
Frustrate their knavish tricks,
On Thee our hopes we fix,
God save us all!

. . .

Lord grant that Marshal Wade
May by thy mighty aid
Victory bring.
May he sedition hush,
And like a torrent rush,
Rebellious Scots to crush.
God save the Queen!

There is also evidence for the fourth criterion, the involvement of low-level, preconscious mechanisms. For instance, we have specialized systems for detecting faces (Nelson, 2001) that are attuned to low-level perceptual cues, such as skin tone and shape, allowing for very early detection of different racial groups (Balas & Nelson, 2010). Even basic information about body shape and kinematics allows us to detect sex and age categories early and quickly (Johnson & Tassinary, 2005, 2007; Montepare & Zebrowitz, 1993). Much of this is implicit and compulsory and occurs prior to the engagement of top-down reasoning processes; that is, we can and do register the age, sex, and ethnicity (and in some accounts even sexual and political orientation) of an individual before we are even consciously aware of seeing them (Kawakami, Amodio, & Hugenberg, 2017).

The fifth criterion for identifying instincts was whether one could associate specific subcortical or early maturing neural circuits with the behavior. Here, our knowledge is very limited. We do not know the answer yet. However, one can postulate the involvement of Panksepp's RAGE, FEAR, DOMINANCE, ATTACHMENT, and SEEKING systems as components. We understand some of the neural circuitry of these components from the animal literature (chapter 11). With respect to in-group/out-group judgments in humans, neuroimaging studies are generating evidence for the involvement of subcortical structures such as the amygdala and striatum. Cortical regions are also engaged, and there is evidence of different cortical representations of the in-group versus the out-group (Amodio, 2014; Gilbert, Swencionis, & Amodio, 2012; Mitchell, Macrae, & Banaji, 2006). Other studies provide preliminary evidence of the role of oxytocin in enhancing trust and thus in-group biases in humans (Baumgartner, Heinrichs, Vonlanthen, Fischbacher, & Fehr, 2008). I will henceforth use capitalization of In-Group/Out-Group to indicate when I'm referring to the instinct, lowercase otherwise. I will not use Panksepp's all-caps notation because our knowledge of the underlying neural machinery is still very limited.

While human In-Group/Out-Group systems have some evolutionary underpinnings on other branches of the phylogenetic tree, they do far outstrip nonhuman precursors in several important respects. Nonhuman animals form coalitions and alliances with kin and other familiar members of the pack or troupe for specific resource acquisition purposes. Human groupings do not need to have a specific purpose or involve kin or familiar members. They are also not disjoint (i.e., they allow overlapping memberships). One can be a member of group A for one purpose and a member of group C but not D for another purpose, while other members of group A may be members of groups B and D. Finally, the bases for human groupings are much more extensive, ranging from evolutionarily salient properties to the arbitrary and trivial (Tajfel, 1970). A brief, familiar list could include kinship, physical features such as color and shape of the face, sex, age, language, nationality or territory, diet, religion, social or cultural norms, political alliances, income (resources), clothing, hairstyle, jobs and hobbies, sports teams, or even which supermarket line we stand in.

These seemingly unlimited groupings result from the same basic evolutionary rule: "Similar equals safe; different could be dangerous." But how is "similar to us" determined? One possibility is some mechanism that allows us to gauge intrinsic differences between self and others. A second possibility is that it is determined indirectly by gauging what is most common and prevalent in our environment. This is then taken as a marker for the self and used to gauge the other. The data indicate that the latter is the case (Bar-Haim, Ziv, Lamy, & Hodes, 2006; Quinn, Yahr, Kuhn, Slater, & Pascalis, 2002). Let me illustrate this with a personal anecdote. I was born in India of Indian parents but brought to Canada as a child and raised in a small Canadian town. I did not see another non-Caucasian person (apart from my parents and siblings) in my neighborhood, in my school, or in my community until I was 16 years old. So my perceptual face recognition system was fine-tuned on differentiating Caucasian faces, not Indian faces. When I was 15 years old, my family visited India and I was reintroduced to my aunts, uncles, and cousins. I have two lasting memories from this experience. First, everyone was brown . . . in the streets, in the shops, on the buses and trains . . . everywhere. To this day, I remember how uncomfortable I felt in this environment. My second memory is of not being able to tell people apart, particularly in the movies. They all looked the same. I remember asking my mother why they cast the same woman in all the female roles in the movies! We are now beginning to understand the functional neuroanatomy and psychophysics underlying my difficulty in recognizing Indian faces (Balas & Nelson, 2010; Golby, Gabrieli, Chiao, & Eberhardt, 2001; Michel, Rossion, Han, Chung, & Caldara, 2016; Nelson, 2001). The issue here is not just one of speed and accuracy of processing, but also of preference for the familiar and bias against the unfamiliar stimuli.

It is no longer politically correct to make statements about people from different racial or ethnic groups all looking alike or making us uncomfortable, but that does not make such statements any less true. What is, of course, interesting here is that I was born into one racial community but raised in another community. My identity and in-group preferences and biases were established not by my genetic membership but rather by the environment in which I was raised. The two are usually the same, in which case the latter can serve as a proxy for the former. It may be useful to think of the In-Group/Out-Group system along the lines of the imprinting mechanism encountered in chapter 5. The imprinting is an innate disposition, but what the bird imprints on is a function of environmental input during a critical window. In normal circumstances, it is fitness enhancing.[14]

Another issue to consider is the relationship among the numerous differentiating properties. They are not alike. Some properties, such as kinship, physical features (race or color) or sex, and language seem more fundamental (or fitness enhancing) in differentiating "us" and "them," while others, such as "favorite hockey team," seem superficial or trivial (but even such trivial properties can result in considerable violence directed at the out-group). This would also imply greater entrenchment of the more evolutionarily salient properties compared to the superficial properties, an idea supported by involvement of very low-level perceptual systems for detecting differences in race, sex, and age, noted earlier, and some evolutionary arguments made by evolutionary psychologists (Cosmides, Tooby, & Kurzban, 2003).[15]

But how can we account for differential entrenchment of grouping properties if the groupings are a function of the environment? The first thing to do is to broaden the notion of "environment" beyond social and cultural learning. For example, even with something like the taste preference displayed by infants in the studies discussed here, there will be a genetic component, but there are also environmental components. A child's taste preferences are affected by the mother's diet during pregnancy (Prescott, 2012). Second, even if sameness and difference are explicitly learned by association or belief formation, through interaction with the environment, their degree of entrenchment will be a product of maturation windows of the corresponding neural systems. Certain neural systems, such as the visual system, the gustatory system, and language system, mature early. Once these neural systems mature, their capacity for rewiring is severely limited. We have already encountered this issue in chapter 10 with the visual system and will revisit it in greater detail in chapter 14. This may provide a partial explanation for why preferences for physical features, language, and diet occur much earlier (Buttelmann, Zmyj, Daum, & Carpenter, 2013; Howard, Henderson, Carrazza, & Woodward, 2015; Pascalis, de Haan, & Nelson, 2002), and may be more deeply entrenched than preferences for sports teams and cars.

With socialization, these basic predispositions are modulated by learning and culture and we begin to see interaction with the rational mind (Kawakami et al., 2017; Kawakami, Hugenberg, & Dunham, 2020). In an experiment involving children six to eight years old, it was shown that children are more likely to attribute positive behaviors to in-group members and negative behaviors to out-group members, even when these groupings are randomly assigned. Negative information about group members was then introduced into the equation. The impact of the negative information was attenuated when it was attributed to an in-group member and accentuated when it was attributed to an out-group member (Baron & Dunham, 2015). This experiment illustrates interaction between instincts and reason (i.e., socially and culturally engendered beliefs). In this example, the instinct is pushing back against the new negative evidence received about the in-group and hindering belief revision by the reasoning mind.

Every society uses song and poetry to enhance and reinforce in-group similarity and out-group differences. These are often intermingled with territoriality. Here are two examples, one from my country of birth:[16]

> My fellow countrymen,
> Hoist our beloved flag on this auspicious occasion . . .
> But also fill your eyes with tears,
> Remember the brave warriors who did not return home,
> I sing this song so you do not forget. . . .
>
> When the Himalayas were assaulted
> Our freedom was in jeopardy,
> They fought until their last breath,
> And then laid down their lives at the border . . .
>
> The blood that stained the mountain,
> That blood was Hindustani,
> For those who martyred themselves,
> Remember their great sacrifice!

And another from my adopted country (and city), halfway around the world (John McCrae, "In Flanders Fields"):

Take up our quarrel with the foe:
To you from failing hands we throw
The torch; be yours to hold it high.
If ye break faith with us who die
We shall not sleep, though poppies grow
In Flanders fields.

One can also use song and poetry to try to ameliorate out-group differences and promote an all-encompassing unitary vision, as Rabindranath Tagore ("Chitto Jetha Bhayshunyo") did in anticipation of Indian independence:

Where the mind is without fear and the head is held high;
Where knowledge is free;
Where the world has not been broken up into fragments
By narrow domestic walls;
Where words come out from the depth of truth;
Where tireless striving stretches its arms towards perfection;

Where the clear stream of reason has not lost its way;
Into the dreary desert sand of dead habit;
Where the mind is led forward by thee;
Into ever-widening thought and action;
Into that heaven of freedom,
My Father, let my country awake.

But as the reality of the intergroup violence in the aftermath of Indian independence and partition revealed, this often remains an unrealized intellectual aspiration. There are very real limits to the control the reasoning mind has over In-Group/Out-Group systems.

Reason through education can modulate but not eradicate in-group/out-group bias. Most North American universities consciously and explicitly strive to be bastions of diversity and acceptance. Rarely is the ideal achieved. Let me relate one specific instance from my own university. Some years ago, when I served on the Committee on Examination and Academic Standards in the Faculty of Pure and Applied Science, there was one particular meeting in which we dealt with several cases of academic dishonesty. The committee members consisted of four Caucasian colleagues and myself. The first student to come before the committee was charged with cheating on an exam in a particular course. The evidence was strong, the case considered serious, and the committee unanimously agreed on a severe punishment.

The second student to appear was also charged with cheating in the same course, on the same exam, and the evidence was equally compelling. However, in this case my four colleagues thought that perhaps the fright and trauma of being charged with cheating and appearing before the committee might be sufficient punishment for the student and perhaps no penalty was necessary or warranted. The facts in the two cases were identical . . . same course, same professor, same test, same evidence, and it was the first offense for both students. So why the different conclusions? The first student had the misfortune of being brown, whereas the second was a nice Caucasian girl. I genuinely believe that my colleagues harbored no ill intent against the visible minority student or even had any conscious awareness of the unfairness in the treatment of the two students. When I pointed out that the facts in the two cases were identical but there was a major discrepancy in the penalties imposed, my colleagues caught their mistake and reversed their decision such that both students were penalized equally. This example shows both the value and limitations of the reasoning mind in modulating not only In-Group/Out-Group systems but all instinctive behaviors.

Tethered Rationality and Failure of Belief Revision

Let's return to the various examples of failure of belief revision with which we began the chapter. When considering reluctance to update beliefs that are based on scientific evidence, it is worth differentiating between science, scientists, and the lay public. In selecting the scientific examples for this chapter, I intentionally included an example where scientists initially got the answer wrong (excess acidity causes peptic ulcers). I selected another example where individual scientists may have cheated and presented false results (MMR vaccine causes autism). This was to emphasize the reality that scientists are human. Some are incompetent. Some are brilliant. Some are fraudulent. Some have great integrity. Some are strongly attached to a particular theory and will defend it in the face of increasing counterevidence. In these latter cases, the motivated reasoning model (figure 13.1c) is often perfectly adequate to capture their reasoning. But with the preponderance of evidence, the reasoning machinery does get turned in on itself and the false beliefs are revised. The scientific methodology (for reasons noted earlier) overcomes the shortcomings of individuals and revises and self-corrects as new information comes to light. Over the long term, it has the potential to converge upon the ideal reasoning model in figure 13.1a.

This is not the case when empirical false beliefs are held by the lay public in the service of their policy preferences and political motives. In these cases, the issues of veridicality and coherence are secondary. Instinctual factors such as group membership come to the forefront. While groups may be formed on the basis of social and economic policy preferences, group bias itself is not a social construct based on beliefs, so it cannot be reasoned away. Anyone who shares your views is an in-group member, and anyone who disagrees with you is an out-group member. In-group members are friends; out-group members are evil and wish to do you harm. If the scientific evidence supports the policy preferences of the other group (or neither group), then science itself gets relegated to the out-group and subject to attack and ridicule:

> Like some suckers still do, I once believed that "science" was a rigorous process where you tested theories and revised those theories in response to objective evidence. But in today's shabby practice, "science" is just a package of self-serving lies buttressing the transnational liberal elite's preferred narrative. Our alleged betters hope that labeling their propaganda "science" will science-shame you into silence about what everyone knows is a scam. (Schlichter, 2019)

There are powerful feelings of pleasure associated with membership in the in-group (perhaps through activation of the ATTACHMENT system) and equally powerful feelings associated with FEAR, RAGE, and POWER/ DOMINANCE directed at the out-group. Striving to destroy the out-group activates the SEEKING system. These feelings override any pleasurable feelings associated with coherence relations between evidence and conclusion. They can be intoxicating. We seek them out and maintain them by forming bubbles or echo chambers that reinforce the false beliefs that allow for the contemptuous dismissal of the out-group scientists, along with their evidence. There need not be a great deal of reasoning involved in these processes. They are automatic. They can be rationalized by the cognitive mind after the fact and articulated in terms of beliefs and desires and reason, but these are *not* the driving mechanisms. The machinery of instincts is driving the tethered mind in these situations.

The activation of these instinctive systems not only explains antivaccination advocates but also may be a missing piece of the puzzle in getting the skeptics to believe the science of global warming. If the participants refuse to believe the facts (i.e., as articulated by the current best science), there is no way of diffusing or diminishing the dilemma of the tragedies of the commons. If this persists, we will all go the way of the Canadian Maritimes fishermen and for the same tragic reasons. It is not inevitable, but unless people can derive greater pleasure from coherence relations than from activation of In-Group/Out-Group systems, dismissal of the science is an unfortunate consequence of the tethered mind. One cognitive strategy to minimize activation of In-Group/Out-Group systems is the formation of superordinate groups that encompass both subgroups (Sherif, 1988), but as was the case in the aftermath of Indian independence and partition, it does not always work.

In the case of political beliefs, where no scientific evidence enters the picture, the role of the In-Group/Out-Group system is very explicit. The source of the specific belief in the example may be a statement such as "a more liberal immigration policy is good for the country." This statement is being made by a member of an out-group. This is sufficient to trigger suspicion and FEAR and RAGE. This particular example also involves the territoriality instinctive system. The interaction of In-Group/Out-Group, FEAR, RAGE, DOMINANCE, and territoriality is a toxic mix. There is also the pleasure associated with the ATTACHMENT system as a function of bonding with the in-group, and the pleasure in the activation of the SEEKING system to crush the out-group. Once these instinctive systems are triggered, they are going to activate the relevant action tendencies, which in turn will amplify the rhetoric, which will further accentuate the instincts, which will further amplify the rhetoric, resulting in a dangerous escalating cycle leading to the cry that "the only good Democrat is a dead Democrat."[17]

A purely reason-based analysis in terms of motivated reasoning and sloppy reasoning cannot explain these phenomena. The motivated reasoning strategy can explain much of scientific reasoning. It can explain why someone may consciously press someone else's instinctive buttons to further their agenda, as noted in the discussion of the White House impeachment defense in chapter 8, but it cannot explain why the rational mind falls into the trap. The model of tethered rationality along with particular instincts provides a much more compelling explanation. Once the buttons triggering the In-Group/Out-Group system are activated, the pleasure generated by this system may be much greater than any pleasure (or indeed displeasure) associated with any evidence and coherency of the argument for impeachment. In the case of President Trump's supporters, the overpowering feelings of ATTACHMENT to the in-group and FEAR and RAGE directed at the out-group (Democrats) totally drowned out any pleasure associated with coherency of arguments for impeachment and completely drove their behavior. In the case of the Democrats, the activation of the same In-Group/Out-Group system directed FEAR and RAGE at the Republicans and triggered the SEEKING, fairness, and cheater detection systems, which hypersensitized them to any evidence of wrongdoing, motivated them to be selective, and enhanced their adherence to coherency relations (because the evidence seemed to support wrongdoing by the president). There are no satisfactory purely rational explanations of these behaviors.

These issues are also germane to completing the discussion of my American friend's aversion to universal healthcare. Apart from the issues discussed in chapter 9, there is the fear and loathing of the ever-present label of "socialism." One might form reasoned views on different systems of social and economic organization after reading Adam Smith, John Locke, John Stuart Mill, Karl Marx, John Maynard Keynes, Kenneth Galbraith, and even Ayn Rand, among others. This exploration of the space of possibilities may lead to reasoned preferences for certain socioeconomic systems, but this is not the case for my American friend. For him it is an uninterpreted label for the outgroup: un-American, unpatriotic, subhuman. He belongs to the in-group: American, patriotic, exceptional. Once this system gets activated, it provides a decisive blow to the other systems discussed in chapter 9, even self-interest.

At the risk of repetition, let me be clear about the distinction between reason and instincts. Reason involves coherence relations among propositional contents. This machinery was articulated in chapter 6. Instincts are characterized in terms of specific valence-laden appetitive states, pent-up arousal reservoirs, specific action tendencies, the properties of the triggering stimuli, the release and satisfaction derived from the consummatory response, and the specific subcortical neural pathways and neurotransmitters underwriting the system, as discussed in chapters 4, 11, and 12. The engagement of these instinct-specific mechanisms is necessary to discharge these examples.

Belief 3: Religious Beliefs

The case of religious belief is different still and not addressed by the motivated reasoning or sloppy reasoning models. I think that tethered rationality can give us some traction. There is no evidentiary basis for the belief that (the Judeo-Christian) God created the world in six days, some 6,000 years ago, or indeed for the existence of this or any other god. In the context of modern scientific knowledge, such statements from religious texts seem arbitrary and disconnected from reality. The natural sciences tell us that the Earth is approximately 3 billion years old and that the universe has been expanding for approximately 20 billion years. New evidence may result in revision of these beliefs. Religious beliefs differ from other false beliefs in that the belief in some god-deity is universal, found in every culture and society (Brown, 2004). Only the local particulars differ. Man is not only the rational animal; he is also the "praying animal" (Jenson, 1983).

If we look for the existence of God in the actual world, we are disappointed and left with the puzzle of explaining how such beliefs are formed, maintained, and propagated by a rational mind. Perhaps we are looking in the wrong place to verify the existence of God. We typically assume that God created the world and then look for evidence of this God in the world. But what if *we* created God? Then it would be appropriate to look inside our heads for his existence. There are roughly three types of explanations that have been advanced for religious belief systems: (1) belief in God is a primary instinct with a genetic basis; (2) belief in God is a cognitive construct of other primal instincts; and (3) belief in God is a sociocultural construct of the rational mind. Each of these positions has certain strengths and weaknesses that have been discussed in the literature (Voland & Schiefenhövel, 2009). I believe the second is the most plausible, given the available evidence.

We have recently come to understand that feelings of religiosity can be manipulated by pathological or artificial modulation of brain activity. Temporal lobe epileptic seizures can result in feelings of mysticism and hyper-religiosity (Ramachandran & Blakeslee, 1998). Such feelings can be re-created through various brain manipulation techniques, including transcranial magnetic stimulation (Booth, Koren, & Persinger, 2005; Persinger, Saroka, Koren, & St-Pierre, 2010). Schizophrenia often results in the hearing of voices that may be interpreted as messages from God (Frith & Johnstone, 2003).[18] Other neurological impairments, such as Parkinson's disease, can result in a decrease of religious feelings in patients (Butler, Mcnamara, & Durso, 2010).

Universality and the fact that neurological and neuropharmacological interventions, often in the absence of cognitive deficits, can affect feelings of religiosity speak against a strictly sociocultural model of religion. This leaves the first two possibilities. While I am making a strong case for instincts throughout the book, I am loath to unnecessarily multiply the number of primary instincts, particularly where evidence is sparse. The second account remains. Religious beliefs may be cognitive manifestations to accommodate interaction among some of the primal instincts that we have already encountered. Insofar as power, hierarchy, servile submission to authority, need for a father figure, and being loved and taken care of as in childhood, for example, are important components of religions, POWER, FEAR, and ATTACHMENT instincts may be relevant:

Isaiah 1:19–20: "If you are willing and obedient, you will eat the best from the land; but if you resist and rebel, you will be devoured by the sword."

Religious beliefs may be cognitive constructs (beliefs with propositional content subject to coherence relations) that have no correspondence with the external world but are constructed to satisfy the needs and feelings associated with these primal instinctive systems. The pleasure derived from activating these systems outweighs the pleasure associated with maintaining coherence and veridicality with the external world.

Consistent with this account, there is some neurological evidence that beliefs involving political and religious issues engage frontal orbital cortex and limbic (subcortical) regions while "normal" evidence-based beliefs (e.g., that apples are fruit) involve the left prefrontal cortex (Gozzi, Zamboni, Krueger, & Grafman, 2010; Harris et al., 2009; Kaplan, Gimbel, & Harris, 2016; Kapogiannis et al., 2009; Moutsiana, Charpentier, Garrett, Cohen, & Sharot, 2015). Furthermore, inferences based on world knowledge and semantic/conceptual connections, along with simple logical and causal connections, preferentially activate the left prefrontal cortex (among other areas), and where there is a conflict between the believability of the premises and/or conclusion and the logic of the inference, the right prefrontal cortex is engaged in the detection of the conflict and/or inhibition of the belief-based response (Goel, 2007; Goel et al., 2000; Goel & Dolan, 2003; Stollstorff et al., 2012).

Are False Beliefs and Bubbles Sustainable: Do Facts Matter?

The most natural explanation for the persistence of false beliefs in the face of counterevidence is offered by tethered rationality. This suggests that the phenomenon is not primarily a reason-based issue. It is not a conservative or liberal issue; it is a human issue. We all share these primal instinctual systems, and once they are activated, they have certain action tendencies associated with them that the reasoning mind may have limited control over. They result in the formation of self-sustaining echo chambers or bubbles that can use reason to ratchet up the activation of the instinctual systems, resulting in an ever-escalating spiral.

To continue experiencing the intoxicating feelings of intense pleasure, fear, rage, and animosity associated with activation of the primal In-Group/Out-Group system, most people will happily exist in these echo chambers or bubbles. It is the equivalent of indulging in chocolate cake. But even too much chocolate cake will eventually kill you. If our beliefs are disassociated from the world, the resulting actions will be also. Is this sustainable and adaptive? Do veridicality and consistency no longer matter? The whole point of the reasoning mind is to ensure that our mental representations are veridical and consistent. While the world is not black and white, certain descriptions of the world are more accurate than others. Certain characterizations of the world are more conducive to our survival and thriving than others.

Mischaracterizing the world results in false beliefs. Beliefs affect actions. Creatures with false beliefs will suffer consequences ranging from the failure to maximize opportunities to immediate death. For example, humans survived for tens of thousands of years believing that the Earth was flat and at the center of the universe. Correcting this erroneous belief led to numerous other scientific and technological advancements that could not have occurred otherwise. False beliefs about vaccination have led to declining vaccination rates in the United States and United Kingdom, resulting in increased rates of measles, mumps, and rubella among children (Iacobucci, 2019; Pilkington, 2019). Failure to believe the global warming evidence has not stopped the Earth from warming. Failure to believe the science of gender identity does not make it a socially constructed choice. Intentionally triggering primal instinctual systems in response to political disagreements, instead of engaging in rational argumentation, will end in physical violence and societal fragmentation, irrespective of what we may choose to believe.

While self-selected echo chambers are seductively pleasurable and may harbor us for short periods of time, in the long run, when false beliefs collide with the world, the world will *always* win. A blunt reminder is provided by the 2020 COVID-19 pandemic. In an echo chamber where you believe the coronavirus, COVID-19, is a common cold "weaponized" by Democrats to bring down President Trump (Chiu, 2020), or it is a flu confined to 15 cases that will quickly go down to zero (The White House, 2020), or that "science should not stand in the way of [school openings]" (Wade, 2020), or that the requirement of wearing a mask is a political ploy to stomp on our freedoms and usher in a communist dictatorship, it may be adequate to ignore the experts and thump your chest and shout at your political opponents to make it go away. But if, in the actual world, the COVID-19 virus is a new, highly contagious, poorly understood pathogen, with a mortality rate estimated (in mid-2020) at 10 to 30 times that of influenza, then it is probably wiser to follow the advice of infectious disease experts and take recommended remedial actions such as large-scale testing, wearing face masks, social distancing, and ultimately vaccination. The virus does not have a group affiliation. It does not care what you believe. It is an equal opportunity killer. Denial of the facts by many Americans, including the president, resulted in the richest and most technologically advanced country in the world having the highest number of infections and deaths in the world (more than 650,000 as of this writing). A more veridical and consistent belief network about the nature and structure of the world, *continuously revised as new information becomes available*, will always result in more fitness-enhancing actions. Veridicality and consistency matter. The world eventually bites back. As I often tell my children, "Stupid is stupid." Or the reader may prefer the version attributed to Albert Einstein: "The only things that are infinite are the universe and human stupidity. And I'm not sure about the universe."

14 Global Belief Revision Is Constrained by Neural Maturation

A worldview is a commitment, a fundamental orientation of the heart, that can be expressed as a story or in a set of presuppositions (assumptions which may be true, partially true or entirely false) which we hold (consciously or subconsciously, consistently or inconsistently) about the basic constitution of reality, and that provides the foundations on which we live and more and have our being.

—James W. Sire

Most people catch their presuppositions [worldview] from their family and surrounding society the way a child catches measles.

—Francis Schaeffer

Kids don't have the ruts yet that adults have carved into their minds.

—James P. Hogan

At this point in our story, we have completed the explanation of the tethered rationality model and successfully applied it to the various problems introduced throughout the volume. There remains one outstanding issue to address: our ability to engage in large-scale belief revisions, such as required for changing "worldviews," seems to be a function of age. For instance, in a poll conducted in June 2014, 78% of Democrats under the age of 40 expressed support for policies to limit greenhouse gas emissions compared to 62% of Democrats over the age of 65. While the numbers were lower for Republicans, the age gap remained (Nuccitelli, 2016). Why should age be an important factor in revising beliefs about climate change and other widely held beliefs that are part of the fabric of our worldview? In this chapter, we turn to brain development for a possible answer to this question.

I begin by differentiating beliefs from worldviews, followed by an overview of brain development and maturation, organized around a sculpting analogy: brains are largely shaped by removing material rather than adding it. This is important in that once materials are "chipped away" they are no longer available. The emphasis of this chapter will be on the pre- and postmaturation properties of neural systems. In the prematuration phase, there is considerable (even excessive) plasticity in all neural systems. Once a system matures, opportunities for plasticity diminish precipitously. I raised this issue and reviewed some of the properties of pre- and postmaturation visual system neurons in chapter 10. In this chapter, I want to extend the discussion to the association cortex and suggest that similar developmental principles and trajectories, with a shifted and protracted window into early adulthood, apply and may account for our inability to change worldviews as adults. I will conjecture that once we reach adulthood there may not be enough neural resources left to undertake large-scale architectural reorganization of neural networks encoding worldviews.

Local Belief Revision versus Worldview Revision

The reasoning mind is concerned with beliefs. Beliefs are acquired continuously over a lifetime. However, they are not a random collection of information or facts. The mind actively structures and organizes them into hierarchical belief networks, where specific content nodes are connected to other nodes by relations (e.g., part/whole, causal, logical), not unlike the little fragment depicted in figure 5.2 but much more vast and complex. Explicit in our discussion of reasoning is not only the idea that as we encounter new information we can add it to our belief network, but also that if it is inconsistent with our existing beliefs, we have the ability to revise those beliefs. This seems obviously correct (once qualified with the insights of tethered rationality). For example, I used to believe that all reptiles were cold-blooded. Recently, I learned that leatherback turtles are reptiles but are not cold-blooded. They can maintain core body temperature at 26°C even when diving into near freezing waters (Davenport, Jones, Work, & Balazs, 2015). My original belief was incorrect, so now it has been updated to match the facts in the world. This is an example of what I refer to as local belief revision. I changed my beliefs about reptiles being cold-blooded to incorporate an exception. It did not have any impact on my beliefs about rattlesnakes, tuna fish, or whether it is appropriate to mow the lawn on Sunday. It was a local, isolated change to a small part of my belief network.

I want to contrast such local belief changes with more extensive changes that essentially require revisions in our "worldview," where worldviews are "an underlying, hidden level of culture that is highly patterned—a set of unspoken, implicit rules of behavior and thought that controls everything we do. . . . It is particularly resistant to manipulative attempts to change from the outside" (Edward Hall quoted in Hiebert, 2008). Worldviews include not just our explicit beliefs but also the unarticulated presuppositions[1] of our beliefs and even the parameter settings of our sensory systems that enable us to easily make certain perceptual, phonological, and taste distinctions (and give preference to them) over others. Interestingly, while we effortlessly acquire and modify worldviews while we are young, we are often unable to undertake large-scale revisions of worldviews once we have attained adulthood. There is surprisingly little data to address this issue. There is a body of literature on "attitudinal changes," which are smaller-scale changes than the worldviews I'm considering here, but nonetheless, these data do suggest an inverse relationship between age and attitudinal change. It is much easier to change attitudes in children than in adults. This is known as the "impressionable years hypothesis" (Alwin & Krosnick, 1991; Krosnick & Alwin, 1989).

As an example of a worldview and resilience to revision, consider the role of *varna* in the organization of Indian society. *Varnas* constitute an ordering of society determined by "essential natures" (i.e., birth) that is divided into four hierarchical classes: Brahmins, Kshatriyas, Vaishyas, and Sudras.[2] The concept appears in various texts as early as the Rig Veda. In the Bhagavad-Gita, it is stated as follows (18:41–45):

41. Of Brâhmanas and Kshatriyas and Vaishyas, as also of Sudras, O scorcher of foes, the duties are distributed according to the Gunas born of their own nature.

42. The control of the mind and the senses, austerity, purity, forbearance, and also uprightness, knowledge, realisation, belief in a hereafter,— these are the duties of the Brâhmanas, born of (their own) nature.

43. Prowess, boldness, fortitude, dexterity, and also not fleeing from battle, generosity and sovereignty are the duties of the Kshatriyas, born of (their own) nature.

44. Agriculture, cattle-rearing and trade are the duties of the Vaishyas, born of (their own) nature; and action consisting of service is the duty of the Sudras, born of (their own) nature.

45. Devoted each to his own duty, man attains the highest perfection. How, engaged in his own duty, he attains Perfection, listen.

The original concept of *varna* has since been further subdivided into hundreds of *jatis* (Beteille, 1996; Rajesh, 2018). The *varna/jati* system is known throughout the world by the Portuguese word "caste." While its origins and initial purpose are still debated by historians, it has served to provide hierarchical order, structure, and meaning to Indian life (Sharma, 1990), as did the Great Chain of Being to medieval European life. *Varna* permeates every aspect of Hindu life from religious, to social, to cultural. Your *varna* traditionally determined who you could marry, what job you could do, what you ate, who you ate it with, if and how you were educated, what you wore, and many other aspects of life.

After being in place for thousands of years, the system was officially outlawed in 1950, when the modern Indian Constitution came into effect. Passage of the law meant that it had widespread support at the political and governmental levels and presumably among the population, at least in the sense that they could entertain the belief that the *varna* system was discriminatory and should be abolished. Seventy years later, look at any Indian matrimonial website and one of the first pieces of information that will be offered (and expected) will be *varna/jati*, even if it is only to say that it does not matter. In fact, the information is already encoded in surnames. It continues to have enormous impact on the lives and worldviews of Indians (Deshpande, 2008). This is not unusual. Worldviews are notoriously difficult to revise (Galperti, 2019; Hiebert, 2008).

First-generation Indians who migrated to North America as adults insist on marrying their children along *varna* lines, even though they may never have taught their children what *varna* they belong to! Contrast these adult immigrants with children born in India but brought to North America as teenagers. They were also educated and socialized into the *varna* system from birth, but they are able to successfully abandon this worldview and adopt the (diametrically opposed) North American worldview after socialization. So the question of interest is, why can the children modify their worldviews but those who migrate as adults cannot?

This phenomenon is very common. It partially explains the difficulty older adults encounter in revising their beliefs on climate change. For most individuals born in the previous 35 years, man-made climate change is an idea that is intertwined with their global belief structures. However, those born more than 60 years ago grew up in a world where it was not part of our general social, economic, and intellectual/scientific framework. For this latter group to accept the idea now involves not only local changes to their belief networks but widespread global belief revision.

As a third example, consider John Locke, Thomas Jefferson, and Charles Darwin, some of the most intelligent and well-educated men of European ancestry who have ever lived. Their dim views on the intellectual capabilities (and even humanity) of those not descended from Europeans are documented. If we were to transport them to the twenty-first century, at the height of their intellectual powers, could we convince them with evidence and reason (i.e., the tools of rationality) that non-Europeans are intellectually equal to Europeans? Despite Thomas Jefferson's aspirations (*Letters of Thomas Jefferson*) that

> Nobody wishes more than I do to see such proofs as you exhibit, that nature has given to our black brethren, talents equal to those of the other colors of men, and that the appearance of a want of them is owing merely to the degraded condition of their existence, both in Africa & America,

I'm skeptical that we would succeed. However, if we transplanted them when they were children, they would have no difficulty accepting the proposition. This example is a little different from the previous two in that it requires revision of not only large-scale belief structures and their underlying presuppositions but also low-level sensorimotor parameter settings that inform the beliefs and presuppositions.

The question for this chapter is, why is it difficult for healthy, mature adult brains to engage in large-scale revision of an established worldview? Why can children do this effortlessly, while older adults struggle? We do not know the answer to this question. I advance a hypothesis based on data from neuronal development: worldview revision requires neuronal reorganization at multiple levels, including large-scale belief networks, presuppositions, and sensorimotor parameters. Once brain systems have matured, there are very few neuronal resources left to underwrite such significant degrees of reorganization. We know that this is the case for subcortical and primary cortex systems. I'm postulating that the same principles may apply to the association cortex, which constitutes the neural basis of our belief networks. Before laying out my argument, I once again ask the reader's indulgence and patience while I explain the basics of neural development. This exercise is necessary for the same reason as it was in the previous cases of gender, economic decision-making, reciprocity and cheater detection, lust, energy management, and in-group/out-group formation—details and nuance matter. The reader can only evaluate my conclusions in the context of these details.

Brain Development: Overview

Brain development is a rapid, intricate process, where 86 billion neurons are created on the order of months (approximately 4.6 million per hour) and an estimated 140 trillion synapses are created on the order of years

(2.5 billion per hour) during the critical periods specified in figure 14.1 (Silbereis, Pochareddy, Zhu, Li, & Sestan, 2016; Tang, Nyengaard, Groot, & Gundersen, 2001). It is guided by genetic and environmental factors: "Brains do not develop normally in the absence of critical genetic signaling *and* they do not develop normally in the absence of essential environmental input" (Stiles & Jernigan, 2010, p. 345). These two aspects of brain development are captured nicely in the terminology of "experience expectant" and "experience dependent" (Greenough et al., 1987) that was introduced in chapter 10. Experience expectancy captures the fact that the genetic program unfolds with certain innate, genetically encoded expectations about the environment of the organism. Experience dependency allows for the actual environment of the organism to impact neural growth and development.

The basic process of brain development is one of first generating excess neural resources arranged according to genetic instructions and then fine-tuning and shaping the neural system, via elimination of resources, in response to internal and external environmental factors. Once eliminated, resources are forever gone and cannot be resurrected, even if needed at a later date. As we often talk about "sculpting" neural systems, let's take a minute to consider the analogy. There are at least two forms of sculpting. One can sculpt in clay. This involves shaping the material by pressing and stretching at different points. If you have created a bird and later want to change it to a tiger, you just knead the clay and start over. However, if you are sculpting in a material such as marble, you shape it by chipping away and removing material. Once a piece has been removed, it can never be put back. It is gone. In this case, if you sculpted a bird and wanted to change it into another similar-looking bird, you could probably continue chipping away and do so. However, if you wanted to change it into something very different, such as a tiger, you may need to find yourself another piece of marble. I think this marble analogy captures some aspects of the sculpting of neural networks.

The particulars of neural development and maturation will be discussed in terms of the following nine steps (see figure 14.1): (1) formation of neural tube (neurulation); (2) neurogenesis (neural cell generation); (3) neural migration; (4) axon and dendrite growth; (5) apoptosis (cell death); (6) synaptogenesis (synaptic growth); (7) gliogenesis (glial cell generation and migration); (8) myelination; and (9) experience-dependent synaptic pruning. As the reader can see from figure 14.1, many of the processes occur prenatally and others postnatally. The characteristics of each of these

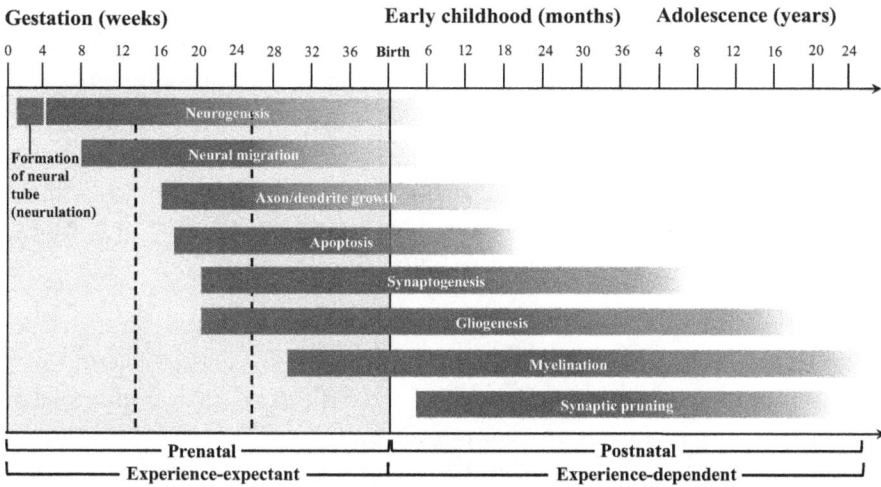

Figure 14.1
Important stages in neural development and maturation. Graphed from data from Andersen (2003) and Stiles and Jernigan (2010).

phases are important for our purposes, as is the fact that development extends to early adulthood but then terminates.

Prenatal (Experience-Expectant) Brain Development

Neural development begins with the formation of the neural tube by the third week after conception. By the fourth week, the anterior end of the neural tube begins to show three subdivisions: the forebrain (prosencephalon), midbrain (mesencephalon), and hindbrain (rhombencephalon). By the end of the seventh week, the forebrain further differentiates into the telencephalon (which will become the cerebral hemispheres) and the diencephalon (thalamus and hypothalamus). The hindbrain differentiates into the metencephalon (cerebellum and pons) and myelencephalon or medulla. The reader should recognize some of these regions from figure 10.2.

The differentiation of these basic regions begins an ongoing process of neuronal patterning within the central nervous system that results in gradual differentiation, organization, and refinement of the hindbrain and spinal column (Lumsden & Keynes, 1989), major components of the diencephalon and midbrain regions (Nakamura, Katahira, Matsunaga, & Sato, 2005), and sensory and motor neocortex regions (Sur & Rubenstein, 2005).

These processes begin during the embryonic stage and progress along different timelines, from the inside out (figure 10.6), such that hindbrain and spinal column segmentation is completed before subcortical region differentiation, followed by primary neocortex differentiation, which—because of its need for environmental experience—extends into the early postnatal period (Campbell, 2005; Stiles & Jernigan, 2010). This is followed by differentiation of the more general-purpose association cortex, which continues into early adulthood. Let us briefly examine each step.

Neurogenesis

Neurogenesis (creation of new neurons) through mitosis begins with formation of the neural tube. Initially, the mitosis process simply multiplies the population of neural progenitor cells (i.e., cells that give rise to neurons) in the ventricular zone. Once a sufficient number of progenitor cells have accumulated, further mitosis generates one progenitor cell and one (undifferentiated) neural cell. This process begins during the sixth week and continues through midgestation.

Initially, the cells are all alike and are called neuroblasts, but there are several different types of neurons in a brain. Early in the neurogenesis process, progenitor cells can receive signals to generate any type of neuronal cell, but toward the end of the process, they become restricted to generating the types of neural cells still needed. There is a species-specific "birthdate" for each part of an animal's brain, determined by a highly stereotyped chronological development program conserved across species. Cell type seems to be determined at the time of cell division.

Neural Migration

Once generated, the neural cells migrate to distal predetermined locations. This migration is concentrated within an 18-week period but continues until birth. Brains develop in an orderly fashion from the inside out. More recently generated cells migrate past and layer on top of earlier-migrating cells (Campbell, 2005; Stiles & Jernigan, 2010). This recapitulates the evolutionary development outlined in chapter 10. Migration is followed by the maturation process, which involves both prenatal and postnatal components. Prenatal maturation consists largely of axon and dendrite growth, apoptosis, synaptogenesis, and gliogenesis. These processes do continue postnatally. Myelination and pruning are largely postnatal processes requiring environmental input. By the time these processes have completed, in early adulthood, neural resources for further development and reorganization have been largely expended (as in the marble sculpting analogy).

Axon and Dendrite Growth

Once the cells reach their target destination in the brain, they undergo a process of axonal and dendritic growth and acquire the distinctive appearance of their particular cell type. There is an exuberance associated with this growth process. Again, while this process is concentrated prenatally, it does continue postnatally for approximately 18 months (Stiles & Jernigan, 2010).

Apoptosis

Neurogenesis generates approximately 50% more cells than will actually be utilized. This is a design feature of brains whereby natural preprogrammed cell death, known as apoptosis, is used to sculpt or shape the developing nervous system. A number of factors influence cell death. One factor is the size of the field of the body surface that needs to be connected to a region of the central nervous system. For example, the removal or grafting of a leg in tadpoles, prior to the formation of spinal cord connections, respectively increases or decreases the number of spinal motor neurons that will die. Another factor is numerical matching between cell populations that need to interconnect. For instance, given two clusters of cells, A and B, where $A = 100$ cells and $B = 50$ cells, once the interconnections have been made, the cells in A that do not participate in the connections will die. Apoptosis is critical to brain development. It suggests that the brain is being shaped not by the growth of new neurons as needed but by the elimination of neurons not needed. Not only are neuronal cells programmed for apoptosis but so are the progenitor cells, resulting in their gradual elimination. This means that no new neurons can be generated (with perhaps the exception of the hippocampus) for any further reorganization (Stiles & Jernigan, 2010).

Synaptogenesis

Synaptogenesis is the process of proliferation of synapses at the terminal end of axons. It is the synapses that will connect the neuron to other neurons (via dendritic spines). As with neurogenesis, synaptogenesis is also profuse, with many more connections made than needed. It begins during midgestation and continues for some years postnatally. Different brain structures have different timelines for synaptogenesis, as discussed later.

Gliogenesis

Neurons constitute one type of cell found in the brain. The other type is glial cells. They provide various critical support functions for neurons. The process of gliogenesis involves the proliferation and migration of glial cells. It starts prenatally, also around midgestation, but continues for extended periods postnatally. As with neurons, there is an overproduction of glial cells. Glial apoptosis follows the delayed time course of gliogenesis, with the number of surviving oligodendrocyte cells corresponding to the axonal area available to be myelinated (McTigue & Tripathi, 2008). Myelination continues postnatally into early or perhaps even middle adulthood (Coupé, Catheline, Lanuza, & Manjón, 2017).

Prenatal Neural Maturation in Sensorimotor Systems

As already noted, neural organization is guided by two distinct types of mechanisms. The first is activity independent, whereby initial axon growth and pathfinding is guided by chemical gradients. This largely occurs prenatally and results in a rough overall configuration. The second mechanism relies on sensory and motor experience (i.e., action potential activity) to fine-tune the neural machinery. For most cortical systems, this will have to wait until after birth.

Certain basic systems that need to come online prenatally to maintain the fetus will need to go through this neural fine-tuning process much earlier. In fact, there is evidence that this process even begins in the sensory systems prior to the availability of any sensory information. For example, development of the visual system begins with formation of the neurons connecting the retina to the lateral geniculate nucleus (LGN). This is followed by LGN neurons innervating layer 4 of the visual cortex. Both the LGN and primary visual cortex are organized to preserve retinotopic mappings. The LGN is organized into six layers alternatively innervated by each eye. The initial invasion of the LGN by retinal ganglion neurons is somewhat coarse and sloppy. In a series of seminal experiments on nonhuman mammals, Carla Shatz and her colleagues have shown asynchronous neural activity (action potentials) in the retinal-LGN system even before the formation of rods and cones (i.e., before vision would be possible). This simulated neural activity is used to sculpt and refine innervation patterns and segregate the LGN into eye-specific layers (through growth and pruning of synapses and dendritic spines).

The primary visual cortex is organized into alternating striped configurations of ocular dominance columns, corresponding to each eye. We know from the pioneering experiments of Hubel and Wiesel (chapter 10) that this segregated mapping for each eye requires visual input (because the banding can be disrupted by obscuring light input to one or both eyes), but data from Shatz and her colleagues show that prenatal action potential activity in the LGN-cortical connections (prior to exposure to any light) accounts for this organization (Ghosh et al., 1990; Katz & Shatz, 1996). The stimulation and organization continue postnatally with actual light.

Postnatal (Experience-Dependent) Brain Development and Behavior

It is important to appreciate that a number of developmental processes continue postnatally and a number do not. Most importantly, with a few exceptions, there is no postnatal neurogenesis (Stiles & Jernigan, 2010). Once the 86 billion neurons are generated and migrate to their assigned location (the obsolete ones having died), the only possibilities for neuronal change are axonal and dendritic outgrowth (continues for approximately 18 months after birth), synaptogenesis, synaptic pruning, and gliogenesis and myelination. These processes account for the fourfold increase in brain size that occurs from birth to early adolescence (Silbereis et al., 2016).

Gliogenesis and myelination follow a structure- and function-specific timetable, tracking neural growth and maturation, and continuing into early to middle adulthood. The generation of one type of glial cell, called oligodendrocyte glial, seems to persist indefinitely and can be activated in response to injury. The oligodendrocyte cells are responsible for myelination in the central nervous system. The process of myelination involves glial cells wrapping themselves around nearby axons to provide a protective coating. Myelination contributes to axonal health, integrity, and survival, and even influences neuronal size and axon diameter (McTigue & Tripathi, 2008). There is even evidence that a subset of oligodendrocyte cells participates in excitatory and inhibitory connections with neurons and may contribute to neuronal signaling (Lin & Bergles, 2004). Astrocytes, another type of glial cell, constitute the majority of glial cells in the brain, and there is evidence that they participate in and support synapse formation, elimination, and functioning (Eroglu & Barres, 2010).

Synaptogenesis and pruning are experience-dependent shaping mechanisms of neural systems, particularly cortical systems, that begin prenatally but continue into late adolescence, even early adulthood. *They are perhaps the final opportunity to undertake large-scale architectural reorganization of neural structures.* The timetable varies greatly across region-specific cortical structures and functions. This is depicted for sensorimotor systems (primary cortex), language systems (Broca's area and angular gyrus), and general cognition (prefrontal cortex) in figure 14.2. These staggered developmental stages are closely aligned with the need to use the corresponding systems.

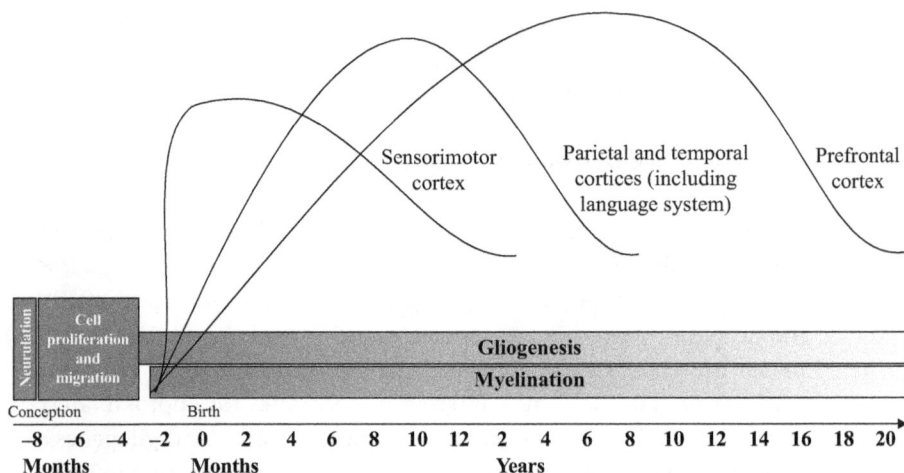

Figure 14.2
Time course of synaptogenesis and synaptic pruning for sensorimotor, language, and higher cognitive systems. The general time course of gliogenesis and myelination is also indicated. Figure based on Casey, Tottenham, Liston, & Durston (2005), modified based on data from Gogtay et al. (2004).

The maturation of neuronal and glial cells permanently structures and sculpts brain systems and associated behaviors. Once these processes are complete, there simply are no neural resources left for any large-scale architectural reorganization of the system. Again, I refer the reader to the marble sculpting analogy. As pieces are chipped away, they cannot be put back and reconfigured into a different shape. This is best understood in the cases of subcortical structures and the primary cortex, but my conjecture is that the same principles should apply to all cortical systems.

Postnatal Neural Maturation in Sensorimotor Systems
Synaptogenesis in motor and sensory systems, such as vision, peaks approximately one month prior to birth, followed by rapid decline as a function of pruning in response to environmental input that plateaus around two years of age (in humans). This ∩-shaped curve corresponds to a critical window during which the system is in an experience-dependent mode and requires external stimuli to complete neural configuration. The completion of the pruning (and myelination) processes corresponds to maturation of the system, beyond which it is not responsive to reconfiguration. We saw in chapter 10 that the experiments of Hubel and Wiesel and of Blakemore and Cooper violated the developing brain's experience expectation during the

critical receptive window through sensory deprivation, resulting in neural (and behavioral) reorganization. Hubel and Wiesel's eye suturing experiments, by creating an unexpected environment, sharply reduced neural connections in the lateral geniculate nucleus and dramatically altered banding in ocular dominance columns in the primary visual cortex as the connections from the open eye innervated the neural cortex of the closed eye. The behavioral experiments by Blakemore and Cooper, which selectively placed light-naive kittens in rooms with either vertical stripes or horizontal stripes, noted permanent neuronal rewiring and behavioral manifestations of the artificial environmental manipulation.

In fact, the flexibility during the critical window (prior to maturation) is so great that the auditory cortex can be used to see! In normal development, because of the excess production of neurons and synapses, there are some transitory input pathways from the retina to the primary auditory cortex that are eliminated in a competitive process during the course of normal sensory input. If the normal auditory input pathways to the primary auditory cortex are surgically destroyed in one-day-old ferrets, there is an absence of competitive auditory input. In this situation, the auditory cortex is recruited for vision. It acquires the organization of the visual cortex and behaves like the visual cortex, albeit imperfectly (Pallas, Roe, & Sur, 1990; Sur, Garraghty, & Roe, 1988).

These examples speak to plasticity as a fundamental feature of immature neural systems. In this chapter, I want to play the devil's advocate and highlight the equally real fact that Hubel and Wiesel's eye suturing manipulation and Blakemore and Cooper's behavioral manipulations had no impact on adult animals with fully mature visual systems. If the ferrets in the preceding experiment had been two years old, perhaps even two weeks old, the primary auditory cortex would not have adapted and reorganized as visual cortex. If the receptive window of opportunity is missed, no amount of sensory stimulation is going to change the organization of the mature visual or auditory cortex. Once this window passes, the organization of the primary cortex and associated subcortical structures is largely fixed. They cannot be reorganized or repurposed to deal with a radically different environment. This would be part of the underlying explanation of why my visual system fine-tuned to differentiate Caucasian faces more accurately than Indian faces (chapter 13). These subconscious perceptual biases—even though they may be an accidental feature of the environment—will feed into the In-Group/Out-Group system and infiltrate up into presuppositions and belief systems.

It is important to emphasize why the system cannot be reorganized. What made possible the reorganization of the immature auditory cortex to "see" was the initial overabundance of neurons and synapses and their random connections from the retina to the primary auditory cortex. These excess neurons and connections are eliminated via a competitive process during normal development. However, when the actual auditory input connections were surgically eliminated, the random connections from the visual system were able to dominate and reorganize the auditory cortex. Once the windows for generating new neurons and synaptogenesis have closed and the elimination of neural and synaptic connections from the visual system to the auditory cortex has been completed via neuronal death and synaptic pruning, there are no neural resources left to reorganize the auditory system to "see." No new neurons can be grown or new synapses generated. I'm suggesting the same logic should apply to other cortical systems.

Postnatal Neural Maturation in Language Systems

Neural systems for language also mature reasonably early, with corresponding behavioral consequences. The auditory and motor cortices necessary for input and output of language would follow the developmental trajectory of sensorimotor systems. The phonological system is associated with the posterior temporal gyrus (Brodmann area 22); the lexical-semantic processing is generally thought to involve left temporal parietal regions, including the angular gyrus and supramarginal gyrus, and syntactic processing is generally associated with Broca's area and the left lateral premotor cortex (Sakai, 2005).

During its premature phase, the phoneme detection system is able to differentiate phonemes from any natural human language. With neural maturation, this ability quickly recedes and the system becomes attuned to the phonemes of the child's native language. This begins to occur as early as six months, long before language acquisition itself. Once the system matures, it cannot be retrained to perfectly differentiate phonemes from the nonnative language (Kuhl, Williams, Lacerda, Stevens, & Lindblom, 1992; Ventureyra, Pallier, & Yoo, 2004). Evidence of this inability is very common among immigrant families. Many second-generation Indians born in the United Kingdom and North America and given traditional Indian names cannot correctly pronounce their own names. A similar story can be told for the syntactic processing system, with some shift in the window.

Babbling and single-word production begin around one year of age, followed by multiple words, and then sentences at around three years of age. The window for syntax acquisition is thought to close at around 12 years (Sakai, 2005), though some studies are suggesting it may be as long as 17.5 years (Hartshorne, Tenenbaum, & Pinker, 2018). Synaptogenesis in language-related systems (angular gyrus and Broca's area) peaks at around six to eight months, followed by exposure-dependent pruning, which plateaus at around 8 to 10 years of age (Thompson & Nelson, 2001).

As with the sensory systems, the maturation of phonological and syntactic neural systems will lay down preferences for the sounds and structure of native languages over the "other" languages that again feed into In-Group/Out-Group systems. The windows in the case of language are not as tight as in the case of sensorimotor systems, but there is every reason to believe that the same general principles will apply.

Postnatal Neural Maturation in Association Cortex

Other cortical areas also mature in an ordered sequence. Phylogenetically older cortical areas (like the inferior medial surface of temporal lobes) are the first to complete synaptogenesis and pruning. Phylogenetically newer cortical areas, consisting of higher-order association cortex, such as superior temporal gyrus, posterior parietal cortex, and prefrontal cortex, follow later. The dorsolateral prefrontal cortex, a region specifically associated with higher-level cognitive processes such as reasoning and problem solving, is the last to mature. It continues to experience exuberant synaptogenesis until six to eight years of age, followed by synaptic pruning that continues into the mid-twenties (Gogtay et al., 2004). Maturation of the higher-order association cortex must also have some behavioral consequences analogous to the maturation of the sensorimotor cortex and language-specific cortex. Or is there something special about the association cortex?

I did refer to the association cortex as "softwired" and contrasted it with the "hardwired" brain stem, diencephalon, subcortical systems (excluding the hippocampus), and primary cortex in chapter 10. But the main differences are in terms of the specificity of the processing they undertake and the relative contributions of experience-expectant and experience-dependent factors in ontological development, resulting in different maturation timelines. The association cortex deals with more general and abstract information combined from various sensory systems and is of necessity much more experience dependent, resulting in a protracted maturation timeline. But otherwise, in terms of basic units and principles of organization, all neural systems should be similar. Once the systems have been sculpted and excess resources eliminated, they should be similarly resistant to massive change, because of lack of raw materials for reorganization.

One function of the association cortex is to maintain veridical and coherent belief systems. We continue to add new beliefs over a lifetime. For example, I recently learned that the male platypus has ankle spurs containing venom. We can also revise specific beliefs that we discover to be incorrect, as in the earlier example of the leatherback turtle. Every change in beliefs will presumably require some local strengthening and/or weakening of synaptic connections in the context of the current configuration of the neural network. This is clearly possible throughout our lifetime, but the phenomenon of interest in this chapter is not local belief revision but rather global revision requiring large-scale architectural reorganization of the system.

What are the functional consequences of neural maturation in the association cortex for global belief revision? This is not a question that is often asked. The examples with which we began the chapter—deeply held worldviews such as those involving the *varna* system, climate change, and racial biases—suggest some limits to reorganization. Once large-scale belief networks are sculpted and the underlying neural substrate has matured, the lack of neural resources may hamper global belief revision. In the case of revising beliefs on things such as racial equality, it is even more difficult because of the additional involvement of perceptual systems. These latter systems mature very early, and once parameters and preferences are set, they are very resistant to change. Again, think of the marble sculpting analogy: no neurons can be generated, shaping through synaptogenesis and synaptic pruning is not available, and neither are gliogenesis and myelination. Without these resources, how are belief systems to undergo large-scale architectural reorganization? This is a conjecture, a hypothesis. Currently, there is little evidence one way or the other. Because belief formation and revision are uniquely human phenomena, it is not possible to explore the issue with animal models. Obvious ethical concerns preclude invasive neural exploration in humans. It may be some time before we are able to address this issue directly with data.

But there are at least three reasons to suggest that it is a hypothesis worth exploring. First, the anecdotal behavioral evidence is overwhelming. We all form worldviews roughly corresponding to the world in which we "came of age" in our teens. This determines the music we like, the ideas and people we admire, our social norms, our religious norms, and our economic norms and expectations. In short, it determines our worldview. We acquire this worldview effortlessly, like a first language . . . or measles. When people think of the "good old days," they are thinking of the world experiences that sculpted their neural systems from early adolescence to early adulthood. This period corresponds very closely to synaptogenesis, pruning, and myelination in the association cortex, particularly the prefrontal cortex.

The second piece of evidence is the actual measurable data regarding sensorimotor systems and language systems reviewed here. In these cases, there are clear-cut measurable relationships between neural maturation and function. The relationship is much tighter in the case of the sensorimotor systems, with clearly demarcated "critical periods" in animal models. It is more relaxed in the case of language, with more flexible "sensitivity periods," but nonetheless the pattern is there. As far as we currently understand, there is nothing special about the neurons in the more general association cortex. If we look at the graph in figure 14.2, the curve for the association cortex looks identical to the curves for sensorimotor systems and language systems, except that it is protracted and shifted to the right. That is, the association cortex undergoes the same synaptic generation, pruning, gliogenesis, and myelination processes, except that they are delayed and extend into young adulthood, so most of the shaping is occurring via environmental interaction. The response profile to environmental expectation is also similar. Where expectations are violated prior to maturation, there is enormous scope for compensation, and where they are violated after maturation, recovery potential is much more limited.[3]

A third piece of evidence is provided by lessons from building computational neural networks. The way to build and train neural networks is by imposing an overall architecture and then connecting nodes to other nodes through a random assignment of weights (chapter 5 appendix). A learning algorithm then operates on this network, and over the course of experience with data, the network is sculpted to transform one set of patterns into another set of patterns. The more interesting of these networks are not necessarily computing specified functions. They are behaving in accordance with their training data set and learning algorithm. For example, it is difficult to say what function a self-driving car is computing as it makes its way down the highway. The way the car will respond to any situation will be a function of the data set it was trained on and the learning algorithm that was used. Training physically resculpts the network by strengthening certain connections and weakening or eliminating others.

Suppose we have access to the Google deep learning network trained to recognize human faces and cat faces from YouTube videos (Le, 2013). It has been fine-tuned for recognizing human faces and cat faces, but now we want it to differentiate between rabbit faces and dog faces. This is a different task. Should we retrain the network or start from scratch with a new one? The two tasks are very similar, so some local retraining or resetting of weights in one or two layers of the network with the new data set may allow it to differentiate between rabbit and dog faces. If we want the network to do a very different task, such as learning to detect credit card fraud, play the game Go, or drive a car down city streets, it will be much faster to start from scratch, because it will be necessary to change not only the local weights in the different layers but also the overall (global) architecture. This will require a total rewiring of the system. It may not be possible to rewire the system simply through training with additional data because local weight changes may not provide sufficient resources to facilitate the global architectural revisions necessary for the new tasks.[4]

Consequences for Models of Rationality

If it is indeed the case that the scope for global belief revision is limited after neural maturation of the association cortex, it has interesting implications for our standard cognitive and social science Platonic models of mind, where rationality is divorced from biology. Even if in some ideal world the machinery of reason may allow for extensive and perpetual belief revision, the actual biology that supports the machinery may not. If correct, this suggests that certain global belief revisions required to maintain veridicality and/or coherence of belief networks may not be undertaken, not because of failure of reason or even intervention by nonreasoning systems but because the brain may lack the neural resources for large-scale architectural reorganization. Like tethered rationality, this constraint arises from taking the neurobiology seriously but is an independent issue.[5]

If someone has been brought up in a particular worldview, they cannot after a few lectures replace it with a different one. My neighbors sometimes complain that "the problem with immigrants is that they do not conform to our Canadian (more generally Western) worldview. It would be desirable if they did so." One solution that is sometimes proposed to facilitate this is mandatory training and education for immigrants in "Canadian studies." If what I'm suggesting is correct, these programs may teach them to sing our national anthem, "O Canada," but will not bring about deep structural changes in beliefs, behaviors, and thought processes of adult immigrants. The only solution is to let first-generation adult immigrants live out their lives fixed in their old worldviews. The second generation will be as Canadian as anyone else. It is wiser to set one's immigration policy knowing this rather than harboring false assumptions about changing worldviews through cultural assimilation and then being disappointed. Similar constraints may apply to cases such as climate change.

The implications for revising worldviews in which sensorimotor systems play a part in the original formulation are even more dire. For example, if the sensorimotor systems play a critical role in the entrenchment of certain

In-Group/Out-Group properties, such as race, no amount of sensitivity training (beyond the maturation window) is going to reorganize the parameters and preferences wired into the perceptual system.

More generally, these arguments highlight the issue of the relative roles of biology and environment (nature vs. nurture) in determining human behavior. Any discussion of nature and nurture usually devolves into a shouting match between the relative roles of genes and social environment. For the last hundred years, academic theories have emphasized how socialization (through belief formation) dominates human behavior. One reason for tenaciously holding onto this view is the assumption that social norms can be revised; biological constraints less so—leaving one the illusion of being able to shape the world in any arbitrary manner. I have challenged this all-too-common misconception. My argument for tethered rationality was an argument for equal time for nonreasoning factors. The argument in this chapter is a reminder that biological constraints are not simply limited to genetic factors. Neural maturation reduces plasticity, which in turn limits the ability of the environment (social or otherwise) to shape neural systems.

Notice that this is not an argument against the importance of environmental factors, including socialization. My point is to highlight that these factors are most effective during certain developmental windows and less so beyond them. We saw the importance of timing of pharmacological, hormonal, and other chemical environmental factors in gender determination. Timing was also a critical factor in the environmental input that allowed my visual system to fine-tune to better recognize Caucasian faces than Indian faces and my language system to differentiate and articulate certain phonemes over others. Once the window passed, my visual and language systems, along with their encoded preferences and biases, were set. In this chapter, I have proposed that neural developmental considerations are also a critical factor in laying down and revising worldviews. These too congeal after neural maturation. These data and arguments are reminders that it is the brain—not some disembodied mind—that is being socialized. We will not fully understand socialization and its underlying limitations until we understand how the brain develops and matures.

*   *   *

In previous chapters, I have argued that the failure to recognize the tethered nature of the reasoning mind has led us to models of human behavior that do not even have the resources to explain our basic choices regarding food, sex, and politics. The solution offered is one where the rational mind is tethered to the associative, instinctive, and autonomic minds and behavior is a blended response of all these systems. The considerations in this chapter point to possible biological limits on the reasoning mind in adult brains to undertake large-scale, global belief revision. This is another biological constraint on human rationality, independent of the tethering. It likewise has some unwelcome implications for changing behaviors.

VI What Follows from the Tethered Mind?

An ideology that tacitly appeals to biological equality as a condition for human emancipation crops the idea of freedom. Moreover, it encourages decent men to tremble at the prospect of "inconvenient" findings that may emerge in future scientific research.

—Marvin Bressler

The title of this book is not *Less than Reason*, or even *Reason or Less*; it is *Reason and Less*. This means that I'm acknowledging and embracing the role of reason in human affairs, but within obvious biological constraints. The reasoning mind is not disembodied. It is tethered to evolutionarily older systems and subject to the basic laws of neurobiology.

I cannot imagine a single colleague who would claim that reason is *literally* independent of biology. Yet in discussing some of the everyday but sensitive examples that I have raised in these pages, many balked at the uncomfortable implications of the tethered mind—but rather than engaging in a discussion of how tight or loose the tethering might be in any given situation, cautioned me about using the example; best not to go there. This advice—however well-intentioned—does not make for good science.

I think there are two reasons for this reaction: fear and hope. The fear is that where behaviors that involve harm to others are involved—and someone needs to be held accountable—any reference to biology may serve to absolve the perpetuator from responsibility. This is a genuine concern, but largely dependent on the specifics of the model under consideration. I touch upon this issue in my closing comments, but it requires a separate volume to do it justice. I think one thing that can be said with considerable confidence is that models that place us on a very tight biological leash or an infinitely long leash are nonstarters. We need to meet in the space of actual possibilities—as guided by disinterested science—for any meaningful dialogue and progress.

For individuals concerned with changing societal behaviors, it is hopeful to believe that behaviors can be changed by arbitrarily changing beliefs (through untethered learning and reason). (This is actually inconsistent with the data presented in part IV, but hope reigns eternal.) While the mechanisms are different, this is exactly the same position the behaviorists arrived at 80 years earlier, and for the same reason: they were obsessed with changing behavior, and found it easier to manipulate environmental factors (through operant conditioning) than neurobiological factors (for technological and ethical reasons). This approach is reminiscent of the drunkard's streetlight fallacy: looking for our car keys under the lamp post on the street, where the light is better, even though we dropped them in the parking lot. We can of course continue doing this for another 80 or 800 years, but that will not change the facts in the world. Ultimately, how effective we are at bringing about lasting societal changes will be a function of the accuracy of the underlying model of human behavior that we utilize.

15 Concerns, Consequences, and Conclusions

Human action can be modified to some extent, but human nature can not be changed.
—Abraham Lincoln

Human nature is complex. Even if we do have inclinations toward violence, we also have an inclination to empathy, to cooperation, to self-control.
—Steven Pinker

Three hundred years after Alexander Pope memorably versified the two prongs of human behavior—reason and "animal passions"—constituting the Western-Christian model, our best current theories are choosing to deal with the resulting dilemma by ignoring one prong or the other. The standard social and cognitive science model embraces reason and ignores (or denies) the "animal passions," while the evolutionary psychology massive modularity model embraces the "animal passions" and ignores (or denies) reason. Both options are untenable. While we cannot accept the details of the Western-Christian account without giving up the natural sciences, we must accept the reality of the underlying intuition. The most cursory look at the data demands it: human behavior consists of both reason and "animal passions."

Tethered rationality is an obvious, commonsense alternative model. It recognizes the deeply held intuitions, incorporated into the Western-Christian model, that human behavior is a function of both reason and evolutionarily older systems. It utilizes the insights and data concerning human behavior, and knowledge of underlying biology, generated during the twentieth century to recast the dichotomy into the following four systems: the autonomic mind, the instinctive mind, the associative mind, and the reasoning mind. Each of these "minds" has been studied and elaborated by hundreds of distinguished scientists. I largely accept the characterizations of these various systems offered by these scientists. Their common mistake was to conclude that only they were correct and those pursuing other programs were misguided. My contribution in this volume is fivefold: (1) to remind readers of the obvious—that while each of these research programs captures some important facet of human behavior, a full model will need to consider how all of these systems contribute to behavior; (2) to provide data demonstrating tethering among the different systems resulting in a blended behavioral response; (3) to show how these hierarchically organized tethered systems are a natural consequence of brain evolution; (4) to propose feelings of pleasure and displeasure as the common currency that allows the different systems to communicate and interact; and (5) to provide a control structure consistent with the underlying biology.

As developed in this volume, tethered rationality does a more convincing job of explaining real-world human choices and decisions than either the standard social and cognitive science model or the evolutionary psychology massive modularity model. Many of the examples used in this book concerned choices involving food, sex, and politics (power relations). This is not an accident. These are among the most consequential decisions we make. They are also the very decisions that the standard rationality models stumble on. In each case, we have seen tethered rationality offer more convincing explanations.

Concerns and Consequences

Every model of human behavior will have social and legal consequences. Western social and legal frameworks have long been based on the Western-Christian model of behavior, with increasing input from the social and cognitive science model. Legal scholars assume that (Jones, 1997, p. 168):

> behavior (excluding that caused by reflexes, chemical imbalances, and the like) is simply what the mind tells the body to do. This, in turn, suggests that behavior can be shifted as easily (or at least with equal difficulty) in any direction. To properly shift behavior, it appears, law need only alter those sociocultural influences that lead a mind to direct it. (italics added)

That is, behavior can be changed simply by changing beliefs and desires and/ or trying harder. So, if we have a rule that "thou shall not do X," where X can be anything from rolling through a stop sign, to cheating on taxes, to infidelity, to racial profiling, the behavior can be equally easy (or difficult) for an individual to control or change to conform to the social or legal norm.

I am proposing two important caveats to this model of human behavior. First, it is true that social environments play a massive role in sculpting behavior. The association cortex is shaped largely by environmental interaction, but, as we have seen, the shaping is most effective prior to full maturation of neural systems. Once neural systems mature, there may not be sufficient resources remaining for large-scale belief revision. This means that even socialized behaviors, particularly those ingrained not just in belief systems but also presuppositions and sensorimotor systems, such as racial biases, cannot be altered equally easily at any time point.

Second, the model of tethered rationality says that the Platonic, disembodied conception of reason is a fiction. The reasoning mind is tethered to the associative, instinctive, and autonomic minds. All these systems modulate behavior. We considered the workings of tethered rationality in some detail in the example concerning my propensity to overindulge in pizza and chocolate cake despite being overweight. I suspect most readers would be content with this explanation and the underlying implication that losing weight is not *just* a matter of changing beliefs or "trying hard enough." The rational mind and the decision to lose weight *is* part of the story, but only one part. The rational mind gets to decide that it wants to lose 20 pounds. The rational mind can take proactive steps to organize life to reduce the temptation of chocolate cake. Insofar as the realization of this desire results in greater pleasure than consuming chocolate cake, the intake of chocolate cake will be curbed and some weight loss will follow, but if the taste of chocolate cake results in greater pleasure than being 20 pounds slimmer, the desired outcome is much less likely. By recognizing the involvement of multiple systems, we allow for the possibility and consequences of individual differences in each system, so, given 10 individuals with "equally strong desires" to lose 20 pounds, the results will vary because of individual differences in homeostatic, instinctive, and associative systems and the underlying reward (wanting and liking) systems. A number of factors specific to energy management were identified in chapter 12. In my case, some of these factors have a greater impact on my eating behavior than my rational mind.

By contrast, the social and cognitive science model and Western-Christian model must rely on changing beliefs and trying harder, respectively. The latter leads to "fat-shaming." James Corden, host of *The Late Late Show* on American TV, responded to fat-shaming as follows (CBS, 2019):

> There's a common and insulting misconception that fat people are stupid and lazy and we're not. We get it, we know. We know that being overweight isn't good for us and I've struggled my entire life trying to manage my weight and I suck at it. . . . Let's be honest, fat-shaming is just bullying. And bullying just

> makes the problem worse. . . . [He then identifies and implicates socioeconomic and genetic factors that contribute to obesity.] . . . Defects in the leptin gene are directly linked to obesity. . . . If making fun of fat people made them lose weight there would be no fat kids in school and I would have a six pack by now.

He is correct. The conscious desire to lose weight is only one of multiple factors that were identified in chapter 12.

Many readers may also be content with the tethered rationality explanation of John Edwards's decision to have an affair during his Democratic presidential nomination campaign (chapter 12). His rational mind knew that, given his circumstances and goals, this was not a good idea. If the news of the affair leaked, the US voting public would never forgive him and it would end his candidacy. But at certain critical moments his noncognitive systems had greater pleasurable arousal associated with them than could be inhibited by the negative consequence signaled by the reasoning system. The reader may even deem the explanations implicating tethered rationality in the White House impeachment defense, climate change denial, and the behavior of teenage daughters as perfectly acceptable.

Once we have a model of any phenomenon, it should be applied consistently across the board until it fails and needs to be modified or replaced. If we keep applying the tethered rationality model to human behaviors, we quickly encounter a challenging issue that needs to be confronted and debated at a societal level. This issue arises when we consider behaviors that result in harm to others. If tethered rationality says that the reasoning mind is not in complete control, does it absolve us of behavior harmful to others?

One such behavior is sexual harassment, which can range from persistent romantic overtures, to unwanted sexual advances, to coercion and intimidation, to violent rape and abuse. This is an extremely sensitive and complex topic, but it does serve to highlight the critical issues in play. In 2015, the Canadian Armed Forces were undertaking a program review to eliminate sexual harassment in the military. General Tom Lawson, the top general of the Canadian Armed Forces, gave an interview to the Canadian Broadcasting Corporation on the topic and the progress being made (CBC, 2015). Lawson stated that the behavior "disturbs the great majority of everyone in uniform and yet, we're still dealing with it. It would be a trite answer but it is because we are biologically wired in a certain way and there will be those who believe that it is a reasonable thing to press themselves and their desires on others. It's not the way it should be." He added, "We are going to tackle that. We've been successful in tackling other cultures."

There is some ambiguity as to the range of behaviors that were under discussion in the interview. Sexual attraction may be in play at one end of the sexual harassment spectrum, but at the other end the behavior may be more about dominance, power, and rage (Phipps, Ringrose, Renold, & Jackson, 2018; Sundaram & Jackson, 2018). For the sake of the discussion, let's assume he was addressing situations where sexual attraction does come into play. In referring to sexual attraction as being "biologically wired in a certain way," the general is stating a scientific fact. St. Paul knew as much. Today we understand much of the underlying neurobiology (chapter 11). If Lawson had been talking about mice or monkeys, no one would have thought twice about it, but his acknowledgment that sexual attraction is biologically hardwired in *humans* resulted in public uproar. Canadian prime minister Stephen Harper noted that "the comments made here are offensive, they are inappropriate, they are inexplicable." The general went on to apologize for the remark and stressed that "sexual misconduct in any form, in any situation is clearly unacceptable. . . . My reference to biological attraction being a factor in sexual misconduct was by no means intended to excuse anyone from responsibility for their actions. . . . I am committed, alongside Canadian Armed Forces leadership, to addressing the issue of sexual misconduct." He was roundly condemned and would have been fired if his term was not to end within a month.

Lawson's original statement and subsequent clarifications seem consistent with the tethered rationality model being proposed here. He was not suggesting that we are driven *only* by our autonomic, instinctive, and associative systems. I assume he would recognize that we are rational agents and can choose to do otherwise. He was reaffirming the social unacceptability of sexual harassment behavior and reaffirming commitments to stamp it out. However, his remarks did suggest—in accordance with the tethered rationality model—a blended response incorporating input from both lower-level systems and the reasoning system, complicating the issue of controlling the behavior. I assume it was fear of his blended response assumption that led to the universal condemnation.

We can all agree that men or women engaged in sexual harassment are behaving in an unacceptable manner. The behavior needs to be curtailed and they need to be held accountable. The question is how best to proceed. What tools do we have at our disposal? The only solution that the cognitive and social science model recognizes involves reconstruction of societal norms, including gender roles and patriarchal hierarchies (McCarry, 2010; Phipps et al., 2018; Sundaram & Jackson, 2018). Socialization is one factor that should be considered. It may well be part of the solution, particularly if applied early, prior to maturation of the association cortex. However, to view it as the *only* solution is to be unnecessarily naive. We have examined the science behind a number of behaviors—gender identities, reciprocal cooperation and cheating, in-group/out-group bias, sexual arousal and mating, and weight management. There is no reasonable interpretation of the data in which they are *just* social constructs. The biology will not be denied. The underlying science matters. Details and nuance matter.

Admitting what is not only intuitive but biologically obvious—that the reasoning mind is tethered to simpler, nonreasoning systems—may be part of the solution here. Choosing to deny this and accept a disembodied mind floating somewhere above the body, totally unconstrained by biological principles that govern every other living organism on the planet, is no different than denying the value of vaccination or denying climate change and is driven by similar arational processes (chapter 13). You can deny the science, but it doesn't change the facts in the world. After all, the model does not cause the behavior. It just offers an explanation for it. A more accurate explanation will result in more effective remedies for unacceptable behaviors.

Why not accept the science and also accept that, unlike mice, we also have a reasoning mind, which is an equally real part of our biology? It is not the CEO in charge of all behavior, but it does have an important input into overall behavior. It can utilize various strategies to control and dampen the evolutionarily older systems. Some of these strategies were mentioned in chapter 12. In normal circumstances, with normal levels of early long-term socialization to norms and threat of potential punishment, most men and women do not engage in sexual harassment. But what of those who won't (or can't) adhere to the social norms? As long as we are committed only to changing beliefs—and it is less than effective—the only recourse is ever-greater punishment.

If we embrace the tethered rationality model and accept the underlying biological processes, we have a larger repertoire of tools at our disposal to deal with transgressors. These tools include early social conditioning (i.e., changing of beliefs) but also include classical and associative conditioning, pharmaceutical manipulation of hormones, and targeted brain stimulation techniques, among others, as our scientific knowledge advances. How, when, and under what conditions these various remedies are to be applied are societal decisions. The model has no direct input into these decisions, though insofar as these are moral decisions, and morality is a cognitive construction based on our instinctive, intuitive notions, the model could also have input into setting standards (Haidt, 2003; Hauser, 2006). As long as this is done through informed societal debate, it is probably a healthy exercise.

We must pull our heads out of the sand and face up to who and what we are—and are not—as a species.

Concluding Notes to My Colleagues

This book does not offer a better or different theory of reasoning. I use the existing cognitive accounts in the literature, albeit with the following important conceptual modifications: (1) the reasoning mind is tethered to phylogenetically older systems; (2) coherence is about feelings of rightness; and (3) neural maturation may constrain large-scale global belief revision. Rather, my purpose has been to step back and raise a number of metatheoretical issues. What is our subject matter? Is the goal to explain real-world human behavior or to discover the types of errors people make during formal logical reasoning? Surely the latter is only of interest insofar as it informs the former. For many years, I have stood up at conferences, in front of my most distinguished colleagues, and given the following example:

> My young son and daughter were squabbling and I said to the elder (my son):
> "If you want dinner tonight, then you need to stop tormenting your sister."

I would then conclude that "given he wants dinner, and draws the correct conditional inference, order will be restored." Not a single colleague has ever stood up and pointed out that not only is this false, it is absurd. Only someone who has never met human children could possibly draw such a conclusion. Drawing the logical inference may be necessary, but it is not sufficient for the desired behavior. Not even close. Whether order is restored will also involve input from lower-level systems. For example, is my son hungry? Did someone torment him at school and he's taking it out on his sister? Did his sister do something to initiate the behavior? My children taught me this.

My colleagues may point out that this is all well and fine, but science advances by simplification to bare-bones elements and the experimental manipulation of those simple elements. In the case of reason, this involves isolating coherence relations between propositions. This is exactly what the preceding example and those in chapter 7 are exploring. Once we understand the basic elements and simple connecting relations, we can use them to construct explanations of more complex phenomena (i.e., actual real-world behavior). This is a seductive but problematic argument for reasons succinctly captured by Jonathan Swift ([1726] 2012, p. 214) when he wrote that "those people suppose, that because the smallest circle hath as many degrees as the largest, therefore the Regulation and Management of the World require no more Abilities than the handling and turning of a Globe."

We are assuming that these mechanisms will scale up linearly to account for the more complex phenomena. This is possible but not inevitable or even probable. If we have a priori evidence to believe that only reason drives human behavior, then the existing cognitive science reasoning research program would be a plausible strategy. Common sense and data indicate otherwise. Reason is one of several determinants of human behavior. If this is the case, one can spend numerous lifetimes studying the minutiae of coherence relations and yet make very little headway in understanding human behavior. This has been my experience. After more than 20 years of studying the neural basis of reasoning, I realize that there is very little consequential human behavior that I can explain.

The challenge is that there is no a priori way of knowing which ideas and route will lead to deeper insights. But we have been working the existing cognitive reasoning models for the past 50 years and are no closer to explaining the behavior of teenage daughters, MAGA (Make America Great Again) neighbors, and my propensity for eating chocolate cake despite being overweight. New graduate students entering the field might take this as one indicator that the limits of the standard cognitive reasoning program have been reached and try to select different approaches for the next 50 years. The point here is not to discourage the study of coherence relations but rather to note that the subject matter of coherence relations may need to be transformed once the larger context in which it occurs is understood and incorporated into the theoretical framework. I would encourage new students entering the field to take up Jaak Panksepp's challenge and accept "one grand but empirically robust premise—that higher aspects of the human mind are still strongly linked to the basic neuropsychological processes of 'lower' animal minds" (2011, p. 1792).

In some form or another, this must be a truism. The laws of biology did not end or change with *Homo sapiens*. But many of my senior colleagues will find reasons to reject the proposition because we have been taught otherwise since graduate school. Many of us were explicitly taught that the biology does not matter, just get the computer program right and everything else will follow.[1] This determined our coursework, research methodologies, and how we thought about problems. Many subfields in cognitive science have abandoned this misguided advice, but in the reasoning world, we continue to adhere to it. There are at least three reasons for this. First, reasoning is about coherence relations between propositions, and it is easier to see propositions in high-level computer languages than in dopamine receptors and ion channels. Second, given our commitment to computer modeling, we have become enamored by the independence of the hardware and software. Reason is about software. Third, nonhuman animals do not reason, so what could we possibly learn from the neuroscience work on mice and rats? Having been educated in this tradition, I am belatedly discovering that the biology does matter. It affects how we frame questions and think about solutions.

I agree that there is something qualitatively different about reason. I agree that it is uniquely human. But from this it does not follow that it floats over the biology untethered, like the Holy Ghost. Biology, evolutionary theory, and behavioral data demand that reason be integrated into autonomic, instinctive, and associative systems. But even if we are prepared to make the effort, another difficulty is in comprehending how propositional attitudes and coherence relations can coexist and communicate with these phylogenetically older systems.

I have offered the conjecture that this may be done through feelings. There are feelings associated with believing (or not believing) propositions. There are feelings associated with coherence and incoherence relations. These are not unlike the feelings associated with hunger, the taste of chocolate cake, and lust. They provide a key or common currency for integrating reasoning with the lower-level systems and give us a more complete account of human behavior. The tethered rationality account proposed here is one attempt in this direction. It warrants detailed empirical exploration.

My colleagues in evolutionary psychology, working on massive modularity models of mind, have made substantive contributions by reminding us that instincts are real and play a central role in guiding human behavior, but it makes no sense to deny the existence of reason. If one cannot account for reason in a particular construal of evolutionary theory, one must modify the theory, not banish what needs to be explained. Ironically, despite approaching the problem from the theory of evolution, a part of biology, I believe these colleagues have been misled by an unquestioned commitment to computational ideas from the 1980s regarding levels of analysis and independence of hardware and software. This commitment is inconsistent with more modern understanding of neural computation and encourages an undifferentiated, accumulative view of modules or behavioral repertoire, disconnected from neurobiology.

I would encourage new graduate students entering the field of evolutionary psychology to take the literature on coherence relations (i.e., reasoning), comparative neuroanatomy, and neurobiology seriously and use these insights to reconceptualize evolutionary psychology so that it can accommodate not just instincts but also autonomic systems, associative systems, and reasoning systems.

To colleagues in the related fields of sociology, politics, law, and economics, I would say that the phenomena that you are trying to explain are far too rich and interesting to be captured by the standard cognitive reasoning model and the ubiquitous concept of heuristics, or the massive modularity models. These models are holding you back and preventing you from saying what you need to say. Tethered rationality provides you with more comprehensive machinery to build your explanatory models.

Concluding Notes to the General Reader

Your intuitions tell you that human behavior is a function of the reasoning mind and "animal passions." The data support your intuitions. Be wary when told that we are just reasoning minds and that the lower-level systems are illusory or irrelevant. Be equally wary when told that reason is illusory and our behavior is just a function of lower-level systems. Minimally, you could adhere to the Western-Christian model, despite its incoherent dualism. It will provide better predictive power than the current academic models. Or preferably, you could recast your intuitions in the form of tethered rationality. This will provide a more comprehensive model of human behavior, consistent with your intuitions and embedded in the neurosciences. With a more comprehensive model comes a larger repertoire of tools to explain human behavior. But tethered rationality also has some (perhaps unwelcome) far-reaching social and legal implications. Dealing with them will require difficult, intelligent, open-minded discussions about the extent to which we can and cannot use reason to construct or structure the world in any arbitrary manner we wish. Embrace this challenge knowing that having an accurate model of human behavior will always be beneficial in shaping that behavior.

Notes

Chapter 1

1. Originally, utility was postulated as a measure of pleasure or satisfaction by Jeremy Bentham and John Stuart Mill (Troyer, 2003), but modern economics has redefined it in terms of a series of alternatives correlated with relative desire or want. While the *Homo economicus* model is widely criticized in the literature, alternatives are hard to find.

2. The quality of healthcare is measured in terms of number of physicians per person (2.4 per thousand in the United States compared to 3.1 on average for OECD countries), number of hospital beds (2.6 per 1,000 population in the United States compared to 3.4 on average for OECD countries), and life expectancy at birth, which also lags behind the OECD average. However, the cancer survival rate in the United States is higher than in OECD countries. The United States does lead the world in health research, but that is a very different business than delivering healthcare (Kane, 2012).

Chapter 2

1. But to question Darwin's characterization of the higher mental faculties of man and animals is not to question the theory of evolution. It remains our best account of the origin of species and a major bulwark of much modern biology and psychology.

2. An even more recent formulation of the theory of evolution, referred to as the integrated synthesis (Noble, 2015), further incorporates epigenetic insights, that many factors not explicitly encoded in the genome, such as development, environmental chemicals, drugs, aging, and diet, affect gene *expression*. This is nicely illustrated by the following experiment involving cross-species cloning. Nuclear genomes from common carp were transplanted into nucleated eggs of goldfish (Sun et al., 2005). According to the gene-centric view of the modern synthesis, the result should be an organism determined by the species from which the genome

was taken. But the resulting fish contained characteristics of both carps (long body shape, two pairs of barbels, normal tail, and normal eyes) and goldfish (number of vertebrae), illustrating that phenotype expression is not simply determined by the genotype but is an interaction between genotype and environment. Genomes are not isolated from the organism and the environment. To date, there are very few examples of epigenetic transgenerational inheritance in plants and nonmammalian organisms but none in complex organisms.

Part II

1. The reader may wish to compare my use of the phrase with that of Dennett (1996).

Chapter 4

1. In fact, it has morphed into the term *module*, which we will discuss in chapter 9.

2. This remains a contentious claim. See Tomasello (1995) and Sampson (2005) for alternative views.

3. McDougall here is actually committing to much more than affect. He is also committing to the conscious purposefulness or goal-directedness of instinctive behavior, which was the main point of contention between him and Lorenz.

4. Deviation rates are estimated at 1:10,000 for male to female and 1:30,000 for female to male (Swaab, 2007).

5. Though focused on instincts, the ethologists were not blind to the need for learning. We have already encountered a few examples of learning. Lorenz viewed the phylogenetic history of an organism as a chain, where some links constituted instinctively specified behavior patterns and other links learning faculties. Much the same as links in a chain, instincts and learning remained distinct. The number of different types of links and their dispersion along the chain would vary among organisms (Richards, 1974).

Chapter 5

1. Hume offered the three overlapping principles of resemblance, contiguity, and cause-and-effect.

2. Interestingly, the composition of the saliva differs in the two cases.

3. This is Thorndike's law of effect (Thorndike, 1927).

4. This is a highly contentious point widely debated in the literature (Chomsky, 1959; Pinker, 1994; Putnam, 1981; Skinner, 1957).

5. For example, to say that I'm thirsty, where thirst is normally understood as a desire to drink, is just shorthand for saying that if there were a glass of water sitting in front of me, I would reach out and grasp it, bring it to my lips, and drink (Hempel, 1980).

6. This is not to say that we have full or veridical access to our mental states; but we certainly have *some* access. We can usually say with some confidence whether we are angry, in love, or believe that it is raining outside.

7. In terms of mental states, this claim can be rephrased as follows: if mental states exist, they are malleable or waxlike and can be shaped in arbitrary ways by environmental interaction.

8. If we express this in the vocabulary of mental states, we would say that our mental states are *not* waxlike and cannot be arbitrarily shaped.

9. These links also need to include logical, conceptual, and causal relations. But these take us beyond the strictly associative mechanisms being discussed here. See chapter 6.

10. David Over, personal communication. Notice that the counterfactual still is not adequate. For example, "if January had not been, then February never existed" does not imply a causal relation between January and February.

Chapter 6

1. See Goel (1995) for an extensive discussion of this issue.

2. *Intending* itself is an intentional state.

3. There is a debate in the philosophical literature as to whether only human mental states have intentionality or whether external representations such as sentences and pictures can also be intentional. This raises the distinction between "intrinsic intentionality" and "derived intentionality." Mental states are usually considered to be intrinsically intentional and external representations are thought to derive their intentionality from the mental states of their creators or observers. See Searle (1983) for a further discussion of this issue.

4. I will argue in chapter 11 that basic mental states such as fear, desire, and lust are the common heritage of at least all mammals, though the availability of more complex states such as jealousy, guilt, or shame will be restricted, perhaps just to humans.

5. I'm using the terms *propositions* and *sentences* interchangeably. Technically this is incorrect. Sentences belong to specific languages and are analyzed in terms of a subject-predicate structure. Propositions are considered more abstract constructs that can grasp the meaning of any sentence in any language and are often referred

to as having an object-concept structure. The distinction is not material for my purposes.

6. Notice that this is the same structural format used in figure 5.2 to indicate that human knowledge is not organized simply by co-occurrence but rather in terms of structured object-concept relations, so "dog" and "cat" are related by the relation "chase," where "dog" is the agent, "cat" is the object, and "chase" is a two-term relation. Similarly, the two-term relation "eat" connects the agent "dog" with the object "meat." The same pair of objects can be related by many different relations. This is not the case for contiguity or co-occurrence.

7. It was this capacity of language that convinced Descartes that man could not be explained as just a machine, unlike nonhuman animals (Descartes, [1637] 2008).

8. All these criteria need to be taken with a grain (or maybe pillar) of salt when applied to natural languages. For example, in the sentences "George loves his wife," "George loves his daughter," "George loves his mother," and "George loves his dog," the word *loves* has slightly different meanings, suggesting that the ultimate story will need to be more complex (or very different) to accommodate these nuances.

9. For a criticism of this view see, for example, Lakoff (1987).

10. The reader is referred to Penn et al. (2008) and Premack (2007) for discussion of how some of the other differences are explained by this apparatus.

11. Indeed, they cannot, as we have no mathematical construct for causation. The closest might be asymmetry.

12. By including causal relations as part of coherence relations, I'm referring to the representation of causal relations in thoughts and propositions via ordering, use of closed-form terms such as *because*, and causal inference from semantic and contextual (world) knowledge (Mulder, 2008). An example of the latter might be that when told that "the cat became frightened and ran away," you might make the causal inference that it was chased by a dog.

13. It is like being asked to prove a Euclidean postulate, but the whole point of postulates is that they are self-evident.

14. There is both behavioral and neural data to suggest that we have general-purpose inconsistency detection systems (Goel, 2019; Marinsek, Turner, Gazzaniga, & Miller, 2014).

15. I have given logical examples, but one can easily give examples from probability theory; the same point applies with respect to the intuitive axioms.

16. There are at least three major issues. First, the biggest shortcoming of the model is that the intentionality of computational systems is derived (from the intentionality of the programmer), while human intentionality is intrinsic (Searle,

1983). Second, Turing-type computation is local, being sensitive only to the syntax. It cannot consider the world and therefore cannot deal with conceptual inference (Fodor, 2000). Third, Turing computation is restricted to representational states that are syntactically disjoint and differentiated. Physical symbol systems further require representational states that are both syntactically and semantically disjoint and differentiated, and unambiguous. Human cognition relies on a much richer notion of representation, as noted by Cassirer. These issues are discussed at length in Goel (1995).

Chapter 7

1. There is a movement in the reasoning literature, referred to as the "New Paradigm," that argues that all arguments involve probabilities and that Bayesian probability theory is the most appropriate formal model for reasoning (Elqayam & Evans, 2013; Elqayam & Over, 2012).

2. Allen Newell, Cliff Shaw, and Herbert Simon (1959) were among the first to introduce the term "heuristics" to the cognitive science literature. Their usage was meant to distinguish between formal computational procedures that involved blind, systematic search of a problem space (universal methods) and procedures that exploited task-specific knowledge to circumvent the search space (heuristic methods). On this construal, heuristics are a function of an individual's world knowledge. The reader will note that this is very different than the usage of the term by Tversky and Kahneman as a structural feature of the cognitive machinery shared by everyone. The term has been used by Gerd Gigerenzer in a third way, to refer to any nonoptimal procedure for solving a problem. Gigerenzer (1996, 2015) has criticized the Tversky and Kahneman program as vacuous, and has taken his heuristics research program in a very different direction by arguing that heuristics are a feature of the system, not a flaw; identifying many specific heuristics and implementing them in computer programs; and focusing on the adaptive nature of these heuristics rather than on rationality.

3. These accounts have various origins. A historical perspective specific to logical reasoning is offered by Evans (2004, 2016) and Frankish and Evans (2009). For a more general perspective, see Chaiken and Trope (1999).

4. I am not the only one to have doubts about the importance of these problems; see, for example, Charness, Karni, & Levin (2010).

5. There may be a coherent case to make for a Sloman (1996) type model to reach down into the associative mind. It may also be necessary to appeal to the associative mind to give a full account of conceptual inference.

6. For critiques of dual mechanism theories within the cognitive framework, the reader may consult Osman (2004), Kruglanski (2013), and Melnikoff & Bargh (2018).

Chapter 8

1. It is important to separate the use of the term *induction* here from its use in mathematics. Mathematical induction, despite the name, is a species of deduction.

2. For example, Mount Everest and my neighbor share the properties of being located less than 100,000,000 miles from the sun, less than 100,000,001 miles from the sun, and so on. See also Murphy & Medin (1985).

3. Hume identified three necessary components of causation: (1) causes and effects are contiguous in space and time, (2) causes precede effects, and (3) there is a necessary link or connection between cause and effect. Hume argued that it was impossible to find the "necessary connection" in the world. One can only find a sequence of events or a constant conjunction between events. Nonetheless, we all accept the following two principles in our reasoning about the world: (1) every event has a cause and (2) like causes have like effects. But how do we know these principles are true? There is no way to establish these principles through observations. We can only establish them through inductive inference, hence the circularity in trying to explain induction via causation.

4. There's an old saying that science has grown more by what it has learned to ignore than by what it has taken into account.

Chapter 9

1. There is another important account in the literature due to Gerd Gigerenzer and his colleagues (Gigerenzer, 2015; Todd & Gigerenzer, 2000), which proposes that the mind is an "adaptive toolbox." It is sometimes lumped in with the heuristics and biases literature but is more accurately compared with the massive modularity account. It takes its cues from evolutionary considerations and focuses on adaptation rather than rationality. I have always found it difficult to assimilate because it does not differentiate between representational and nonrepresentational processes, much less between representations with and without propositional content. Because of space considerations, I omit any substantive discussions of this account.

2. This is still an unsettled issue in evolutionary biology. See, for example, Wilson and Wilson (2007).

3. The rule can be illustrated by the following story. Two brothers (of similar age and both with no offspring) are swimming when one gets caught in an undertow and is dragged underwater. Does the other one help him, even at the risk of getting caught in the undertow himself?

Given some assumptions, we can use Hamilton's rule to make a prediction. Assumptions:

$r=.5$ (siblings have a .5 coefficient of relatedness)

$B=2$ (the number of children that the beneficiary might have)

$C=2\times.05$ (two is the number of projected children for the altruist; .05 is the risk of the altruist perishing while trying to save the brother)

$rB>C=1>.1$

In this case, the rule predicts the seemingly altruistic action would be genetically profitable (for selfish, evolutionarily beneficial reasons). Notice that as the coefficient of relatedness weakens—that is, genetic distance increases (e.g., uncle/nephew $=.25$, first cousins $=.125$)—altruistic assistance will be less likely. This indeed seems to be the case. No reciprocity is required for kin-based altruism.

4. One proposal is to argue for "indirect reciprocity," whereby we help those who have helped others in the past. This can confer a valuable reputation on the cooperator, which may lead others to help them in the future. Insofar as the cost of punishing or rewarding is less than the value of the reputation, the system can be made consistent with the theory of evolution (Nowak & Sigmund, 1998, 2005).

5. For example, in 2015, Volkswagen pleaded guilty to using software to suppress emissions of nitrogen oxide during tests to get around emission pollution laws. In Canada, no criminal charges were filed (Maher, 2020). The 2008 financial crisis saw many financial institutions make risky or illegal choices, which ended up costing the US taxpayer (not the corporations) $498 billion (Harbert, 2019). Only one person was jailed in relation to the financial crisis (Eisinger, 2014).

6. Tragedy of commons problems have been characterized mathematically as follows (Dawes, 1980). Two choices are possible, D (defecting) or C (cooperating). My payoff is a function of whether I choose D or C and the number of other participants who choose D or C. There are a total of N players. D (m) is the payoff for the defectors, where m people cooperate. C (m) is the payoff to the cooperators when m players (including themselves) cooperate.

    1. D $(m)>C(m+1)$

The payoff is always higher for a defector when m other people cooperate than for the individual who becomes the $m+1$ cooperator (m ranges from zero to $N-1$).

    2. D $(0)<C(N)$

Universal cooperation results in a higher payoff than universal defection.

7. Notice that the dispute was not ultimately about fairness or cheating, because the quotas would apply to everyone.

8. Beliefs with propositional content belong to the reasoning mind, but they can be arrived at through perceptual, associative, and cognitive processes.

9. I am grateful to Larry Fiddick for discussions of this issue.

Chapter 10

Epigraph: Quoted in (DeFelipe, 2011).

1. There are some exceptions, such as kinesis and taxis in bacteria that result from metabolic changes in organisms.

2. Multicellular animals without nervous systems include sponges and trichoplax.

3. Seventh-century Indian mathematician credited with first using zero as a number (rather than just as a placeholder), incorporating the decimal point, and specifying operational rules close to modern understanding.

4. There are multiple divisions useful for various purposes.

5. EQ=actual brain size/expected size, where expected size=body size$^{2/3}\times k$, where $k=0.12$ (average index of cephalization for living mammals).

Chapter 11

1. The reader may wish to consult Daniel Dennett's (2003) notion of "heterophenom-enology."

2. It is also important to distinguish efferent autonomic motor control nerves from the afferent viscerasensory nerves.

3. The issue is whether this system can capture the obvious fact that different feelings *feel* different, not only in terms of valence and arousal but in their individual quality. Both the satisfaction of hunger pangs and sexual arousal result in feelings that are positive and high in intensity but nonetheless feel very different from each other, even unique. No one could confuse one with the other. One solution to this shortcoming is to postulate a third component that gives specific qualities to the valence. Another type of solution is to dispense with this particular framework and argue for specific neurological generators for specific feelings.

4. Moods are another type of affective state. They differ from emotions in that they are not directed, they do not typically have a specific trigger, and are long lasting.

5. In some cases, the specific nature of the emotion and/or its object may be ambiguous. For example, if I'm humiliated in front of others, I may feel anger at the person who humiliates me or shame before the audience. If a parent gives a toy to one child but not to the other, the latter may feel jealousy toward the sibling and/or anger at the parent (Elster, 1998).

6. Jaak Panksepp and Lucy Biven (2012) note that electrical stimulation in the "lower regions of the brain, such as brainstem and periaqueductal gray nucleus (PAG), induces more intense feelings with less electrical current than in other areas such as amygdala."

7. Could a circuit have evolved that results in behavior without feeling like anything? Yes. Many homeostatic processes, such as modulation of blood glucose levels, provide examples. The behavior of organisms such as the capricorn beetle probably also falls into this category. But the data suggest that much of the behavior of mammals (and perhaps all vertebrates) does not.

8. There are disputed claims that certain meditation techniques allow some individuals to "reach down" and modulate autonomic systems (Benson et al., 1982; Heathers et al., 2018; Kox et al., 2014).

9. One alternative would be to use a neutral term such as "survival circuits," but this term has already been used by LeDoux (2012) as an alternative for emotions, to signal noncommitment to feelings. My intention is simply to differentiate between full-blown human emotions involving propositional content and emotions available to creatures without propositional attitudes.

10. Sexual arousal is not only a primal emotion or instinct but also involves interoceptive homeostasis systems and exteroceptive sensory systems (Panksepp & Biven, 2012).

11. There are some who argue that sexual arousal is a drive rather than an appetite. An appetite requires an external stimulus for activation. A drive, by contrast, has an internal source, and ejaculation would be considered necessary to maintain homeostasis (Singer & Toates, 1987).

12. Females reported experiencing sexual desire on average nine times per week.

13. The story of sexual maturity is more complex in females. It begins with the maturation of the ovaries and production of estrogen and progesterone. The main subcortical neural system involved is the ventromedial hypothalamus (Panksepp & Biven, 2012; Pfaus, 2009).

14. See commentary in Searle (1983).

15. For other arguments, see Kriegel (2003) and Searle (1992).

16. For instance, in the above tiger example, if my tiger-belief is veridical, my engagement in tiger-avoiding behaviors will be appropriate and conducive to my survival. If there is a mismatch between my beliefs and the facts in the world, my actions will be generally inappropriate. If there is no tiger under my desk, but I believe there to be one, I will run away unnecessarily. If there is a tiger under my desk, but I do not believe that there is one, I will be eaten. Beliefs that are not veridical may be harmful.

17. As noted earlier, not all aspects of the autonomic system (for example, reflex arcs, glucose monitoring in normal circumstances) are associated with feelings.

Chapter 12

1. This is, of course, the American dream. Thousands of years from now, when some future historians dig up the remnants of the American empire and ask, "What did it mean to be American? What was their unique contribution to the world?," the answer may well be the belief that "anyone can be president . . . if you try hard enough." This is truly a novel idea in the intellectual and social history of the world. It does have the merit of allowing everyone to achieve their maximum potential, but taking it literally can lead not only to disappointment but to a skewed science of human behavior. Thomas Edison would've been closer to the mark if he had said "genius is 50% inspiration and 50% perspiration."

2. These models are concerned with computational limitations, but this is a separate issue.

Chapter 13

1. Trump supporter, personal communication.

2. Unfortunately, most of the lay public does not or cannot directly access or comprehend the scientific literature. They must rely on media sources, which vary greatly in competence.

3. The argument here is that prior beliefs affect rational reasoning by priming belief-biased responses. For example, a study involving artificial syllogistic reasoning reported that participants' prior ideological beliefs impaired their ability to recognize the validity of logical arguments where the conclusion was inconsistent with their beliefs, and invalidity of logical arguments where the conclusion was consistent with their beliefs (Gampa et al., 2019). This is the phenomenon of belief bias discussed in examples 3a and 3b in chapter 7, extended to ideological beliefs. Similar types of considerations and explanations will apply. See also discussions in Duckitt & Sibley (2009), Kraft et al. (2015), and Petersen, Skov, Serritzlew, & Ramsøy (2013).

4. Chapter 7 concluded that the only valid distinction that the primary literature pointed to was between deduction and induction. All reasoning examples of concern here involve induction.

5. There are of course many documented failures of the legal system attributable to judges (and juries) also being goal-directed, motivated reasoners.

6. Analytical ability was measured using the CRT task (Frederick, 2005) rather than IQ scores.

7. One "science-based" argument we are sometimes presented with is the assertion that carbon dioxide is not a pollutant but rather an essential component of the atmosphere. This is true but irrelevant. Water is essential for life—nothing lives without it—but fill your lungs with water and see what happens.

8. There are many examples to select from. Earlier, I mentioned the widely reported "death panels" supposedly mandated by the Affordable Health Care Act (Gonyea, 2017). Some Americans see parallels between the impeachment of Trump and the struggles of Jesus: "Chris . . . said he thought Democrats' pursual of Trump was 'evil'—akin to those who 'bore false witness' against Christ in order to crucify him. He doesn't watch TV, he said: He gets his news from conservative YouTube channels, especially Fox's" (Hensley-Clancy, 2019).

9. Or alternatively, we could utilize the reasoning machinery to try and sustain our current desire in light of the new evidence. For example, in the case of global warming, if our denial of the scientific evidence is motivated by the fact that we enjoy our current lifestyle and do not wish to change it, then why don't we turn the reasoning machinery to the problem of maintaining the enjoyable lifestyle while reducing carbon emissions?

10. Signs at British establishments in India during the Raj.

11. Notably, bonobos, which are as closely related to us as chimpanzees (Gibbons, 2012), do not engage in out-group violence (Wrangham, 1999).

12. Occasional coalitional violence is also reported in spotted hyenas and cheetahs (Wrangham, 1999).

13. For similar conclusions using different experimental manipulations, see Buttelmann et al. (2013), Howard et al. (2015), and Powell & Spelke (2013). See Salvadori et al. (2015) for failure to replicate Mahajan and Wynn (2012).

14. Some data suggest genetic components to biases such as political leanings (Hatemi et al., 2010).

15. Interestingly, there seem to be no good data regarding the differential entrenchment of grouping properties.

16. A few lines translated from the popular Indian patriotic song "Aye Mere Wotan Ke Logo."

17. Tweet by Trump supporter, retweeted by Trump himself (@realDonaldTrump, May 28, 2020).

18. Joan of Arc is often cited as an example.

Chapter 14

1. I'm not sure what "presuppositions" are. On the one hand, they consist of beliefs that support other beliefs, but as we keep moving downward we eventually come to what some philosophers have called the background (Searle, 1983). The background does not consist of propositions or articulated beliefs but rather nonrepresentational assumptions necessary for beliefs to be evaluated. One example that Searle offers

is that when you order a hamburger, you do not want it to be made of petrified beef or beef from cows raised on Saturn or defecated on by *Tyrannosaurus rex*. These requirements are unstated. Most of us have not even thought of these possibilities, and it would be odd to consider them part of our belief network. However, if we were presented with petrified hamburger, we would object and say that is not what we ordered. For my purposes, I will simply consider background presuppositions to be encoded into the architectural features of neural networks.

2. The word *varna* means color. There is a plausible (but contested) account supported by genetic studies in which fair-skinned Aryans from the Caucasian mountain regions entered the Indus Valley some 5,000 years ago. The region was inhabited by dark-skinned natives. The invaders (or immigrants) established the color-based *varna* system to maintain racial purity by discouraging interbreeding with the dark-skinned natives and maintain the higher-status societal positions for themselves. A recent, large-scale genetic study indicates that Brahmins in northern India have maintained a higher percentage of Aryan genetic ancestry than other *varnas* (Narasimhan et al., 2018).

3. In undertaking brain scans of normal healthy adults for studies of logical reasoning, I once scanned an adult individual only to discover that one hemisphere of their cortex was atrophied. It never developed. Yet this individual was in every way—intellectually, emotionally, socially—normal. The surviving hemisphere had fully compensated for the missing one. This speaks to the enormous plasticity of all types of immature neural systems. But there is also considerable data to show that lesions to the association cortex, even the prefrontal cortex, acquired in adulthood do leave permanent deficits (Goel et al., 2017; Goel, Marling, Raymont, Krueger, & Grafman, 2019; Goel, Grafman, Tajik, Gana, & Danto, 1997; Goel & Grafman, 2000).

4. I'm grateful to Stefan Heck for discussion of neural networks.

5. Notice that this constraint is very different from the "bounded rationality" constraint articulated by Herbert Simon. Bounded rationality is a recognition that cognitive and computational systems do not have the time and memory resources to do optimization; a "good enough" solution must suffice (Simon, 1997).

Chapter 15

1. Herbert Simon made such comments to the entering class of graduate students at Carnegie Mellon University in 1986.

Bibliography

ABC News. (2019, June 17). President Trump: 30 hours, interview with George Stephanopoulos, part 1 (No. 39). *20/20*. https://www.youtube.com/watch?v=K_IR9OFthXQ

Abizaid, A., Liu, Z.-W., Andrews, Z. B., Shanabrough, M., Borok, E., Elsworth, J. D., . . . Horvath, T. L. (2006). Ghrelin modulates the activity and synaptic input organization of midbrain dopamine neurons while promoting appetite. *The Journal of Clinical Investigation, 116*(12), 3229–3239. https://doi.org/10.1172/JCI29867

Alexander, G. M. (2003). An evolutionary perspective of sex-typed toy preferences: Pink, blue, and the brain. *Archives of Sexual Behavior, 32*(1), 7–14. https://link.springer.com/article/10.1023%2FA%3A1021833110722

Alexander, G. M. (2006). Associations among gender-linked toy preferences, spatial ability, and digit ratio: Evidence from eye-tracking analysis. *Archives of Sexual Behavior, 35*(6), 699–709. https://doi.org/10.1007/s10508-006-9038-2

Alexander, G. M., & Hines, M. (2002). Sex differences in response to children's toys in nonhuman primates (*Cercopithecus aethiops sabaeus*). *Evolution and Human Behavior, 23*(6), 467–479. https://doi.org/10.1016/S1090-5138(02)00107-1

Alexander, G. M., Wilcox, T., & Woods, R. (2009). Sex differences in infants' visual interest in toys. *Archives of Sexual Behavior, 38*(3), 427–433. https://doi.org/10.1007/s10508-008-9430-1

Allen, C. (2006). Transitive inference in animals: Reasoning or conditioned associations? In S. L. Hurley & M. Nudds (Eds.), *Rational animals?* (pp. 175–185). Oxford University Press.

Alwin, D. F., & Krosnick, J. A. (1991). Aging, cohorts, and the stability of sociopolitical orientations over the life span. *American Journal of Sociology, 97*(1), 169–195. https://doi.org/10.1086/229744

American Psychological Association. (2014, December 1). *Answers to your questions about transgender people, gender identity, and gender expression*. American Psychological Association. https://www.apa.org/topics/lgbt/transgender

Amodio, D. M. (2014). The neuroscience of prejudice and stereotyping. *Nature Reviews Neuroscience, 15*(10), 670–682. https://doi.org/10.1038/nrn3800

Amstislavskaya, T. G., & Popova, N. K. (2004). Female-induced sexual arousal in male mice and rats: Behavioral and testosterone response. *Hormones and Behavior, 46*(5), 544–550. https://doi.org/10.1016/j.yhbeh.2004.05.010

Andersen, S. L. (2003). Trajectories of brain development: Point of vulnerability or window of opportunity? *Neuroscience & Biobehavioral Reviews, 27*(1), 3–18. https://doi.org/10.1016/S0149-7634(03)00005-8

Atran, S. (2001). A cheater-detection module? Dubious interpretations of the Wason selection task and logic. *Evolution and Cognition, 7*(2), 187–192.

Austen, J. (2013). *Sense and sensibility: An annotated edition* (P. Meyer Spacks, Ed.). The Bellknap Press of Harvard University Press.

Axelrod, R., & Hamilton, W. D. (1981). The evolution of cooperation. *Science, 211*(4489), 1390–1396. https://doi.org/10.1126/science.7466396

Balas, B., & Nelson, C. A. (2010). The role of face shape and pigmentation in other-race face perception: An electrophysiological study. *Neuropsychologia, 48*(2), 498–506. https://doi.org/10.1016/j.neuropsychologia.2009.10.007

Ballinger, A., McLoughlin, L., Medbak, S., & Clark, M. (1995). Cholecystokinin is a satiety hormone in humans at physiological post-prandial plasma concentrations. *Clinical Science, 89*(4), 375–381. https://doi.org/10.1042/cs0890375

Bar-Haim, Y., Ziv, T., Lamy, D., & Hodes, R. M. (2006). Nature and nurture in own-race face processing. *Psychological Science, 17*(2), 159–163. https://doi.org/10.1111/j.1467-9280.2006.01679.x

Baron, A. S., & Dunham, Y. (2015). Representing "us" and "them": Building blocks of intergroup cognition. *Journal of Cognition and Development, 16*(5), 780–801. https://doi.org/10.1080/15248372.2014.1000459

Barrett, L. F. (2006). Are emotions natural kinds? *Perspectives on Psychological Science, 1*(1), 28–58. https://doi.org/10.1111/j.1745-6916.2006.00003.x

Barrett, L. F., Quigley, K. S., & Hamilton, P. (2016). An active inference theory of allostasis and interoception in depression. *Philosophical Transactions of the Royal Society B: Biological Sciences, 371*(1708), 20160011. https://doi.org/10.1098/rstb.2016.0011

Bartoshuk, L. M. (2000). Comparing sensory experiences across individuals: Recent psychophysical advances illuminate genetic variation in taste perception. *Chemical Senses, 25*(4), 447–460. https://doi.org/10.1093/chemse/25.4.447

Batterham, R. L., Cowley, M. A., Small, C. J., Herzog, H., Cohen, M. A., Dakin, C. L., . . . Bloom, S. R. (2002). Gut hormone PYY 3–36 physiologically inhibits food intake. *Nature, 418*(6898), 650–654. https://doi.org/10.1038/nature00887

Baumgartner, T., Heinrichs, M., Vonlanthen, A., Fischbacher, U., & Fehr, E. (2008). Oxytocin shapes the neural circuitry of trust and trust adaptation in humans. *Neuron, 58*(4), 639–650. https://doi.org/10.1016/j.neuron.2008.04.009

Beach, F. A., Jr. (1937). The neural basis of innate behavior. I. Effects of cortical lesions upon the maternal behavior pattern in the rat. *Journal of Comparative Psychology, 24*(3), 393–440. https://doi.org/10.1037/h0059606

Benson, H., Lehmann, J. W., Malhotra, M. S., Goldman, R. F., Hopkins, J., & Epstein, M. D. (1982). Body temperature changes during the practice of g Tum-mo yoga. *Nature, 295*(5846), 234–236. https://doi.org/10.1038/295234a0

Bentham, J. ([1789] 1823). *An introduction to the principles of morals and legislation.* Clarendon Press.

Berenbaum, S. M., & Hines, M. (1992). Early androgens are related to childhood sex-typed toy preferences. *Psychological Science, 3*(3), 203–206. https://doi.org/10.1111/j.1467-9280.1992.tb00028.x

Berg, J., Dickhaut, J., & McCabe, K. (1995). Trust, reciprocity, and social history. *Games and Economic Behavior, 10*(1), 122–142. https://doi.org/10.1006/game.1995.1027

Bermudez, J. L. (2002). Rationality and psychological explanation without language. In J. L. Bermudez & A. Millar (Eds.), *Reason and nature: Essays in the theory of rationality* (pp. 233–264). Oxford University Press.

Berridge, K. C. (1991). Modulation of taste affect by hunger, caloric satiety, and sensory-specific satiety in the rat. *Appetite, 16*(2), 103–120. https://doi.org/10.1016/0195-6663(91)90036-R

Berridge, K. C. (2009). "Liking" and "wanting" food rewards: Brain substrates and roles in eating disorders. *Physiology & Behavior, 97*(5), 537–550. https://doi.org/10.1016/j.physbeh.2009.02.044

Berridge, K. C., Ho, C.-Y., Richard, J. M., & DiFeliceantonio, A. G. (2010). The tempted brain eats: Pleasure and desire circuits in obesity and eating disorders. *Brain Research, 1350*, 43–64. https://doi.org/10.1016/j.brainres.2010.04.003

Berridge, K. C., & Kringelbach, M. L. (2013). Neuroscience of affect: Brain mechanisms of pleasure and displeasure. *Current Opinion in Neurobiology, 23*(3), 294–303. https://doi.org/10.1016/j.conb.2013.01.017

Berridge, K. C., & Kringelbach, M. L. (2015). Pleasure systems in the brain. *Neuron, 86*(3), 646–664. https://doi.org/10.1016/j.neuron.2015.02.018

Berthoud, H.-R., Münzberg, H., & Morrison, C. D. (2017). Blaming the brain for obesity: Integration of hedonic and homeostatic mechanisms. *Gastroenterology, 152*(7), 1728–1738. https://doi.org/10.1053/j.gastro.2016.12.050

Beteille, A. (1996). Varna and Jati. *Sociological Bulletin, 45*(1), 15–27. https://doi.org /10.1177/0038022919960102

Bindra, D. (1969). A unified interpretation of emotion and motivation. *Annals of the New York Academy of Sciences, 159*(3), 1071–1083. https://doi.org/10.1111/j.1749 -6632.1969.tb12998.x

Bindra, D. (1974). A motivational view of learning, performance, and behavior modification. *Psychological Review, 81*(3), 199–213. https://doi.org/10.1037/h0036330

Blake, P. R., McAuliffe, K., & Warneken, F. (2014). The developmental origins of fairness: The knowledge–behavior gap. *Trends in Cognitive Sciences, 18*(11), 559–561. https://doi.org/10.1016/j.tics.2014.08.003

Blakemore, C., & Cooper, G. F. (1970). Development of the brain depends on the visual environment. *Nature, 228*(5270), 477–478. https://doi.org/10.1038/228477a0

Bohnet, I., & Frey, B. S. (1999). Social distance and other-regarding behavior in dictator games: Comment. *American Economic Review, 89*(1), 335–339. https://doi.org /10.1257/aer.89.1.335

Boldog, E., Bakken, T. E., Hodge, R. D., Novotny, M., Aevermann, B. D., Baka, J., ... Tamás, G. (2018). Transcriptomic and morphophysiological evidence for a specialized human cortical GABAergic cell type. *Nature Neuroscience, 21*(9), 1185–1195. https://doi.org/10.1038/s41593-018-0205-2

Bolhuis, J. J., & Macphail, E. M. (2001). A critique of the neuroecology of learning and memory. *Trends in Cognitive Sciences, 5*(10), 426–433. https://doi.org/10.1016 /S1364-6613(00)01753-8

Bond, A. R., Kamil, A. C., & Balda, R. P. (2003). Social complexity and transitive inference in corvids. *Animal Behaviour, 65*(3), 479–487.

Booth, J. R., Koren, S. A., & Persinger, M. A. (2005). Increased feelings of the sensed presence and increased geomagnetic activity at the time of the experience during exposures to transcerebral weak complex magnetic fields. *International Journal of Neuroscience, 115*(7), 1053–1079. https://doi.org/10.1080/00207450590901521

Boyd, R., Gintis, H., Bowles, S., & Richerson, P. J. (2003). The evolution of altruistic punishment. *Proceedings of the National Academy of Sciences, 100*(6), 3531–3535. https://doi.org/10.1073/pnas.0630443100

Bozarth, M. (1994). Pleasure systems in the brain. In *Pleasure: The politics and the reality* (pp. 5–14). John Wiley & Sons.

Braine, M. D. S. (1978). On the relation between the natural logic of reasoning and standard logic. *Psychological Review, 85*(1), 1–21.

Brand, S., & Rakic, P. (1979). Genesis of the primate neostriatum: [3H]thymidine autoradiographic analysis of the time of neuron origin in the rhesus monkey. *Neuroscience, 4*(6), 767–778. https://doi.org/10.1016/0306-4522(79)90005-8

Brasted, P. J., Bussey, T. J., Murray, E. A., & Wise, S. P. (2003). Role of the hippocampal system in associative learning beyond the spatial domain. *Brain, 126*(5), 1202–1223. https://doi.org/10.1093/brain/awg103

Breedlove, S. M. (1994). Sexual differentiation of the human nervous system. *Annual Review of Psychology, 45*, 389–418. https://doi.org/10.1146/annurev.ps.45.020194 .002133

Brentano, F. ([1874] 2012). *Psychology from an empirical standpoint.* Routledge.

Briscoe, S. D., & Ragsdale, C. W. (2018). Homology, neocortex, and the evolution of developmental mechanisms. *Science, 362*(6411), 190–193. https://doi.org/10.1126 /science.aau3711

Brosnan, S. F., Silk, J. B., Henrich, J., Mareno, M. C., Lambeth, S. P., & Schapiro, S. J. (2009). Chimpanzees (*Pan troglodytes*) do not develop contingent reciprocity in an experimental task. *Animal Cognition, 12*(4), 587–597. https://doi.org/10.1007/s10071 -009-0218-z

Brosnan, S. F., de Waal, F. B. M., & Proctor, D. (2014). Reciprocity in primates. In S. D. Preston (Ed.), *The interdisciplinary science of consumption* (pp. 3–32). MIT Press.

Brown, C. M., Greenwood, D. R., Kalyniuk, J. E., Braman, D. R., Henderson, D. M., Greenwood, C. L., & Basinger, J. F. (2020). Dietary palaeoecology of an early Cretaceous armoured dinosaur (Ornithischia; Nodosauridae) based on floral analysis of stomach contents. *Royal Society Open Science, 7*(6), 200305. https://doi.org/10.1098/rsos.200305

Brown, D. E. (2004). Human universals, human nature & human culture. *Daedalus, 133*(4), 47–54. https://doi.org/10.1162/0011526042365645

Bucciol, A., & Piovesan, M. (2011). Luck or cheating? A field experiment on honesty with children. *Journal of Economic Psychology, 32*(1), 73–78. https://doi.org/10.1016/j .joep.2010.12.001

Buchan, N. R., Croson, R. T. A., & Dawes, R. M. (2002). Swift neighbors and persistent strangers: A cross-cultural investigation of trust and reciprocity in social exchange. *American Journal of Sociology, 108*(1), 168–206. https://doi.org/10.1086/344546

Buller, D. J. (2006). *Adapting minds: Evolutionary psychology and the persistent quest for human nature.* MIT Press.

Burgdorf, J., Knutson, B., & Panksepp, J. (2000). Anticipation of rewarding electrical brain stimulation evokes ultrasonic vocalization in rats. *Behavioral Neuroscience, 114*(2), 320–327. https://doi.org/10.1037/0735-7044.114.2.320

Burgdorf, J., Kroes, R. A., Moskal, J. R., Pfaus, J. G., Brudzynski, S. M., & Panksepp, J. (2008). Ultrasonic vocalizations of rats (*Rattus norvegicus*) during mating, play, and aggression: Behavioral concomitants, relationship to reward, and self-administration of playback. *Journal of Comparative Psychology, 122*(4), 357–367. https://doi.org/10 .1037/a0012889

Burgdorf, J., Panksepp, J., & Moskal, J. R. (2011). Frequency-modulated 50kHz ultra-sonic vocalizations: A tool for uncovering the molecular substrates of positive affect. *Neuroscience & Biobehavioral Reviews*, *35*(9), 1831–1836. https://doi.org/10.1016/j.neubiorev.2010.11.011

Burgdorf, J., Wood, P. L., Kroes, R. A., Moskal, J. R., & Panksepp, J. (2007). Neuro-biology of 50-kHz ultrasonic vocalizations in rats: Electrode mapping, lesion, and pharmacology studies. *Behavioural Brain Research*, *182*(2), 274–283. https://doi.org/10.1016/j.bbr.2007.03.010

Butler, P. M., Mcnamara, P., & Durso, R. (2010). Deficits in the automatic activation of religious concepts in patients with Parkinson's disease. *Journal of the International Neuropsychological Society*, *16*(2), 252–261. https://doi.org/10.1017/S1355617709991202

Buttelmann, D., Zmyj, N., Daum, M., & Carpenter, M. (2013). Selective imitation of in-group over out-group members in 14-month-old infants. *Child Development*, *84*(2), 422–428. https://doi.org/10.1111/j.1467-8624.2012.01860.x

Cabrera, A. (1919). *Genera mammalium*. Museo Nacional de ciencas naturales; Biodiversity Heritage Library. https://doi.org/10.5962/bhl.title.46757

Cahill, L. (2006). Why sex matters for neuroscience. *Nature Reviews Neuroscience*, *7*(6), 477–484. https://doi.org/10.1038/nrn1909

Campbell, K. (2005). Cortical neuron specification: It has its time and place. *Neuron*, *46*(3), 373–376. https://doi.org/10.1016/j.neuron.2005.04.014

Carlisle, E., & Shafir, E. (2005). Questioning the cheater-detection hypothesis: New studies with the selection task. *Thinking & Reasoning*, *11*(2), 97–122. https://doi.org/10.1080/13546780442000079

Carney, S. (2017). *What doesn't kill us: How freezing water, extreme altitude, and environmental conditioning will renew our lost evolutionary strength*. Rodale. https://books.google.ca/books?id=cjp8DQAAQBAJ&lpg=PP1&pg=PR4#v=onepage&q&f=false

Carter, C. S., Witt, D. M., Kolb, B., & Whishaw, I. Q. (1982). Neonatal decortication and adult female sexual behavior. *Physiology & Behavior*, *29*(4), 763–766. https://doi.org/10.1016/0031-9384(82)90254-2

Carter, G., & Wilkinson, G. (2013). Does food sharing in vampire bats demonstrate reciprocity? *Communicative & Integrative Biology*, *6*(6), e25783-1–e25783-6. https://doi.org/10.4161/cib.25783

Casey, B. J., Tottenham, N., Liston, C., & Durston, S. (2005). Imaging the developing brain: What have we learned about cognitive development? *Trends in Cognitive Sciences*, *9*(3), 104–110. https://doi.org/10.1016/j.tics.2005.01.011

Cassirer, E. (1944). *An essay on man: An introduction to a philosophy of human culture*. Yale University Press.

CBC. (2015, June 16). Gen. Tom Lawson interview. CBC. https://www.cbc.ca/player/play/2669648128

CBS. (2019, September 12). James Corden responds to Bill Maher's fat shaming take [Interview]. *The Late Late Show*. CBS. https://www.youtube.com/watch?v=Ax1U04c4gaw

Chaiken, S., & Trope, Y. (1999). *Dual-process theories in social psychology*. Guilford Press.

Charness, G., Karni, E., & Levin, D. (2010). On the conjunction fallacy in probability judgment: New experimental evidence regarding Linda. *Games and Economic Behavior, 68*(2), 551–556. https://doi.org/10.1016/j.geb.2009.09.003

Cheney, D. L., & Seyfarth, R. M. (1987). The influence of intergroup competition on the survival and reproduction of female vervet monkeys. *Behavioral Ecology and Sociobiology, 21*(6), 375–386. https://doi.org/10.1007/BF00299932

Chiu, A. (2020, February 25). Rush Limbaugh on coronavirus: "The common cold" that's being "weaponized" against Trump. *The Washington Post*. https://www.washingtonpost.com/nation/2020/02/25/limbaugh-coronavirus-trump/

Chomsky, N. (1959). A review of B. F. Skinner's verbal behavior. *Language, 35*(1), 26–58.

Chomsky, N. (1972). *Language and mind*. Harcourt Brace Jovanovich.

Clark, J. T. (2013). Component analysis of male sexual behavior. In P. M. Conn (Ed.), *Paradigms for the study of behavior* (pp. 32–53). Academic Press.

Clark, R. E. (2004). The classical origins of Pavlov's conditioning. *Integrative Physiological & Behavioral Science, 39*(4), 279–294. https://doi.org/10.1007/BF02734167

Clausing, K. A. (2016). *The effect of profit shifting on the corporate tax base in the United States and beyond* (SSRN Scholarly Paper 2685442). Social Science Research Network. https://papers.ssrn.com/abstract=2685442

Clemens, L. G., & Gladue, B. A. (1978). Feminine sexual behavior in rats enhanced by prenatal inhibition of androgen aromatization. *Hormones and Behavior, 11*(2), 190–201. https://doi.org/10.1016/0018-506X(78)90048-X

Clemens, L. G., Gladue, B. A., & Coniglio, L. P. (1978). Prenatal endogenous androgenic influences on masculine sexual behavior and genital morphology in male and female rats. *Hormones and Behavior, 10*(1), 40–53. https://doi.org/10.1016/0018-506X(78)90023-5

Clutton-Brock, T. H. (1999). Selfish sentinels in cooperative mammals. *Science, 284*(5420), 1640–1644. https://doi.org/10.1126/science.284.5420.1640

Cochrane, E., Lipton, E., & Cameron, C. (2020, January 16). G.A.O. report says Trump administration broke law in withholding Ukraine aid. *The New York Times*. https://www.nytimes.com/2020/01/16/us/politics/gao-trump-ukraine.html

Colibazzi, T., Posner, J., Wang, Z., Gorman, D., Gerber, A., Yu, S., . . . Peterson, B. S. (2010). Neural systems subserving valence and arousal during the experience of induced emotions. *Emotion, 10*(3), 377–389. https://doi.org/10.1037/a0018484

Collins, C. E., Turner, E. C., Sawyer, E. K., Reed, J. L., Young, N. A., Flaherty, D. K., & Kaas, J. H. (2016). Cortical cell and neuron density estimates in one chimpanzee hemisphere. *Proceedings of the National Academy of Sciences, 113*(3), 740–745. https://doi.org/10.1073/pnas.1524208113

Congregation for Catholic Education for Educational Institutions. (2019). *"Male and female he created them": Toward a path of dialogue on the question of gender theory in education*. Catholic Truth Society. https://www.vatican.va/roman_curia/congregations/ccatheduc/documents/rc_con_ccatheduc_doc_20190202_maschio-e-femmina_en.pdf

Connor, J. M., & Serbin, L. A. (1977). Behaviorally based masculine- and feminine-activity-preference scales for preschoolers: Correlates with other classroom behaviors and cognitive tests. *Child Development, 48*(4), 1411–1416. https://doi.org/10.2307/1128500

Considine, R. V., Sinha, M. K., Heiman, M. L., Kriauciunas, A., Stephens, T. W., Nyce, M. R., . . . Caro, J. F. (1996). Serum immunoreactive-leptin concentrations in normal-weight and obese humans. *New England Journal of Medicine, 334*(5), 292–295. https://doi.org/10.1056/NEJM199602013340503

Coolidge, F. L., Thede, L. L., & Young, S. E. (2002). The heritability of gender identity disorder in a child and adolescent twin sample. *Behavior Genetics, 32*(4), 251–257. https://doi.org/10.1023/A:1019724712983

Cosmides, L. (1989). The logic of social exchange: Has natural selection shaped how humans reason? Studies with the Wason selection task [see comments]. *Cognition, 31*(3), 187–276.

Cosmides, L., & Tooby, J. (1994a). Better than rational: Evolutionary psychology and the invisible hand. *American Economic Review, 84*(2), 327–332. https://www.jstor.org/stable/2117853

Cosmides, L., & Tooby, J. (1994b). Origins of domain specificity: The evolution of functional organization. In L. Hirschfeld & S. Gelman (Eds.), *Mapping the mind: Domain specificity in cognition and culture*. Cambridge University Press.

Cosmides, L., & Tooby, J. (2013). Unraveling the enigma of human intelligence: Evolutionary psychology and the multimodular mind. In R. J. Sternberg & J. C. Kaufman (Eds.), *The evolution of intelligence* (pp. 145–198). Psychology Press.

Cosmides, L., Tooby, J., & Kurzban, R. (2003). Perceptions of race. *Trends in Cognitive Sciences, 7*(4), 173–179. https://doi.org/10.1016/S1364-6613(03)00057-3

Costa, R., & Goldstein, A. (2017, January 15). Trump vows "insurance for everybody" in Obamacare replacement plan. *The Washington Post*. https://www.washingtonpost

.com/politics/trump-vows-insurance-for-everybody-in-obamacare-replacement-plan /2017/01/15/5f2b1e18-db5d-11e6-ad42-f3375f271c9c_story.html

Coupé, P., Catheline, G., Lanuza, E., & Manjón, J. V. (2017). Towards a unified analysis of brain maturation and aging across the entire lifespan: An MRI analysis. *Human Brain Mapping*, *38*(11), 5501–5518. https://doi.org/10.1002/hbm.23743

Craig, A. D. (2002). How do you feel? Interoception: The sense of the physiological condition of the body. *Nature Reviews Neuroscience*, *3*(8), 655–666. https://doi.org/10 .1038/nrn894

Craig, A. D. (2009). How do you feel—now? The anterior insula and human aware-ness. *Nature Reviews Neuroscience*, *10*(1), 59–70. https://doi.org/10.1038/nrn2555

Craig, W. (1917). Appetites and aversions as constituent instincts. *Proceedings of the National Academy of Sciences USA*, *3*(12), 685–688. https://doi.org/10.1073/pnas.3.12 .685

Damasio, A., & Carvalho, G. B. (2013). The nature of feelings: Evolutionary and neurobiological origins. *Nature Reviews Neuroscience*, *14*(2), 143–152. https://doi.org /10.1038/nrn3403

Daniszewski, J. (2002, September 5). Hanging with the cranes. *Los Angeles Times*. https://www.latimes.com/archives/la-xpm-2002-sep-05-fg-cranes5-story.html

Danziger, S., Levav, J., & Avnaim-Pesso, L. (2011). Extraneous factors in judicial deci-sions. *Proceedings of the National Academy of Sciences*, *108*(17), 6889–6892. https://doi .org/10.1073/pnas.1018033108

Darwin, C. ([1871] 1896). *The descent of man and selection in relation to sex*. D. Appleton.

Darwin, C. ([1859] 1995). *The origin of species* (new ed.). Gramercy.

Dauphin, Y. N., Fan, A., Auli, M., & Grangier, D. (2017). Language modeling with gated convolutional networks. *Proceedings of the 34th International Conference on Machine Learning, Vol. 70*, ICML'17 (pp. 933–941). JMLR.org.

Davenport, J., Jones, T. T., Work, T. M., & Balazs, G. H. (2015). Topsy-turvy: Turning the counter-current heat exchange of leatherback turtles upside down. *Biology Let-ters*, *11*(10), 20150592. https://doi.org/10.1098/rsbl.2015.0592

Davids, K., & Baker, J. (2007). Genes, environment and sport performance. *Sports Medicine*, *37*(11), 961–980. https://doi.org/10.2165/00007256-200737110-00004

Davidson, D. (2004). *Problems of rationality*. Clarendon Press.

Dawes, R. M. (1980). Social dilemmas. *Annual Review of Psychology*, *31*(1), 169–193. https://doi.org/10.1146/annurev.ps.31.020180.001125

DeBernardo, F. (2019, June 10). New ways ministry responds to new Vatican docu-ment on gender identity. New Ways Ministry. https://www.newwaysministry.org

/2019/06/10/new-ways-ministry-responds-to-new-vatican-document-on-gender
-identity/

Deer, B. (2011, January 11). How the vaccine crisis was meant to make money. *BMJ,*
342, c5258.

DeFelipe, J. (2011). The evolution of the brain, the human nature of cortical circuits,
and intellectual creativity. *Frontiers in Neuroanatomy, 5.* https://doi.org/10.3389/fnana
.2011.00029

Defelipe, J., Alonso-Nanclares, L., & Arellano, J. (2002). Microstructure of the neocortex: Comparative aspects. *Journal of Neurocytology, 31,* 299–316. https://doi.org/10
.1023/A:1024130211265

Delius, J. D., & Siemann, M. (1998). Transitive responding in animals and humans:
Exaptation rather than adaption? *Behavioral Processes, 42,* 107–137.

Denault, L. K., & McFarlane, D. A. (1995). Reciprocal altruism between male vampire
bats, *Desmodus rotundus. Animal Behaviour, 49*(3), 855–856. https://doi.org/10.1016
/0003-3472(95)80220-7

De Neys, W. (2006a). Automatic–heuristic and executive–analytic processing during
reasoning: Chronometric and dual-task considerations. *The Quarterly Journal of Experimental Psychology, 59*(6), 1070–1100. https://doi.org/10.1080/02724980543000123

De Neys, W. (2006b). Dual processing in reasoning: Two systems but one reasoner. *Psychological Science, 17*(5), 428–433. https://doi.org/10.1111/j.1467-9280.2006.01723.x

De Neys, W. (2017). Bias, conflict and fast logic: Towards a hybrid dual process
future? In W. De Neys (Ed.), *Dual Process Theory 2.0* (pp. 47–65). Routledge.

De Neys, W., & Pennycook, G. (2019). Logic, fast and slow: Advances in dual-process
theorizing. *Current Directions in Psychological Science, 28*(5), 503–509. https://doi.org
/10.1177/0963721419855658

Denis, R. G. P., Joly-Amado, A., Webber, E., Langlet, F., Schaeffer, M., Padilla, S. L., . . .
Luquet, S. (2015). Palatability can drive feeding independent of AgRP neurons. *Cell
Metabolism, 22*(4), 646–657. https://doi.org/10.1016/j.cmet.2015.07.011

Dennett, D. C. (1996). *Kinds of minds: Toward an understanding of consciousness.* Basic
Books.

Dennett, D. C. (2003). Who's on first? Heterophenomenology explained. *Journal of
Consciousness Studies, 10*(9–10), 19–30.

Descartes, R. ([1637] 2008). *A discourse on the method: Of correctly conducting one's
reason and seeking truth in the sciences.* Oxford University Press.

Deshpande, A. (2008). Quest for equality: Affirmative action in India. *Indian Journal
of Industrial Relations, 44*(2), 154–163.

Dessens, A. B., Cohen-Kettenis, P. T., Mellenbergh, G. J., Poll, N. V. D., Koppe, J. G., & Boer, K. (1999). Prenatal exposure to anticonvulsants and psychosexual development. *Archives of Sexual Behavior, 28*(1), 31–44. https://doi.org/10.1023/A:1018789521375

de Vries, G. J., Fields, C. T., Peters, N. V., Whylings, J., & Paul, M. J. (2014). Sensitive periods for hormonal programming of the brain. In S. L. Andersen & D. S. Pine (Eds.), *The neurobiology of childhood* (pp. 79–108). Springer. https://doi.org/10.1007/7854_2014_286

de Waal, F. B. M. (1997). The chimpanzee's service economy: Food for grooming. *Evolution and Human Behavior, 18*(6), 375–386. https://doi.org/10.1016/S1090-5138(97)00085-8

de Waal, F. B. M., & Brosnan, S. F. (2006). Simple and complex reciprocity in primates. In P. M. Kappeler & C. P. van Schaik (Eds.), *Cooperation in primates and humans: Mechanisms and evolution* (pp. 85–105). Springer. https://doi.org/10.1007/3-540-28277-7_5

Dobzhansky, T. (1950). Genetics of natural populations. XIX. Origin of heterosis through natural selection in populations of Drosophila pseudoobscura. *Genetics, 35*(3), 288–302.

Donahue, C. J., Glasser, M. F., Preuss, T. M., Rilling, J. K., & Van Essen, D. C. (2018). Quantitative assessment of prefrontal cortex in humans relative to nonhuman primates. *Proceedings of the National Academy of Sciences, 115*(22), E5183–E5192. https://doi.org/10.1073/pnas.1721653115

Douglass, A. M., Kucukdereli, H., Ponserre, M., Markovic, M., Gründemann, J., Strobel, C., . . . Klein, R. (2017). Central amygdala circuits modulate food consumption through a positive-valence mechanism. *Nature Neuroscience, 20*(10), 1384–1394. https://doi.org/10.1038/nn.4623

Doyle, A. C. ([1892] 2019). *The adventures of Sherlock Holmes (100th anniversary edition): With 100 original illustrations* (illustrated ed.). SeaWolf Press.

Dreber, A., Rand, D. G., Fudenberg, D., & Nowak, M. A. (2008). Winners don't punish. *Nature, 452*(7185), 348–351. https://doi.org/10.1038/nature06723

Duckitt, J., & Sibley, C. G. (2009). A dual-process motivational model of ideology, politics, and prejudice. *Psychological Inquiry, 20*(2–3), 98–109. https://doi.org/10.1080/10478400903028540

Dufort, R. H., & Kimble, G. A. (1956). Changes in response strength with changes in the amount of reinforcement. *Journal of Experimental Psychology, 51*(3), 185–191. https://doi.org/10.1037/h0041095

Dugatkin, L. A. (1998). A model of coalition formation in animals. *Proceedings of the Royal Society B: Biological Sciences, 265*(1410), 2121–2125. https://doi.org/10.1098/rspb.1998.0548

Dupuis-Desormeaux, M., Kaaria, T. N., Mwololo, M., Davidson, Z., & MacDonald, S. E. (2018). A ghost fence-gap: Surprising wildlife usage of an obsolete fence crossing. *PeerJ*, *6*, e5950. https://doi.org/10.7717/peerj.5950

Durkheim, E. ([1895] 2014). *The rules of sociological method and selected texts on sociology and its method*. Simon and Schuster.

Edwards, D. A., & Burge, K. G. (1971). Early androgen treatment and male and female sexual behavior in mice. *Hormones and Behavior*, *2*(1), 49–58. https://doi.org/10.1016/0018-506X(71)90037-7

Egas, M., & Riedl, A. (2008). The economics of altruistic punishment and the maintenance of cooperation. *Proceedings of the Royal Society B: Biological Sciences*, *275*(1637), 871–878. https://doi.org/10.1098/rspb.2007.1558

Eimontaite, I., Nicolle, A., Schindler, I. C., & Goel, V. (2013). The effect of partner-directed emotion in social exchange decision-making. *Frontiers in Psychology*, *4*. https://doi.org/10.3389/fpsyg.2013.00469

Eimontaite, I., Schindler, I., De Marco, M., Duzzi, D., Venneri, A., & Goel, V. (2019). Left amygdala and putamen activation modulate emotion driven decisions in the iterated prisoner's dilemma game. *Frontiers in Neuroscience*, *13*, 741. https://doi.org/10.3389/fnins.2019.00741

Eisinger, J. (2014, April 30). Why only one top banker went to jail for the financial crisis. *New York Times Magazine*. https://www.nytimes.com/2014/05/04/magazine/only-one-top-banker-jail-financial-crisis.html

Ekman, P. (1993). Facial expression and emotion. *American Psychologist*, *48*(4), 384–392. https://doi.org/10.1037/0003-066X.48.4.384

Eliot, L. (2011). The trouble with sex differences. *Neuron*, *72*(6), 895–898. https://doi.org/10.1016/j.neuron.2011.12.001

Elqayam, S., & Evans, J. St. B. T. (2013). Rationality in the new paradigm: Strict versus soft Bayesian approaches. *Thinking & Reasoning*, *19*(3–4), 453–470. https://doi.org/10.1080/13546783.2013.834268

Elqayam, S., & Over, D. (2012). Probabilities, beliefs, and dual processing: The paradigm shift in the psychology of reasoning. *Mind & Society*, *11*(1), 27–40. https://doi.org/10.1007/s11299-012-0102-4

Elster, J. (1998). Emotions and economic theory. *Journal of Economic Literature*, *36*(1), 47–74.

Epley, N., & Gilovich, T. (2016). The mechanics of motivated reasoning. *Journal of Economic Perspectives*, *30*(3), 133–140. https://doi.org/10.1257/jep.30.3.133

Epstein, L. H., Truesdale, R., Wojcik, A., Paluch, R. A., & Raynor, H. A. (2003). Effects of deprivation on hedonics and reinforcing value of food. *Physiology & Behavior*, *78*(2), 221–227. https://doi.org/10.1016/S0031-9384(02)00978-2

Eroglu, C., & Barres, B. A. (2010). Regulation of synaptic connectivity by glia. *Nature, 468*(7321), 223–231. https://doi.org/10.1038/nature09612

Etcoff, N. (2011). *Survival of the prettiest: The science of beauty.* Knopf Doubleday.

Evans, J. St. B. T. (2004). History of the dual process theory of reasoning. In K. I. Manktelow & Man Cheung Chung (Eds.), *Psychology of reasoning: Theoretical and historical perspectives* (pp. 241–266). Psychology Press. https://doi.org/10.4324/9780203506936-12

Evans, J. St. B. T. (2016). Reasoning, biases and dual processes: The lasting impact of Wason (1960). *The Quarterly Journal of Experimental Psychology, 69*(10), 2076–2092. https://doi.org/10.1080/17470218.2014.914547

Evans, J. St. B. T., Barston, J., & Pollard, P. (1983). On the conflict between logic and belief in syllogistic reasoning. *Memory & Cognition, 11,* 295–306.

Evans, J. St. B. T., & Curtis-Holmes, J. (2005). Rapid responding increases belief bias: Evidence for the dual-process theory of reasoning. *Thinking & Reasoning, 11*(4), 382–389. https://doi.org/10.1080/13546780542000005

Evans, J. St. B. T., & Over, D. E. (1996). *Rationality and reasoning.* Psychology Press.

Evans, J. St. B. T., & Stanovich, K. E. (2013). Dual-process theories of higher cognition: Advancing the debate. *Perspectives on Psychological Science, 8*(3), 223–241. https://doi.org/10.1177/1745691612460685

Eyal, G., Verhoog, M. B., Testa-Silva, G., Deitcher, Y., Lodder, J. C., Benavides-Piccione, R., . . . Segev, I. (2016). Unique membrane properties and enhanced signal processing in human neocortical neurons. *ELife, 5.* https://doi.org/10.7554/eLife.16553

Farooqi, I. S., Jebb, S. A., Langmack, G., Lawrence, E., Cheetham, C. H., Prentice, A. M., . . . O'Rahilly, S. (1999). Effects of recombinant leptin therapy in a child with congenital leptin deficiency. *New England Journal of Medicine, 341*(12), 879–884. https://doi.org/10.1056/NEJM199909163411204

Feagin, J. R. (1972). America's welfare stereotypes. *Social Science Quarterly, 52*(4), 921–933.

Febo, M., Felix-Ortiz, A. C., & Johnson, T. R. (2010). Inactivation or inhibition of neuronal activity in the medial prefrontal cortex largely reduces pup retrieval and grouping in maternal rats. *Brain Research, 1325,* 77–88. https://doi.org/10.1016/j.brainres.2010.02.027

Fehr, E., & Fischbacher, U. (2003). The nature of human altruism. *Nature, 425*(6960), 785–791. https://doi.org/10.1038/nature02043

Fehr, E., & Fischbacher, U. (2004). Third-party punishment and social norms. *Evolution and Human Behavior, 25*(2), 63–87. https://doi.org/10.1016/S1090-5138(04)00005-4

Fehr, E., & Gächter, S. (2000). Cooperation and punishment in public goods experiments. *American Economic Review, 90*(4), 980–994. https://doi.org/10.1257/aer.90.4.980

Festinger, L. (1957). *A theory of cognitive dissonance*. Stanford University Press.

Fiddes, I. T., Lodewijk, G. A., Mooring, M., Bosworth, C. M., Ewing, A. D., Mantalas, G. L., . . . Haussler, D. (2018). Human-specific NOTCH2NL genes affect notch signaling and cortical neurogenesis. *Cell, 173*(6), 1356–1369.e22. https://doi.org/10.1016/j.cell.2018.03.051

Fischbacher, U., Gächter, S., & Fehr, E. (2001). Are people conditionally cooperative? Evidence from a public goods experiment. *Economics Letters, 71*(3), 397–404. https://doi.org/10.1016/S0165-1765(01)00394-9

Fisher, T. D., Moore, Z. T., & Pittenger, M.-J. (2012). Sex on the brain? An examination of frequency of sexual cognitions as a function of gender, erotophilia, and social desirability. *Journal of Sex Research, 49*(1), 69–77. https://doi.org/10.1080/00224499.2011.565429

Fletcher, R. (1957). *Instinct in man; in the light of recent work in comparative psychology*. International Universities Press.

Fodor, J. A. (1975). *The language of thought*. Harvard University Press.

Fodor, J. A. (1980). Methodological solipsism considered as a research strategy in cognitive psychology. *Behavioral and Brain Sciences, 3*(1), 63–73. https://doi.org/10.1017/S0140525X00001771

Fodor, J. A. (2000). *The mind doesn't work that way: The scope and limits of computational psychology*. MIT Press.

Fodor, J. A., & Pylyshyn, Z. W. (1988). Connectionism and cognitive architecture: A critical analysis. *Cognition, 28*(1–2), 3–71.

Forsythe, R., Horowitz, J. L., Savin, N. E., & Sefton, M. (1994). Fairness in simple bargaining experiments. *Games and Economic Behavior, 6*(3), 347–369. https://doi.org/10.1006/game.1994.1021

Fowler, J. H. (2005). Altruistic punishment and the origin of cooperation. *Proceedings of the National Academy of Sciences, 102*(19), 7047–7049. https://doi.org/10.1073/pnas.0500938102

Frankish, K., & Evans, J. St. B. T. (2009). The duality of mind: An historical perspective. In J. Evans & K. Frankish (Eds.), *In two minds: Dual processes and beyond* (pp. 1–30). Oxford University Press. https://doi.org/10.1093/acprof:oso/9780199230167.003.0001

Frederick, S. (2005). Cognitive reflection and decision making. *Journal of Economic Perspectives, 19*(4), 25–42. https://doi.org/10.1257/089533005775196732

Frederick, S., Loewenstein, G., & O'Donoghue, T. (2002). Time discounting and time preference: A critical review. *Journal of Economic Literature, 40*(2), 351–401. https://doi.org/10.1257/002205102320161311

Freeman, W., & Watts, J. W. (1942). Prefrontal lobotomy. *Bulletin of the New York Academy of Medicine, 18*(12), 794–812.

Frisch, K. von. (1962). *About biology.* Oliver & Boyd.

Frith, C., & Johnstone, E. C. (2003). *Schizophrenia: A very short introduction.* Oxford University Press.

Frixione, M., & Lieto, A. (2014). Concepts, perception and the dual process theories of mind. *Baltic International Yearbook of Cognition, Logic and Communication, 9*(1). https://doi.org/10.4148/1944-3676.1084

Gächter, S., & Falk, A. (2002). Reputation and reciprocity: Consequences for the labour relation. *The Scandinavian Journal of Economics, 104*(1), 1–26. https://doi.org/10.1111/1467-9442.00269

Galilei, G. ([1638] 1954). *Two new sciences* (H. Crew & A. de Salvio, Trans.). Dover Publications.

Galperti, S. (2019). Persuasion: The art of changing worldviews. *American Economic Review, 109*(3), 996–1031. https://doi.org/10.1257/aer.20161441

Gampa, A., Wojcik, S. P., Motyl, M., Nosek, B. A., & Ditto, P. H. (2019). (Ideo)logical reasoning: Ideology impairs sound reasoning. *Social Psychological and Personality Science, 10*(8), 1075–1083. https://doi.org/10.1177/1948550619829059

Garcia, J., Ervin, F. R., & Koelling, R. A. (1966). Learning with prolonged delay of reinforcement. *Psychonomic Science, 5*(3), 121–122. https://doi.org/10.3758/BF03328311

Garcia, J., & Koelling, R. A. (1966). Relation of cue to consequence in avoidance learning. *Psychonomic Science, 4*(1), 123–124. https://doi.org/10.3758/BF03342209

Garey, J., Goodwillie, A., Frohlich, J., Morgan, M., Gustafsson, J.-A., Smithies, O., . . . Pfaff, D. W. (2003). Genetic contributions to generalized arousal of brain and behavior. *Proceedings of the National Academy of Sciences, 100*(19), 11019–11022. https://doi.org/10.1073/pnas.1633773100

Georgiadis, J. R., & Kringelbach, M. L. (2012). The human sexual response cycle: Brain imaging evidence linking sex to other pleasures. *Progress in Neurobiology, 98*(1), 49–81. https://doi.org/10.1016/j.pneurobio.2012.05.004

Ghosh, A., Antonini, A., McConnell, S. K., & Shatz, C. J. (1990). Requirement for subplate neurons in the formation of thalamocortical connections. *Nature, 347*(6289), 179–181. https://doi.org/10.1038/347179a0

Gibbons, A. (2012, June 13). Bonobos join chimps as closest human relatives. *Science.* AAAS. https://www.sciencemag.org/news/2012/06/bonobos-join-chimps-closest-human-relatives

Gibbs, J., Young, R. C., & Smith, G. P. (1973). Cholecystokinin elicits satiety in rats with open gastric fistulas. *Nature, 245*(5424), 323–325. https://doi.org/10.1038/245323a0

Gigerenzer, G. (1996). On narrow norms and vague heuristics: A reply to Kahneman and Tversky. *Psychological Review, 103*(3), 592–596. https://doi.org/10.1037/0033-295X.103.3.592

Gigerenzer, G. (2015). *Simply rational: Decision making in the real world.* Oxford University Press.

Gigerenzer, G., Gaissmaier, W., Kurz-Milcke, E., Schwartz, L. M., & Woloshin, S. (2007). Helping doctors and patients make sense of health statistics. *Psychological Science in the Public Interest, 8*(2), 53–96. https://doi.org/10.1111/j.1539-6053.2008.00033.x

Gilbert, S. J., Swencionis, J. K., & Amodio, D. M. (2012). Evaluative vs. trait representation in intergroup social judgments: Distinct roles of anterior temporal lobe and prefrontal cortex. *Neuropsychologia, 50*(14), 3600–3611. https://doi.org/10.1016/j.neuropsychologia.2012.09.002

Gillan, D. J. (1981). Reasoning in the chimpanzee. II. Transitive inference. *Journal of Experimental Psychology: Animal Behavior Processes, 7*(2), 150–164. https://doi.org/10.1037/0097-7403.7.2.150

Gladue, B. A., & Clemens, L. G. (1980). Masculinization diminished by disruption of prenatal estrogen biosynthesis in male rats. *Physiology & Behavior, 25*(4), 589–593. https://doi.org/10.1016/0031-9384(80)90126-2

Goel, V. (1995). *Sketches of thought.* MIT Press.

Goel, V. (2007). Anatomy of deductive reasoning. *Trends in Cognitive Sciences, 11*(10), 435–441.

Goel, V. (2019). Hemispheric asymmetry in the prefrontal cortex for complex cognition. In M. D'Esposito & J. H. Grafman (Eds.), *Handbook of clinical neurology* (Vol. 163, pp. 179–196). Elsevier. https://doi.org/10.1016/B978-0-12-804281-6.00010-0

Goel, V., Buchel, C., Frith, C., & Dolan, R. J. (2000). Dissociation of mechanisms underlying syllogistic reasoning. *NeuroImage, 12*(5), 504–514. https://doi.org/10.1006/nimg.2000.0636

Goel, V., & Dolan, R. J. (2003). Explaining modulation of reasoning by belief. *Cognition, 87*(1), B11–B22.

Goel, V., Gold, B., Kapur, S., & Houle, S. (1997). The seats of reason: A localization study of deductive & inductive reasoning using PET (O15) blood flow technique. *NeuroReport, 8*(5), 1305–1310.

Goel, V., & Grafman, J. (2000). The role of the right prefrontal cortex in ill-structured problem solving. *Cognitive Neuropsychology, 17*(5), 415–436.

Goel, V., Grafman, J., Tajik, J., Gana, S., & Danto, D. (1997). A study of the performance of patients with frontal lobe lesions in a financial planning task. *Brain*, *120*(10), 1805–1822.

Goel, V., Lam, E., Smith, K. W., Goel, A., Raymont, V., Krueger, F., & Grafman, J. (2017). Lesions to polar/orbital prefrontal cortex selectively impair reasoning about emotional material. *Neuropsychologia*, *99*, 236–245.

Goel, V., Makale, M., & Grafman, J. (2004). The hippocampal system mediates logical reasoning about familiar spatial environments. *Journal of Cognitive Neuroscience*, *16*(4), 654–664.

Goel, V., Marling, M., Raymont, V., Krueger, F., & Grafman, J. (2019). Patients with lesions to left prefrontal cortex (BA 9 & 10) have less entrenched beliefs and are more sceptical reasoners. *Journal of Cognitive Neuroscience*, *31*(11), 1674–1688.

Goel, V., Shuren, J., Sheesley, L., & Grafman, J. (2004). Asymmetrical involvement of frontal lobes in social reasoning. *Brain*, *127*(4), 783–790.

Gogtay, N., Giedd, J. N., Lusk, L., Hayashi, K. M., Greenstein, D., Vaituzis, A. C., . . . Thompson, P. M. (2004). Dynamic mapping of human cortical development during childhood through early adulthood. *Proceedings of the National Academy of Sciences*, *101*(21), 8174–8179. https://doi.org/10.1073/pnas.0402680101

Golby, A. J., Gabrieli, J. D. E., Chiao, J. Y., & Eberhardt, J. L. (2001). Differential responses in the fusiform region to same-race and other-race faces. *Nature Neuroscience*, *4*(8), 845–850. https://doi.org/10.1038/90565

Goldberg, S., & Lewis, M. (1969). Play behavior in the year-old infant: Early sex differences. *Child Development*, *40*(1), 21–31. https://doi.org/10.2307/1127152

Goldstein, M. H., King, A. P., & West, M. J. (2003). Social interaction shapes babbling: Testing parallels between birdsong and speech. *Proceedings of the National Academy of Sciences*, *100*(13), 8030–8035. https://doi.org/10.1073/pnas.1332441100

Gonyea, D. (2017, January 10). From the start, Obama struggled with fallout from a kind of fake news. NPR.org. https://www.npr.org/2017/01/10/509164679/from-the-start-obama-struggled-with-fallout-from-a-kind-of-fake-news

Goodman, N. (1955). *Fact, fiction, and forecast*. Harvard University Press.

Goodman, N. (1976). *Languages of art: An approach to a theory of symbols* (2nd ed.). Hackett.

Goodson, J. L., & Kingsbury, M. A. (2013). What's in a name? Considerations of homologies and nomenclature for vertebrate social behavior networks. *Hormones and Behavior*, *64*(1), 103–112. https://doi.org/10.1016/j.yhbeh.2013.05.006

Gould, S. J., & Eldredge, N. (1977). Punctuated equilibria: The tempo and mode of evolution reconsidered. *Paleobiology, 3*(2), 115–151. https://doi.org/10.1017/S009483 7300005224

Gozzi, M., Zamboni, G., Krueger, F., & Grafman, J. (2010). Interest in politics modulates neural activity in the amygdala and ventral striatum. *Human Brain Mapping, 31*(11), 1763–1771. https://doi.org/10.1002/hbm.20976

Gray, J. R., Braver, T. S., & Raichle, M. E. (2002). Integration of emotion and cognition in the lateral prefrontal cortex. *Proceedings of the National Academy of Sciences, 99*(6), 4115–4120.

Greenough, W. T., Black, J. E., & Wallace, C. S. (1987). Experience and brain development. *Child Development, 58*(3), 539–559. https://doi.org/10.2307/1130197

Grice, H. P. (1975). Logic and conversation. In P. Cole and J. Morgan (Eds.), *Speech Acts* (pp. 41–58). Academic Press. https://doi.org/10.1163/9789004368811_003

Griggs, R. A., & Cox, J. R. (1982). The elusive thematic-materials effect in Wason's selection task. *British Journal of Psychology, 73*(3), 407–420. https://doi.org/10.1111/j .2044-8295.1982.tb01823.x

Grosenick, L., Clement, T. S., & Fernald, R. D. (2007). Fish can infer social rank by observation alone. *Nature, 445*(7126), 429–432. https://doi.org/10.1038/nature05511

Guillette, L. M., Scott, A. C. Y., & Healy, S. D. (2016). Social learning in nest-building birds: A role for familiarity. *Proceedings of the Royal Society B: Biological Sciences, 283*(1827), 20152685. https://doi.org/10.1098/rspb.2015.2685

Hadwen, W. (1896). *The case against vaccination.* alternative-doctor.com. http://pro -decizii-informate.ro/wp-content/uploads/2015/07/1896-Dr-Walter-Hadwen-The -case-against-vaccination.pdf

Haidt, J. (2003). The moral emotions. In R. J. Davidson, K. R. Scherer, & H. H. Goldsmith (Eds.), *Handbook of affective sciences* (pp. 852–870). Oxford University Press.

Haier, R. J., Jung, R. E., Yeo, R. A., Head, K., & Alkire, M. T. (2004). Structural brain variation and general intelligence. *NeuroImage, 23*(1), 425–433. https://doi.org/10 .1016/j.neuroimage.2004.04.025

Halford, G. S., Wilson, W. H., & Phillips, S. (2010). Relational knowledge: The foundation of higher cognition. *Trends in Cognitive Sciences, 14*(11), 497–505. https://doi .org/10.1016/j.tics.2010.08.005

Hall, Z. J., Bertin, M., Bailey, I. E., Meddle, S. L., & Healy, S. D. (2014). Neural correlates of nesting behavior in zebra finches (*Taeniopygia guttata*). *Behavioural Brain Research, 264*(100), 26–33. https://doi.org/10.1016/j.bbr.2014.01.043

Hamburg, M. (1971). Hypothalamic unit activity and eating behavior. *American Journal of Physiology-Legacy Content, 220*(4), 980–985. https://doi.org/10.1152/ajplegacy .1971.220.4.980

Hamilton, W. D. (1963). The evolution of altruistic behavior. *The American Naturalist, 97*(896), 354–356. https://doi.org/10.1086/497114

Hamilton, W. D. (1964a). The genetical evolution of social behaviour. I. *Journal of Theoretical Biology, 7*(1), 1–16. https://doi.org/10.1016/0022-5193(64)90038-4

Hamilton, W. D. (1964b). The genetical evolution of social behaviour. II. *Journal of Theoretical Biology, 7*(1), 17–52. https://doi.org/10.1016/0022-5193(64)90039-6

Hamlin, J. K., Mahajan, N., Liberman, Z., & Wynn, K. (2013). Not like me = bad: Infants prefer those who harm dissimilar others. *Psychological Science, 24*(4), 589–594. https://doi.org/10.1177/0956797612457785

Handley, S., Newstead, S. E., & Trippas, D. (2011). Logic, beliefs, and instruction: A test of the default interventionist account of belief bias. *Journal of Experimental Psychology: Learning, Memory, and Cognition, 37*(1), 28–43. https://doi.org/10.1037/a0021098

Hansen, J. E., & Lacis, A. A. (1990). Sun and dust versus greenhouse gases: An assessment of their relative roles in global climate change. *Nature, 346*(6286), 713–719. https://doi.org/10.1038/346713a0

Hanson, V. (2019). Impeachment push against Trump may only strengthen him. *National Review.* https://www.nationalreview.com/corner/why-impeachment-frenzy-may-strengthen-trump/

Harbert, T. (2019, February 21). *Here's how much the 2008 bailouts really cost.* MIT Sloan. https://mitsloan.mit.edu/ideas-made-to-matter/heres-how-much-2008-bailouts-really-cost

Hardie, L. J., Rayner, D. V., Holmes, S., & Trayhurn, P. (1996). Circulating leptin levels are modulated by fasting, cold exposure and insulin administration in lean but not Zucker (fa/fa) rats as measured by ELISA. *Biochemical and Biophysical Research Communications, 223*(3), 660–665. https://doi.org/10.1006/bbrc.1996.0951

Hardin, G. (1968). The tragedy of the commons. *Science, 162*(3859), 1243–1248. https://doi.org/10.1126/science.162.3859.1243

Harmon-Jones, E., & Mills, J. (2019). An introduction to cognitive dissonance theory and an overview of current perspectives on the theory. In E. Harmon-Jones (Ed.), *Cognitive dissonance: Reexamining a pivotal theory in psychology* (2nd ed., pp. 3–24). American Psychological Association. https://doi.org/10.1037/0000135-001

Harris, S., Kaplan, J. T., Curiel, A., Bookheimer, S. Y., Iacoboni, M., & Cohen, M. S. (2009). The neural correlates of religious and nonreligious belief. *PLoS One, 4*(10), e7272. https://doi.org/10.1371/journal.pone.0007272

Harrold, J. A., Dovey, T. M., Blundell, J. E., & Halford, J. C. G. (2012). CNS regulation of appetite. *Neuropharmacology, 63*(1), 3–17. https://doi.org/10.1016/j.neuropharm.2012.01.007

Hartshorne, J. K., Tenenbaum, J. B., & Pinker, S. (2018). A critical period for second language acquisition: Evidence from 2/3 million English speakers. *Cognition, 177*, 263–277. https://doi.org/10.1016/j.cognition.2018.04.007

Hassett, J. M., Siebert, E. R., & Wallen, K. (2008). Sex differences in rhesus monkey toy preferences parallel those of children. *Hormones and Behavior, 54*(3), 359–364. https://doi.org/10.1016/j.yhbeh.2008.03.008

Hatemi, P. K., Hibbing, J. R., Medland, S. E., Keller, M. C., Alford, J. R., Smith, K. B., . . . Eaves, L. J. (2010). Not by twins alone: Using the extended family design to investigate genetic influence on political beliefs. *American Journal of Political Science, 54*(3), 798–814. https://doi.org/10.1111/j.1540-5907.2010.00461.x

Haugeland, John. (1981). Semantic engines: An introduction to mind design. In J. Haugeland (Ed.), *Mind design* (pp. 1–34). MIT Press.

Hauser, M. (2006). *Moral minds: How nature designed our universal sense of right and wrong.* Ecco/HarperCollins.

Heath, R. (1972). Pleasure and brain activity in man: Deep and surface electroencephalograms during orgasm. *The Journal of Nervous and Mental Disease, 154*(1), 3–18.

Heathers, J. A. J., Fayn, K., Silvia, P. J., Tiliopoulos, N., & Goodwin, M. S. (2018). The voluntary control of piloerection. *PeerJ, 6*, e5292. https://doi.org/10.7717/peerj.5292

Hebb, D. O. (1939). Intelligence in man after large removals of cerebral tissue: Report of four left frontal lobe cases. *The Journal of General Psychology, 21*, 73–87.

Heinlein, R. A. (1973). *Time enough for love: The lives of Lazarus Long.* Putnam.

Heisler, L. K., & Lam, D. D. (2017). An appetite for life: Brain regulation of hunger and satiety. *Current Opinion in Pharmacology, 37*, 100–106. https://doi.org/10.1016/j.coph.2017.09.002

Heit, E. (2007). What is induction and why study it? In E. Heit & A. Feeney (Eds.), *Inductive reasoning: Experimental, developmental, and computational approaches* (pp. 1–24). Cambridge University Press.

Heit, E., & Rubinstein, J. (1994). Similarity and property effects in inductive reasoning. *Journal of Experimental Psychology, 20*(2), 411–422.

Hempel, C. G. (1980). The logical analysis of psychology. In N. Block (Ed.), *Readings in philosophy of psychology.* Harvard University Press.

Henle, M. (1962). On the relation between logic and thinking. *Psychological Review, 69*(4), 366–378.

Henrich, J., Boyd, R., Bowles, S., Camerer, C., Fehr, E., Gintis, H., & McElreath, R. (2001). In search of Homo economicus: Behavioral experiments in 15 small-scale societies. *American Economic Review, 91*(2), 73–78. https://doi.org/10.1257/aer.91.2.73

Hensley-Clancy, M. (2019, October 31). Inside the Fox News viewer impeachment bubble: Rage, disgust, and uninterrupted defense of Donald Trump. *BuzzFeed News.* https://www.buzzfeednews.com/article/mollyhensleyclancy/fox-news-impeachment -donald-trump-florida?ref=bfnsplash&utm_term=4ldqpho

Herculano-Houzel, S. (2009). The human brain in numbers: A linearly scaled-up primate brain. *Frontiers in Human Neuroscience, 3,* 31. https://doi.org/10.3389/neuro .09.031.2009

Herculano-Houzel, S. (2012). The remarkable, yet not extraordinary, human brain as a scaled-up primate brain and its associated cost. *Proceedings of the National Academy of Sciences, 109*(Suppl. 1), 10661–10668. https://doi.org/10.1073/pnas.1201895109

Herculano-Houzel, S. (2020). Remarkable, but not special: What human brains are made of. In J. H. Kaas (Ed.), *Evolutionary neuroscience* (2nd ed., pp. 803–813). Academic Press. https://doi.org/10.1016/B978-0-12-820584-6.00033-7

Herculano-Houzel, S., Avelino-de-Souza, K., Neves, K., Porfírio, J., Messeder, D., Mattos Feijó, L., . . . Manger, P. R. (2014). The elephant brain in numbers. *Frontiers in Neuroanatomy, 8.* https://doi.org/10.3389/fnana.2014.00046

Herculano-Houzel, S., Catania, K., Manger, P. R., & Kaas, J. H. (2015). Mammalian brains are made of these: A dataset of the numbers and densities of neuronal and nonneuronal cells in the brain of glires, primates, scandentia, eulipotyphlans, afrotherians and artiodactyls, and their relationship with body mass. *Brain, Behavior and Evolution, 86*(3–4), 145–163. https://doi.org/10.1159/000437413

Herculano-Houzel, S., & Lent, R. (2005). Isotropic fractionator: A simple, rapid method for the quantification of total cell and neuron numbers in the brain. *Journal of Neuroscience, 25*(10), 2518–2521. https://doi.org/10.1523/JNEUROSCI.4526-04.2005

Hewett, F. M. (1965). Teaching speech to an autistic child through operant conditioning. *American Journal of Orthopsychiatry, 35*(5), 927–936. https://doi.org/10.1111 /j.1939-0025.1965.tb00472.x

Hiebert, P. G. (2008). *Transforming worldviews: An anthropological understanding of how people change.* Baker Academic.

Hilbe, C., & Traulsen, A. (2012). Emergence of responsible sanctions without second order free riders, antisocial punishment or spite. *Scientific Reports, 2,* 458. https://doi .org/10.1038/srep00458

Hill, A. J., Magson, L. D., & Blundell, J. E. (1984). Hunger and palatability: Tracking ratings of subjective experience before, during and after the consumption of preferred and less preferred food. *Appetite, 5*(4), 361–371. https://doi.org/10.1016/S0195 -6663(84)80008-2

Hines, M., Golombok, S., Rust, J., Johnston, K. J., Golding, J., & Avon Longitudinal Study of Parents and Children Study Team (2002). Testosterone during pregnancy and gender role behavior of preschool children: A longitudinal, population study. *Child Development, 73*(6), 1678–1687. https://doi.org/10.1111/1467-8624.00498

Hines, M., & Kaufman, F. R. (1994). Androgen and the development of human sex-typical behavior: Rough-and-tumble play and sex of preferred playmates in children with congenital adrenal hyperplasia (CAH). *Child Development, 65*(4), 1042–1053. https://doi.org/10.2307/1131303

Hines, M., Pasterski, V., Spencer, D., Neufeld, S., Patalay, P., Hindmarsh, P. C., . . . & Acerini, C. L. (2016). Prenatal androgen exposure alters girls' responses to information indicating gender-appropriate behaviour. *Philosophical Transactions of the Royal Society B: Biological Sciences, 371*(1688), 20150125. https://doi.org/10.1098/rstb.2015.0125

Holland, D. (2007). Bias and concealment in the IPCC process: The "hockey-stick" affair and its implications. *Energy & Environment, 18*(7), 951–983. https://doi.org/10.1260/095830507782616788

Holtzman, S. G. (1979). Suppression of appetitive behavior in the rat by naloxone: Lack of effect of prior morphine dependence. *Life Sciences, 24*(3), 219–226. https://doi.org/10.1016/0024-3205(79)90222-4

Hommel, J. D., Trinko, R., Sears, R. M., Georgescu, D., Liu, Z.-W., Gao, X.-B., . . . DiLeone, R. J. (2006). Leptin receptor signaling in midbrain dopamine neurons regulates feeding. *Neuron, 51*(6), 801–810. https://doi.org/10.1016/j.neuron.2006.08.023

Horgan, T., & Tienson, J. (2002). The intentionality of phenomenology and the phenomenology of intentionality. In D. J. Chalmers (Ed.), *Philosophy of mind: Classical and contemporary readings* (pp. 520–533). Oxford University Press.

House, B., Henrich, J., Sarnecka, B., & Silk, J. B. (2013). The development of contingent reciprocity in children. *Evolution and Human Behavior, 34*(2), 86–93. https://doi.org/10.1016/j.evolhumbehav.2012.10.001

Howard, B., & Moore, R., dir. (2016). *Zootopia*. Walt Disney Studios.

Howard, L. H., Henderson, A. M. E., Carrazza, C., & Woodward, A. L. (2015). Infants' and young children's imitation of linguistic in-group and out-group informants. *Child Development, 86*(1), 259–275. https://doi.org/10.1111/cdev.12299

Howlin, P. A. (1981). The effectiveness of operant language training with autistic children. *Journal of Autism and Developmental Disorders, 11*(1), 89–105. https://doi.org/10.1007/BF01531343

Hubel, D. H., & Wiesel, T. N. (1970). The period of susceptibility to the physiological effects of unilateral eye closure in kittens. *The Journal of Physiology, 206*(2), 419–436. https://doi.org/10.1113/jphysiol.1970.sp009022

Hubel, D. H., Wiesel, T. N., LeVay, S., Barlow, H. B., & Gaze, R. M. (1977). Plasticity of ocular dominance columns in monkey striate cortex. *Philosophical Transactions of the Royal Society B: Biological Sciences, 278*(961), 377–409. https://doi.org/10.1098/rstb.1977.0050

Huffman, L., & Hendricks, S. E. (1981). Prenatally injected testosterone propionate and sexual behavior of female rats. *Physiology & Behavior, 26*(5), 773–778. https://doi.org/10.1016/0031-9384(81)90097-4

Humboldt, W. von. ([1836] 1999). *On language: On the diversity of human language construction and its influence on the mental development of the human species* (M. Losonsky, Ed.; P. Heath, Trans.). Cambridge University Press.

Hume, D. ([1739] 1888). *Treatise of human nature, Book 1: Of the understanding.* Clarendon Press.

Hume, D. ([1748] 1910). An enquiry concerning human understanding. In C. W. Eliot (Ed.), *Harvard classics, Vol. 37* (pp. 287–420). P. F. Collier & Son.

Huston, J., dir. (1951). *The African Queen.* United Artists.

Iacobucci, G. (2019). Child vaccination rates in England fall across the board, figures show. *BMJ, 366.* https://doi.org/10.1136/bmj.l5773

Ikemoto, S. (2010). Brain reward circuitry beyond the mesolimbic dopamine system: A neurobiological theory. *Neuroscience & Biobehavioral Reviews, 35*(2), 129–150. https://doi.org/10.1016/j.neubiorev.2010.02.001

Ingalls, C. (2011). *Welfare fraud at its finest.* KING 5 News. https://www.youtube.com/watch?v=NVqsQq9XcHE

Isaac, R. M., & Walker, J. M. (1988). Group size effects in public goods provision: The voluntary contributions mechanism. *The Quarterly Journal of Economics, 103*(1), 179–199. https://doi.org/10.2307/1882648

James, W. (1878). Brute and human intellect. *The Journal of Speculative Philosophy, 12*(3), 236–276. JSTOR. https://www.jstor.org/stable/25666088

James, W. (1890). *The principles of psychology.* Fine Editions Press.

Jarvis, E. D. (2009). Evolution of the pallium in birds and reptiles. In M. D. Binder, N. Hirokawa, & U. Windhorst (Eds.), *New encyclopedia of neuroscience* (pp. 1390–1400). Springer-Verlag. https://www.jarvislab.net/Publications/Evolution_of_the_pallium.pdf

Jarvis, E. D., Güntürkün, O., Bruce, L., Csillag, A., Karten, H., Kuenzel, W., . . . Butler, A. B. (2005). Avian brains and a new understanding of vertebrate brain evolution. *Nature Reviews Neuroscience, 6*(2), 151–159. https://doi.org/10.1038/nrn1606

Jenson, R. W. (1983). The praying animal. *Zygon, 18*(3), 311–325. https://doi.org/10.1111/j.1467-9744.1983.tb00517.x

Jerison, H. J. (1976). Paleoneurology and the evolution of mind. *Scientific American, 234*(1), 90–101.

Johnson, K. L., & Tassinary, L. G. (2005). Perceiving sex directly and indirectly: Meaning in motion and morphology. *Psychological Science, 16*(11), 890–897. https://doi.org/10.1111/j.1467-9280.2005.01633.x

Johnson, K. L., & Tassinary, L. G. (2007). Compatibility of basic social perceptions determines perceived attractiveness. *Proceedings of the National Academy of Sciences, 104*(12), 5246–5251. https://doi.org/10.1073/pnas.0608181104

Johnson-Laird, P. N. (2006). *How we reason.* Oxford University Press.

Johnson-Laird, P. N., Legrenzi, P., & Legrenzi, M. S. (1972). Reasoning and a sense of reality. *British Journal of Psychology, 63*, 395–400.

Jones, O. D. (1997). Law and biology: Toward an integrated model of human behavior law, human behavior and evolution. *Journal of Contemporary Legal Issues, 8*, 167–208.

Josso, N. (2008). Professor Alfred Jost: The builder of modern sex differentiation. *Sexual Development, 2*(2), 55–63. https://doi.org/10.1159/000129690

Kahan, D. M. (2016). The politically motivated reasoning paradigm, part 1: What politically motivated reasoning is and how to measure it. In *Emerging Trends in the Social and Behavioral Sciences* (pp. 1–16). https://doi.org/10.1002/9781118900772.etrds0417

Kahan, D. M., Jenkins-Smith, H., & Braman, D. (2011). Cultural cognition of scientific consensus. *Journal of Risk Research, 14*(2), 147–174. https://doi.org/10.1080/13669877.2010.511246

Kahan, D. M., Peters, E., Dawson, E. C., & Slovic, P. (2017). Motivated numeracy and enlightened self-government. *Behavioural Public Policy, 1*(1), 54–86. https://doi.org/10.1017/bpp.2016.2

Kahneman, D. (2003). A perspective on judgment and choice: Mapping bounded rationality. *American Psychologist, 58*(9), 697–720. https://doi.org/10.1037/0003-066X.58.9.697

Kahneman, D. (2012). *Thinking, fast and slow.* Penguin.

Kalikow, T. J. (1975). History of Konrad Lorenz's ethological theory, 1927–1939: The role of meta-theory, theory, anomaly and new discoveries in a scientific "evolution." *Studies in History and Philosophy of Science, 6*(4), 331–341. https://doi.org/10.1016/0039-3681(75)90027-8

Kalikow, T. J. (1976). Konrad Lorenz's ethological theory, 1939–1943: "Explanations" of human thinking, feeling and behaviour. *Philosophy of the Social Sciences, 6*(1), 15–34. https://doi.org/10.1177/004839317600600102

Kane, J. (2012, October 22). Health costs: How the U.S. compares with other countries. *PBS NewsHour.* https://www.pbs.org/newshour/health/health-costs-how-the-us-compares-with-other-countries

Kaplan, J. T., Gimbel, S. I., & Harris, S. (2016). Neural correlates of maintaining one's political beliefs in the face of counterevidence. *Scientific Reports, 6,* 39589. https://doi .org/10.1038/srep39589

Kapogiannis, D., Barbey, A. K., Su, M., Zamboni, G., Krueger, F., & Grafman, J. (2009). Cognitive and neural foundations of religious belief. *Proceedings of the National Academy of Sciences, 106*(12), 4876–4881. https://doi.org/10.1073/pnas.0811717106

Katz, L. C., & Shatz, C. J. (1996). Synaptic activity and the construction of cortical circuits. *Science, 274*(5290), 1133–1138. https://doi.org/10.1126/science.274.5290.1133

Kawakami, K., Amodio, D. M., & Hugenberg, K. (2017). Intergroup perception and cognition: An integrative framework for understanding the causes and consequences of social categorization. *Advances in Experimental Social Psychology, 55,* 1–80. https:// doi.org/10.1016/bs.aesp.2016.10.001

Kawakami, K., Hugenberg, K., & Dunham, Y. (2020). Perceiving others as group members: Basic principles of social categorization processes. In P. van Lange & E. T. Higgins (Eds.), *Social psychology: Handbook of basic principles* (3rd ed.). Guilford Press.

Keesey, R. E., & Powley, T. L. (1975). Hypothalamic regulation of body weight: Experiments suggest that the lateral and ventromedial hypothalamus jointly determine the regulation level or "set point" for body fat. *American Scientist, 63*(5), 558–565. JSTOR. https://www.jstor.org/stable/27845682

Keesey, R. E., & Powley, T. L. (2008). Body energy homeostasis. *Appetite, 51*(3), 442–445. https://doi.org/10.1016/j.appet.2008.06.009

Kelly, R. C. (2005). The evolution of lethal intergroup violence. *Proceedings of the National Academy of Sciences, 102*(43), 15294–15298. https://doi.org/10.1073/pnas.0505 955102

Kimble, G. A. (1956). *Principles of general psychology.* Ronald.

King, S. J., Isaacs, A. M., O'Farrell, E., & Abizaid, A. (2011). Motivation to obtain preferred foods is enhanced by ghrelin in the ventral tegmental area. *Hormones and Behavior, 60*(5), 572–580. https://doi.org/10.1016/j.yhbeh.2011.08.006

Ko, J. (2017). Neuroanatomical substrates of rodent social behavior: The medial prefrontal cortex and its projection patterns. *Frontiers in Neural Circuits, 11.* https://doi .org/10.3389/fncir.2017.00041

Kohler-Hausmann, J. (2007). "The crime of survival": Fraud prosecutions, community surveillance, and the original "welfare queen." *Journal of Social History, 41*(2), 329–354. JSTOR. https://www.jstor.org/stable/25096482

Kox, M., van Eijk, L. T., Zwaag, J., van den Wildenberg, J., Sweep, F. C. G. J., van der Hoeven, J. G., & Pickkers, P. (2014). Voluntary activation of the sympathetic nervous system and attenuation of the innate immune response in humans. *Proceedings of*

the National Academy of Sciences, 111(20), 7379–7384. https://doi.org/10.1073/pnas
.1322174111

Kraft, P. W., Lodge, M., & Taber, C. S. (2015). Why people "don't trust the evidence": Motivated reasoning and scientific beliefs. *The Annals of the American Academy of Political and Social Science, 658*(1), 121–133. https://doi.org/10.1177
/0002716214554758

Krebs, J. R., Sherry, D. F., Healy, S. D., Perry, V. H., & Vaccarino, A. L. (1989). Hippocampal specialization of food-storing birds. *Proceedings of the National Academy of Sciences, 86*(4), 1388–1392. https://doi.org/10.1073/pnas.86.4.1388

Kriegel, U. (2003). Is intentionality dependent upon consciousness? *Philosophical Studies, 116*(3), 271–307. https://doi.org/10.1023/B:PHIL.0000007204.53683.d7

Kringelbach, M. L., & Berridge, K. C. (2009). *Pleasures of the brain.* Oxford University Press.

Kringelbach, M. L., & Berridge, K. C. (2010). The functional neuroanatomy of pleasure and happiness. *Discovery Medicine, 9*(49), 579–587.

Krosnick, J. A., & Alwin, D. F. (1989). Aging and susceptibility to attitude change. *Journal of Personality and Social Psychology, 57*(3), 416–425. https://doi.org/10.1037
/0022-3514.57.3.416

Kruglanski, A. W. (2013). Only one? The default interventionist perspective as a unimodel—commentary on Evans & Stanovich (2013). *Perspectives on Psychological Science, 8*(3), 242–247. https://doi.org/10.1177/1745691613483477

Kuhl, P. K., Williams, K. A., Lacerda, F., Stevens, K. N., & Lindblom, B. (1992). Linguistic experience alters phonetic perception in infants by 6 months of age. *Science, 255*(5044), 606–608. https://doi.org/10.1126/science.1736364

Kunda, Z. (1990). The case for motivated reasoning. *Psychological Bulletin, 108*(3), 480–498. https://doi.org/10.1037/0033-2909.108.3.480

Kuo, Z. Y. (1921). Giving up instincts in psychology. *The Journal of Philosophy, 18*(24), 645–664. https://doi.org/10.2307/2939656

Lakoff, G. (1987). *Women, fire, and dangerous things: What categories reveal about the mind.* University of Chicago Press.

Langer, S. K. (1942). *Philosophy in a new key: A study in the symbolism of reason, rite, and art.* Mentor.

Le, Q. V. (2013). Building high-level features using large scale unsupervised learning. *2013 IEEE International Conference on Acoustics, Speech and Signal Processing,* 8595–8598. IEEE. https://doi.org/10.1109/ICASSP.2013.6639343

LeDoux, J. (2012). Rethinking the emotional brain. *Neuron, 73*(4), 653–676. https://
doi.org/10.1016/j.neuron.2012.02.004

LeDoux, J., & Brown, R. (2017). A higher-order theory of emotional consciousness. *Proceedings of the National Academy of Sciences, 114*(10), E2016–E2025. https://doi.org/10.1073/pnas.1619316114

Leknes, S., & Tracey, I. (2008). A common neurobiology for pain and pleasure. *Nature Reviews Neuroscience, 9*(4), 314–320. https://doi.org/10.1038/nrn2333

Lemmens, S. G. T., Schoffelen, P. F. M., Wouters, L., Born, J. M., Martens, M. J. I., Rutters, F., & Westerterp-Plantenga, M. S. (2009). Eating what you like induces a stronger decrease of "wanting" to eat. *Physiology & Behavior, 98*(3), 318–325. https://doi.org/10.1016/j.physbeh.2009.06.008

Lenz, K. M., Nugent, B. M., & McCarthy, M. M. (2012). Sexual differentiation of the rodent brain: Dogma and beyond. *Frontiers in Neuroscience, 6*, 1–13. https://doi.org/10.3389/fnins.2012.00026

LeVine, R. A., & Campbell, D. T. (1972). *Ethnocentrism: Theories of conflict, ethnic attitudes, and group behavior.* John Wiley & Sons.

Levitt, P., & Rakic, P. (1982). The time of genesis, embryonic origin and differentiation of the brain stem monoamine neurons in the rhesus monkey. *Developmental Brain Research, 4*(1), 35–57. https://doi.org/10.1016/0165-3806(82)90095-5

Lieberman, M. D., & Eisenberger, N. I. (2009). Pains and pleasures of social life. *Science, 323*(5916), 890–891. https://doi.org/10.1126/science.1170008

Lin, S., & Bergles, D. E. (2004). Synaptic signaling between GABAergic interneurons and oligodendrocyte precursor cells in the hippocampus. *Nature Neuroscience, 7*(1), 24–32. https://doi.org/10.1038/nn1162

Lincoln, G. A. (1971). The seasonal reproductive changes in the red deer stag (*Cervus elaphus*). *Journal of Zoology, 163*(1), 105–123. https://doi.org/10.1111/j.1469-7998.1971.tb04527.x

Lindquist, K. A., Wager, T. D., Kober, H., Bliss-Moreau, E., & Barrett, L. F. (2012). The brain basis of emotion: A meta-analytic review. *Behavioral and Brain Sciences, 35*(3), 121–143. https://doi.org/10.1017/S0140525X11000446

Lindzen, R. S. (1994). Climate dynamics and global change. *Annual Review of Fluid Mechanics, 26*(1), 353–378. https://doi.org/10.1146/annurev.fl.26.010194.002033

Lorber, J. (1995). *Paradoxes of gender.* Yale University Press.

Lorenz, K. Z. (1937). The companion in the bird's world. *The Auk, 54*(3), 245–273. https://doi.org/10.2307/4078077

Lorenz, K. Z. (1950). The comparative method in studying innate behavior patterns. In *Physiological mechanisms in animal behavior*, Society for Experimental Biology (pp. 221–268). Academic Press. http://klha.at/papers/1950-InnateBehavior.pdf

Lorenz, K. Z. (1952). *King Solomon's ring: New light on animal ways.* Methuen.

Lorenz, K. Z. (1958). The evolution of behaviour. *Scientific American, 199*(6), 67–82. JSTOR. https://www.jstor.org/stable/24944850

Lorenz, K. Z. (1970). Companions as factors in the bird's environment: The conspecific as the eliciting factor for social behaviour patterns. In R. Martin (Trans.), *Studies in Animal and Human Behaviour* (Vol. 1, pp. 101–258). Harvard University Press. https://doi.org/10.4159/harvard.9780674430389.c4

Lovejoy, A. O. (1968). Buffon and the problem of species. In A. O. Lovejoy, B. Glass, O. Temkin, & W. L. Straus (Eds.), *Forerunners of Darwin, 1745–1859* (pp. 84–113). Johns Hopkins Press.

Lovejoy, A. O. (2011). *The Great Chain of Being: A study of the history of an idea.* Transaction.

Lucas, J. R., Brodin, A., de Kort, S. R., & Clayton, N. S. (2004). Does hippocampal size correlate with the degree of caching specialization? *Proceedings of the Royal Society of London, Series B: Biological Sciences, 271*(1556), 2423–2429. https://doi.org/10.1098/rspb.2004.2912

Lumsden, A., & Keynes, R. (1989). Segmental patterns of neuronal development in the chick hindbrain. *Nature, 337*(6206), 424–428. https://doi.org/10.1038/337424a0

Luquet, S., Perez, F. A., Hnasko, T. S., & Palmiter, R. D. (2005). NPY/AgRP neurons are essential for feeding in adult mice but can be ablated in neonates. *Science, 310*(5748), 683–685. https://doi.org/10.1126/science.1115524

MacLeod, D. J., Sharpe, R. M., Welsh, M., Fisken, M., Scott, H. M., Hutchison, G. R., Drake, A. J., & Driesche, S. V. D. (2010). Androgen action in the masculinization programming window and development of male reproductive organs. *International Journal of Andrology, 33*(2), 279–287. https://doi.org/10.1111/j.1365-2605.2009.01005.x

MacLusky, N. J., Naftolin, F., & Goldman-Rakic, P. S. (1986). Estrogen formation and binding in the cerebral cortex of the developing rhesus monkey. *Proceedings of the National Academy of Sciences, 83*(2), 513–516. https://doi.org/10.1073/pnas.83.2.513

Maguire, E. A., Gadian, D. G., Johnsrude, I. S., Good, C. D., Ashburner, J., Frackowiak, R. S. J., & Frith, C. D. (2000). Navigation-related structural change in the hippocampi of taxi drivers. *Proceedings of the National Academy of Sciences, 97*(8), 4398–4403. https://doi.org/10.1073/pnas.070039597

Maguire, E. A., Woollett, K., & Spiers, H. J. (2006). London taxi drivers and bus drivers: A structural MRI and neuropsychological analysis. *Hippocampus, 16*(12), 1091–1101. https://doi.org/10.1002/hipo.20233

Mahajan, N., & Wynn, K. (2012). Origins of "us" versus "them": Prelinguistic infants prefer similar others. *Cognition, 124*(2), 227–233. https://doi.org/10.1016/j.cognition.2012.05.003

Maher, S. (2020, January 21). How did VW avoid criminal charges in Canada over its emissions cheating? *Maclean's*. https://www.macleans.ca/opinion/how-did-vw-avoid-criminal-charges-in-canada-over-its-emissions-cheating/

Malloy, A. (2019, September 26). Pence: Democrats "keep trying to overturn the will of the American people." CNN. https://edition.cnn.com/politics/live-news/whistleblower-complaint-impeachment-inquiry/h_93dd55c8eb2fd9a6077e94e04e15166e

Malthus, T. R. (1798). *An essay on the principle of population as it affects the future improvement of society, with remarks on the speculations of Mr Godwin, M. Condorcet, and other writers*. Text Creation Partnership, 2011. https://quod.lib.umich.edu/e/ecco/004860797.0001.000?view=toc

Mandelbaum, E. (2020). Associationist theories of thought. In E. N. Zalta (Ed.), *The Stanford encyclopedia of philosophy* (Fall 2020). Metaphysics Research Lab, Stanford University. https://plato.stanford.edu/archives/fall2020/entries/associationist-thought/

Margreiter, R. (2006). Chimpanzee heart was not rejected by human recipient. *Texas Heart Institute Journal, 33*(3), 412.

Marinsek, N., Turner, B. O., Gazzaniga, M., & Miller, M. B. (2014). Divergent hemispheric reasoning strategies: Reducing uncertainty versus resolving inconsistency. *Frontiers in Human Neuroscience, 8*, 839. https://doi.org/10.3389/fnhum.2014.00839

Matsuda, K. I., Mori, H., Nugent, B. M., Pfaff, D. W., McCarthy, M. M., & Kawata, M. (2011). Histone deacetylation during brain development is essential for permanent masculinization of sexual behavior. *Endocrinology, 152*(7), 2760–2767. https://doi.org/10.1210/en.2011-0193

Matsumoto, I. (2013). Gustatory neural pathways revealed by genetic tracing from taste receptor cells. *Bioscience, Biotechnology, and Biochemistry, 77*(7), 1359–1362.

Matsushita, S., Suzuki, K., Murashima, A., Kajioka, D., Acebedo, A. R., Miyagawa, S., . . . Yamada, G. (2018). Regulation of masculinization: Androgen signalling for external genitalia development. *Nature Reviews Urology, 15*(6), 358–368. https://doi.org/10.1038/s41585-018-0008-y

Mayr, E. (2000). The biologica species concept. In Q. D. Wheeler & R. Meier (Eds.), *Species concepts and phylogenetic theory: A debate*. Columbia University Press.

McCarry, M. (2010). Becoming a 'proper man': Young people's attitudes about interpersonal violence and perceptions of gender. *Gender and Education, 22*(1), 17–30. https://doi.org/10.1080/09540250902749083

McCarthy, M. M. (2016). Multifaceted origins of sex differences in the brain. *Philosophical Transactions of the Royal Society B: Biological Sciences, 371*(1688), 20150106. https://doi.org/10.1098/rstb.2015.0106

McCulloch, W. S., & Pitts, W. (1943). A logical calculus of the ideas immanent in nervous activity. *Bulletin of Mathematical Biophysics*, *5*(4), 115–133. https://doi.org /10.1007/BF02478259

McDougall, W. (1923). *An introduction to social psychology* (16th ed.). John W. Luce.

McGonigle, B. O., & Chalmers, M. (1977). Are monkeys logical? *Nature*, *267*(5613), 694–696.

McLeod, P. (2019, September 26). Republicans say the real problem is that Trump's conversation with the Ukrainian president leaked. *BuzzFeed News*. https://www .buzzfeednews.com/article/paulmcleod/republicans-trump-ukraine-call-problem -leaks-white-house

McTigue, D. M., & Tripathi, R. B. (2008). The life, death, and replacement of oligo-dendrocytes in the adult CNS. *Journal of Neurochemistry*, *107*(1), 1–19. https://doi.org /10.1111/j.1471-4159.2008.05570.x

Melis, A. P., Altrichter, K., & Tomasello, M. (2013). Allocation of resources to col-laborators and free-riders in 3-year-olds. *Journal of Experimental Child Psychology*, *114*(2), 364–370. https://doi.org/10.1016/j.jecp.2012.08.006

Melnikoff, D. E., & Bargh, J. A. (2018). The mythical number two. *Trends in Cognitive Sciences*, *22*(4), 280–293. https://doi.org/10.1016/j.tics.2018.02.001

Mercer, J. G., Moar, K. M., Rayner, D. V., Trayhurn, P., & Hoggard, N. (1997). Regula-tion of leptin receptor and NPY gene expression in hypothalamus of leptin-treated obese (ob/ob) and cold-exposed lean mice. *FEBS Letters*, *402*(2–3), 185–188. https:// doi.org/10.1016/S0014-5793(96)01525-6

Merker, B. (2007). Consciousness without a cerebral cortex: A challenge for neuro-science and medicine. *Behavioral and Brain Sciences*, *30*(1), 63–81. https://doi.org/10 .1017/S0140525X07000891

Michel, C., Rossion, B., Han, J., Chung, C.-S., & Caldara, R. (2016). Holistic process-ing is finely tuned for faces of one's own race. *Psychological Science*, *17*(7), 608–615. https://journals.sagepub.com/doi/10.1111/j.1467-9280.2006.01752.x

Michotte, A. ([1946] 2017). *The perception of causality*. Routledge. https://doi.org/10 .4324/9781315519050

Milinski, M. (2016). Reputation, a universal currency for human social interac-tions. *Philosophical Transactions of the Royal Society B: Biological Sciences*, *371*(1687), 20150100. https://doi.org/10.1098/rstb.2015.0100

Minsky, M., & Papert, S. A. (2017). *Perceptrons: An introduction to computational geom-etry*. MIT Press.

Mitani, J. C., Watts, D. P., & Amsler, S. J. (2010). Lethal intergroup aggression leads to territorial expansion in wild chimpanzees. *Current Biology*, *20*(12), R507–R508. https://doi.org/10.1016/j.cub.2010.04.021

Mitchell, J. P., Macrae, C. N., & Banaji, M. R. (2006). Dissociable medial prefrontal contributions to judgments of similar and dissimilar others. *Neuron, 50*(4), 655–663. https://doi.org/10.1016/j.neuron.2006.03.040

Montague, C. T., Farooqi, I. S., Whitehead, J. P., Soos, M. A., Rau, H., Wareham, N. J., . . . O'Rahilly, S. (1997). Congenital leptin deficiency is associated with severe early-onset obesity in humans. *Nature, 387*(6636), 903–908. https://doi.org/10.1038 /43185

Montepare, J. M., & Zebrowitz, L. A. (1993). A cross-cultural comparison of impressions created by age-related variations in gait. *Journal of Nonverbal Behavior, 17*(1), 55–68. https://doi.org/10.1007/BF00987008

Moran, C. N., & Pitsiladis, Y. P. (2017). Tour de France champions born or made: Where do we take the genetics of performance? *Journal of Sports Sciences, 35*(14), 1411–1419. https://doi.org/10.1080/02640414.2016.1215494

Morgan, L. (1903). *An introduction to comparative psychology* (new ed., rev.). Walter Scott.

Morgan, P., Ross, A., Mercer, J., & Barrett, P. (2003). Photoperiodic programming of body weight through the neuroendocrine hypothalamus. *Journal of Endocrinology, 177*(1), 27–34. https://doi.org/10.1677/joe.0.1770027

Morris, J. A., Jordan, C. L., & Breedlove, S. M. (2004). Sexual differentiation of the vertebrate nervous system. *Nature Neuroscience, 7*(10), 1034–1039. https://doi.org/10 .1038/nn1325

Moutsiana, C., Charpentier, C. J., Garrett, N., Cohen, M. X., & Sharot, T. (2015). Human frontal-subcortical circuit and asymmetric belief updating. *Journal of Neuroscience, 35*(42), 14077–14085. https://doi.org/10.1523/JNEUROSCI.1120-15.2015

Mulder, G. (2008). *Understanding causal coherence relations = Het begrijpen van causale coherentierelaties*. LOT.

Mullen, E., & Skitka, L. J. (2006). Exploring the psychological underpinnings of the moral mandate effect: Motivated reasoning, group differentiation, or anger? *Journal of Personality and Social Psychology, 90*(4), 629–643. https://doi.org/10.1037/0022 -3514.90.4.629

Münzberg, H., Qualls-Creekmore, E., Yu, S., Morrison, C. D., & Berthoud, H.-R. (2016). Hedonics act in unison with the homeostatic system to unconsciously control body weight. *Frontiers in Nutrition, 3*. https://doi.org/10.3389/fnut.2016.00006

Murphy, G. L., & Medin, D. L. (1985). The role of theories in conceptual coherence. *Psychological Review, 92*(3), 289–315.

Nabokov, V. V. (1991). *The annotated Lolita* (A. Appel, Jr., Ed.). Vintage Books.

Nakajima, S., & Sato, M. (1993). Removal of an obstacle: Problem-solving behavior in pigeons. *Journal of the Experimental Analysis of Behavior, 59*(1), 131–145. https:// doi.org/10.1901/jeab.1993.59-131

Nakamura, H., Katahira, T., Matsunaga, E., & Sato, T. (2005). Isthmus organizer for midbrain and hindbrain development. *Brain Research Reviews, 49*(2), 120–126. https://doi.org/10.1016/j.brainresrev.2004.10.005

Narasimhan, V. M., Patterson, N., Moorjani, P., Lazaridis, I., Lipson, M., Mallick, S., . . . Reich, D. (2018). The genomic formation of South and Central Asia. *BioRxiv*, 292581. https://doi.org/10.1101/292581

Näslund, E., & Hellström, P. M. (2007). Appetite signaling: From gut peptides and enteric nerves to brain. *Physiology & Behavior, 92*(1), 256–262. https://doi.org/10.1016/j.physbeh.2007.05.017

The National Academy of Sciences & The Royal Society. (2020). *Evidence & causes of climate change: Update 2020.* https://royalsociety.org/topics-policy/projects/climate-change-evidence-causes/

National Aeronautics and Space Administration. (2020, May 19). Climate change evidence: How do we know? *Climate Change: Vital Signs of the Planet.* https://climate.nasa.gov/evidence

Naumann, R. K., Ondracek, J. M., Reiter, S., Shein-Idelson, M., Tosches, M. A., Yamawaki, T. M., & Laurent, G. (2015). The reptilian brain. *Current Biology, 25*(8), R317–R321. https://doi.org/10.1016/j.cub.2015.02.049

Nelson, C. A. (2001). The development and neural bases of face recognition. *Infant and Child Development, 10*(1–2), 3–18. https://doi.org/10.1002/icd.239

Newell, A. (1980). Physical symbol systems. *Cognitive Science, 4*, 135–183.

Newell, A., Shaw, J. C., & Simon, H. (1959). *Report on a general problem solving program.* International Conference on Information Processing, Paris, France. https://bit.ly/3cOWnqT

Newell, A., & Simon, H. A. (1972). *Human problem solving.* Prentice-Hall.

Noble, D. (2015). Evolution beyond neo-Darwinism: A new conceptual framework. *Journal of Experimental Biology, 218*(1), 7–13. https://doi.org/10.1242/jeb.106310

Nolte, J. (2019, April 9). "Nolte: Scientists prove man-made global warming is a hoax." Breitbart. https://www.breitbart.com/politics/2019/04/09/nolte-scientists-prove-man-made-global-warming-is-a-hoax/

Nordenström, A., Servin, A., Bohlin, G., Larsson, A., & Wedell, A. (2002). Sex-typed toy play behavior correlates with the degree of prenatal androgen exposure assessed by CYP21 genotype in girls with congenital adrenal hyperplasia. *The Journal of Clinical Endocrinology & Metabolism, 87*(11), 5119–5124. https://doi.org/10.1210/jc.2001-011531

Northcutt, R. G. (2002). Understanding vertebrate brain evolution. *Integrative and Comparative Biology, 42*(4), 743–756. https://doi.org/10.1093/icb/42.4.743

Nowak, M. A., & Sigmund, K. (1998). Evolution of indirect reciprocity by image scoring. *Nature, 393*(6685), 573–577. https://doi.org/10.1038/31225

Nowak, M. A., & Sigmund, K. (2005). Evolution of indirect reciprocity. *Nature, 437*(7063), 1291–1298. https://doi.org/10.1038/nature04131

Nuccitelli, D. (2016, April 21). The climate change generation gap. *Bulletin of the Atomic Scientists.* https://thebulletin.org/2016/04/the-climate-change-generation-gap/

Nugent, B. M., Wright, C. L., Shetty, A. C., Hodes, G. E., Lenz, K. M., Mahurkar, A., . . . McCarthy, M. M. (2015). Brain feminization requires active repression of masculinization via DNA methylation. *Nature Neuroscience, 18*(5), 690–697. https://doi.org/10.1038/nn.3988

Oakes, B., & Last, J. (2020, September 28). Women on Arctic research mission told not to wear tight-fitting clothing. *CBC News.* https://www.cbc.ca/news/canada/north/mosaic-dress-code-sexism-arctic-research-1.5739547

OECD (Organization for Economic Cooperation and Development). (2020, June 4). Social expenditure—aggregated data. OECD.Stat. https://stats.oecd.org/Index.aspx?datasetcode=SOCX_AGG#

Olds, J., & Milner, P. (1954). Positive reinforcement produced by electrical stimulation of septal area and other regions of rat brain. *Journal of Comparative and Physiological Psychology, 47*(6), 419–427. https://doi.org/10.1037/h0058775

Oliver, J. E., & Wood, T. J. (2018). *Enchanted America: How intuition and reason divide our politics.* University of Chicago Press.

Olmstead, M. C., & Franklin, K. B. J. (1997). The development of a conditioned place preference to morphine: Effects of microinjections into various CNS sites. *Behavioral Neuroscience, 111*(6), 1324–1334. https://doi.org/10.1037/0735-7044.111.6.1324

O'Malley, K. (2018, September 13). Kate Moss says she regrets mantra "Nothing tastes as good as skinny feels." *Elle.* https://www.elle.com/uk/life-and-culture/culture/a23113786/kate-moss-regrets-mantra-nothing-tastes-as-good-as-skinny-feels/

O'Reilly, R. C., & Norman, K. A. (2002). Hippocampal and neocortical contributions to memory: Advances in the complementary learning systems framework. *Trends in Cognitive Sciences, 6*(12), 505–510. https://doi.org/10.1016/S1364-6613(02)02005-3

Ortmann, A., Fitzgerald, J., & Boeing, C. (2000). Trust, reciprocity, and social history: A re-examination. *Experimental Economics, 3*(1), 81–100. https://doi.org/10.1023/A:1009946125005

Ortony, A., Clore, G. L., & Collins, A. (1988). *The cognitive structure of emotions.* Cambridge University Press.

O'Shaughnessy, P. J., & Fowler, P. A. (2011). Endocrinology of the mammalian fetal testis. *Reproduction, 141*(1), 37–46. https://doi.org/10.1530/REP-10-0365

Osman, M. (2004). An evaluation of dual-process theories of reasoning. *Psychonomic Bulletin & Review, 11*(6), 988–1010. https://doi.org/10.3758/BF03196730

Oxfam America. (2020). The missing $1,000,000,000. Featured Online Actions. https://action.oxfamamerica.org/riggedreform/

Palanza, P., Parmigiani, S., Liu, H., & Saal, F. S. V. (1999). Prenatal exposure to low doses of the estrogenic chemicals diethylstilbestrol and o,p'-DDT alters aggressive behavior of male and female house mice. *Pharmacology Biochemistry and Behavior, 64*(4), 665–672. https://doi.org/10.1016/S0091-3057(99)00151-3

Pallas, S. L., Roe, A. W., & Sur, M. (1990). Visual projections induced into the auditory pathway of ferrets. I. Novel inputs to primary auditory cortex (AI) from the LP/ pulvinar complex and the topography of the MGN-AI projection. *Journal of Comparative Neurology, 298*(1), 50–68. https://doi.org/10.1002/cne.902980105

Panksepp, J. (1981). The ontogeny of play in rats. *Developmental Psychobiology, 14*(4), 327–332. https://doi.org/10.1002/dev.420140405

Panksepp, J. (2007). Neurologizing the psychology of affects: How appraisal-based constructivism and basic emotion theory can coexist. *Perspectives on Psychological Science, 2*(3), 281–296. https://doi.org/10.1111/j.1745-6916.2007.00045.x

Panksepp, J. (2011). The basic emotional circuits of mammalian brains: Do animals have affective lives? *Neuroscience & Biobehavioral Reviews, 35*(9), 1791–1804. https://doi.org/10.1016/j.neubiorev.2011.08.003

Panksepp, J., & Biven, L. (2012). *The archaeology of mind: Neuroevolutionary origins of human emotions*. W. W. Norton.

Panksepp, J., Lane, R. D., Solms, M., & Smith, R. (2017). Reconciling cognitive and affective neuroscience perspectives on the brain basis of emotional experience. *Neuroscience & Biobehavioral Reviews, 76*(Part B), 187–215. https://doi.org/10.1016/j.neubiorev.2016.09.010

Panksepp, J., Normansell, L., Cox, J. F., & Siviy, S. M. (1994). Effects of neonatal decortication on the social play of juvenile rats. *Physiology & Behavior, 56*(3), 429–443. https://doi.org/10.1016/0031-9384(94)90285-2

Pascalis, O., de Haan, M., & Nelson, C. A. (2002). Is face processing species-specific during the first year of life? *Science, 296*(5571), 1321–1323. https://doi.org/10.1126/science.1070223

Passingham, R. E., & Smaers, J. B. (2014). Is the prefrontal cortex especially enlarged in the human brain? Allometric relations and remapping factors. *Brain, Behavior and Evolution, 84*(2), 156–166. https://doi.org/10.1159/000365183

Pater, J. (2019). Generative linguistics and neural networks at 60: Foundation, friction, and fusion. *Language, 95*(1), e41–e74. https://doi.org/10.1353/lan.2019.0005

Pavlov, I. P. (1928). The reflex of purpose. In W. H. Gantt (Trans.), *Lectures on conditioned reflexes: Twenty-five years of objective study of the higher nervous activity (behaviour) of animals* (pp. 275–281). Liverwright. https://doi.org/10.1037/11081-025

Paz, Y., Mino, C. G., Bond, A. B., Kamil, A. C., & Balda, R. P. (2004). Pinyon jays use transitive inference to predict social dominance. *Nature, 430*(7001), 778–781.

Pelchat, M. L., Johnson, A., Chan, R., Valdez, J., & Ragland, J. D. (2004). Images of desire: Food-craving activation during fMRI. *NeuroImage, 23*(4), 1486–1493. https://doi.org/10.1016/j.neuroimage.2004.08.023

Penn, D. C., Holyoak, K. J., & Povinelli, D. J. (2008). Darwin's mistake: Explaining the discontinuity between human and nonhuman minds. *Behavioral and Brain Sciences, 31*(2), 109–178.

Pennycook, G., & Rand, D. G. (2019). Lazy, not biased: Susceptibility to partisan fake news is better explained by lack of reasoning than by motivated reasoning. *Cognition, 188*, 39–50. https://doi.org/10.1016/j.cognition.2018.06.011

Persinger, M. A., Saroka, K. S., Koren, S. A., & St-Pierre, L. S. (2010). The electromagnetic induction of mystical and altered states within the laboratory. *Journal of Consciousness Exploration & Research, 1*(7), 808–830.

Peters, R., & Mech, L. D. (1975). Scent-marking in wolves. *American Scientist, 63*(6), 628–637.

Petersen, M. B., Skov, M., Serritzlew, S., & Ramsøy, T. (2013). Motivated reasoning and political parties: Evidence for increased processing in the face of party cues. *Political Behavior, 35*(4), 831–854. https://doi.org/10.1007/s11109-012-9213-1

Pfaff, D. W., Martin, E. M., & Faber, D. (2012). Origins of arousal: Roles for medullary reticular neurons. *Trends in Neurosciences, 35*(8), 468–476. https://doi.org/10.1016/j.tins.2012.04.008

Pfaus, J. G. (2009). Reviews: Pathways of sexual desire. *The Journal of Sexual Medicine, 6*(6), 1506–1533. https://doi.org/10.1111/j.1743-6109.2009.01309.x

Pfaus, J. G., Kippin, T. E., Coria-Avila, G. A., Gelez, H., Afonso, V. M., Ismail, N., & Parada, M. (2012). Who, what, where, when (and maybe even why)? How the experience of sexual reward connects sexual desire, preference, and performance. *Archives of Sexual Behavior, 41*(1), 31–62. https://doi.org/10.1007/s10508-012-9935-5

Phipps, A., Ringrose, J., Renold, E., & Jackson, C. (2018). Rape culture, lad culture and everyday sexism: Researching, conceptualizing and politicizing new mediations of gender and sexual violence. *Journal of Gender Studies, 27*(1), 1–8. https://doi.org/10.1080/09589236.2016.1266792

Phoenix, C. H., Goy, R. W., Gerall, A. A., & Young, W. C. (1959). Organizing action of prenatally administered testosterone propionate on the tissue of mediating mating behaviour in the female guinea pig. *Endocrinology, 65*(3), 369–382. https:// doi.org/10.1210/endo-65-3-369

Pilkey, O. H., & Pilkey-Jarvis, L. (2007). *Useless arithmetic: Why environmental scientists can't predict the future.* Columbia University Press.

Pilkington, E. (2019, November 16). US states saw drop in vaccine rates for children as anti-vaxx theories spread. *The Guardian.* https://www.theguardian.com/us-news /2019/nov/16/vaccines-measles-mumps-polio-hepatitis-b

Pinker, S. (1994). *The language instinct: How the mind creates language.* William Morrow.

Pinker, S. (1997). *How the mind works.* W. W. Norton.

Place, N. J., & Glickman, S. E. (2004). Masculinization of female mammals: Lessons from nature. In L. S. Baskin (Ed.), *Hypospadias and genital development* (pp. 243–253). AEMB, vol. 545. Springer. https://doi.org/10.1007/978-1-4419-8995-6_15

Pohl-Apel, G. (1985). The correlation between the degree of brain masculinization and song quality in estradiol treated female zebra finches. *Brain Research, 336*(2), 381–383. https://doi.org/10.1016/0006-8993(85)90673-0

Pollak, J. B. (2019, September 24). Fox News: IG found "whistleblower" had "political bias" in favor of Trump 2020 rival. Breitbart. https://www.breitbart.com/politics /2019/09/24/fox-news-ig-found-whistleblower-had-political-bias-in-favor-of-trump -2020-rival/

Pomeroy, R. (2011, November 3). Fixed action patterns and their human manifestations. *RealClearScience.* http://www.realclearscience.com/blog/2011/11/fixedactionpatterns .html

Pope, A. ([1739] 1843). *Essay on man; in four epistles.* W. B. Fowle and N. Capen. http://archive.org/details/popesessayonmani00pope

Porte, D., Baskin, D. G., & Schwartz, M. W. (2002). Leptin and insulin action in the central nervous system. *Nutrition Reviews, 60*(Suppl. 10), S20–S29. https://doi.org/10 .1301/002966402320634797

Powell, L. J., & Spelke, E. S. (2013). Preverbal infants expect members of social groups to act alike. *Proceedings of the National Academy of Sciences, 110*(41), E3965–E3972. https://doi.org/10.1073/pnas.1304326110

Premack, D. (2007). Human and animal cognition: Continuity and discontinuity. *Proceedings of the National Academy of Sciences, 104*(35), 13861–13867. https://doi.org /10.1073/pnas.0706147104

Prescott, J. (2012). *Taste matters: Why we like the foods we do.* Reaktion Books.

Prince, A., & Smolensky, P. (1997). Optimality: From neural networks to universal grammar. *Science, 275*(5306), 1604–1610. https://doi.org/10.1126/science.275.5306.1604

Przybla, H., & Edelman, A. (2019). Nancy Pelosi announces formal impeachment inquiry of Trump. NBC News. https://www.nbcnews.com/politics/trump-impeachment -inquiry/pelosi-announce-formal-impeachment-inquiry-trump-n1058251

Putnam, H. (1981). The "innateness hypothesis" and explanatory models in linguistics. In N. Block (Ed.), *Readings in philosophy of psychology, Vol. 2*. Harvard University Press.

Quinn, P. C., Yahr, J., Kuhn, A., Slater, A. M., & Pascalis, O. (2002). Representation of the gender of human faces by infants: A preference for female. *Perception, 31*(9), 1109–1121. https://doi.org/10.1068/p3331

Ragni, M., Kola, I., & Johnson-Laird, P. (2017). The Wason selection task: A meta-analysis. *Proceedings of the 39th Annual Meeting of the Cognitive Science Society 2017*. London.

Rajesh, C. (2018). Varna to Jati: An overview of caste system in Hinduism. *Academic Discourse, 7*(1), 35–43. https://www.indianjournals.com/ijor.aspx?target=ijor:adi &volume=7&issue=1&article=005

Rakic, P. (1974). Neurons in rhesus monkey visual cortex: Systematic relation between time of origin and eventual disposition. *Science, 183*(4123), 425–427. https://doi.org/10.1126/science.183.4123.425

Rakic, P. (1977). Genesis of the dorsal lateral geniculate nucleus in the rhesus monkey: Site and time of origin, kinetics of proliferation, routes of migration and pattern of distribution of neurons. *Journal of Comparative Neurology, 176*(1), 23–52. https://doi.org/10.1002/cne.901760103

Rakoczy, H., Warneken, F., & Tomasello, M. (2008). The sources of normativity: Young children's awareness of the normative structure of games. *Developmental Psychology, 44*(3), 875–881. https://doi.org/10.1037/0012-1649.44.3.875

Ramachandran, V. S., & Blakeslee, S. (1998). *Phantoms in the brain: Probing the mysteries of the human mind*. William Morrow.

Ravussin, Y., Gutman, R., Diano, S., Shanabrough, M., Borok, E., Sarman, B., . . . Leibel, R. L. (2011). Effects of chronic weight perturbation on energy homeostasis and brain structure in mice. *American Journal of Physiology—Regulatory, Integrative and Comparative Physiology, 300*(6), R1352–R1362. https://doi.org/10.1152/ajpregu.00429.2010

Regan, P. C., & Atkins, L. (2006). Sex differences and similarities in frequency and intensity of sexual desire. *Social Behavior and Personality, 34*(1), 95–102. https://doi .org/10.2224/sbp.2006.34.1.95

Reiner, A., Perkel, D. J., Bruce, L. L., Butler, A. B., Csillag, A., Kuenzel, W., . . . Jarvis, E. D. (2004). Revised nomenclature for avian telencephalon and some related brainstem nuclei. *Journal of Comparative Neurology, 473*(3), 377–414. https://doi.org/10 .1002/cne.20118

Reuben, E., Sapienza, P., & Zingales, L. (2010). Time discounting for primary and monetary rewards. *Economics Letters, 106*(2), 125–127. https://doi.org/10.1016/j .econlet.2009.10.020

Revkin, A. C. (2003, August 5). Politics reasserts itself in the debate over climate change and its hazards. *The New York Times.* https://www.nytimes.com/2003/08/05 /science/politics-reasserts-itself-in-the-debate-over-climate-change-and-its-hazards .html

Richards, R. J. (1974). The innate and the learned: The evolution of Konrad Lorenz's theory of instinct. *Philosophy of the Social Sciences, 4*(2–3), 111–133. https://doi.org /10.1177/004839317400400201

Riehl, C., & Frederickson, M. (2016). Cheating and punishment in cooperative animal societies. *Philosophical Transactions of the Royal Society B: Biological Sciences, 371*(1687), 20150090. https://doi.org/10.1098/rstb.2015.0090

Rines, J. P., & vom Saal, F. S. (1984). Fetal effects on sexual behavior and aggression in young and old female mice treated with estrogen and testosterone. *Hormones and Behavior, 18*(2), 117–129. https://doi.org/10.1016/0018-506X(84)90037-0

Rips, L. J. (1994). *The psychology of proof: Deductive reasoning in human thinking.* MIT Press.

Roh, E., Song, D. K., & Kim, M.-S. (2016). Emerging role of the brain in the homeo-static regulation of energy and glucose metabolism. *Experimental & Molecular Medi-cine, 48*(3), e216–e216. https://doi.org/10.1038/emm.2016.4

Rolls, E. T. (1989). Functions of neuronal networks in the hippocampus and neocortex in memory. In J. H. Byrne & W. O. Berry (Eds.), *Neural models of plasticity* (pp. 240–265). Academic Press. https://doi.org/10.1016/B978-0-12-148955-7.50017-5

Rosenblatt, F. (1958). The perceptron: A probabilistic model for information storage and organization in the brain. *Psychological Review, 65*(6), 386–408. https://doi.org /10.1037/h0042519

Roser, M. E., Fugelsang, J. A., Dunbar, K. N., Corballis, P. M., & Gazzaniga, M. S. (2005). Dissociating processes supporting causal perception and causal inference in the brain. *Neuropsychology, 19*(5), 591–602. https://doi.org/10.1037/0894-4105.19.5.591

Ross, N., Piñeyrúa, M., Prieto, S., Arias, L. P., Stirner, A., & Galeano, C. (1965). Con-ditioning of midbrain behavioral responses. *Experimental Neurology, 11*(3), 263–276. https://doi.org/10.1016/0014-4886(65)90047-6

Roth, G., & Dicke, U. (2005). Evolution of the brain and intelligence. *Trends in Cog-nitive Sciences, 9*(5), 250–257. https://doi.org/10.1016/j.tics.2005.03.005

Rumelhart, D. E., & McClelland, J. L. (1986). *Parallel distributed processing: Explora-tions in the microstructures of cognition, Volume 1: Foundations.* MIT Press.

Rupar, A. (2019, September 24). Trump's rationale for withholding aid to Ukraine changed overnight. Vox. https://www.vox.com/policy-and-politics/2019/9/24/2088 2081/trump-ukraine-aid-explanation-whistleblower

Ruse, M. (1985). *Sociobiology: Sense or Nonsense?* (2nd ed.). Reidel.

Russell, J. A. (2003). Core affect and the psychological construction of emotion. *Psychological Review, 110*(1), 145–172. https://doi.org/10.1037/0033-295X.110.1.145

Sabia, C. (2019, September 25). Breaking: Trump releases transcript of Ukrainian phone call, major bust for Democrats. *The Federalist Papers.* https://thefederalistpapers .org/opinion/breaking-trump-releases-transcript-ukrainian-phone-call-major-bust -democrats

Sachs, B. D. (2007). A contextual definition of male sexual arousal. *Hormones and Behavior, 51*(5), 569–578. https://doi.org/10.1016/j.yhbeh.2007.03.011

Sakai, K. L. (2005). Language acquisition and brain development. *Science, 310*(5749), 815–819. https://doi.org/10.1126/science.1113530

Salvadori, E., Blazsekova, T., Volein, A., Karap, Z., Tatone, D., Mascaro, O., & Csibra, G. (2015). Probing the strength of infants' preference for helpers over hinderers: Two replication attempts of Hamlin and Wynn (2011). *PLoS One, 10*(11). https://doi .org/10.1371/journal.pone.0140570

Sam, A. H., Troke, R. C., Tan, T. M., & Bewick, G. A. (2012). The role of the gut/brain axis in modulating food intake. *Neuropharmacology, 63*(1), 46–56. https://doi.org/10 .1016/j.neuropharm.2011.10.008

Sampson, G. (2005). *The "language instinct" debate* (rev. ed.). A. & C. Black.

Sandel, M. J. (2020). *The tyranny of merit: What's become of the common good?* Farrar, Straus and Giroux.

Sandhu, S. (2019, September 6). Boris Johnson asks queen to suspend Parliament from mid-September—leading to no-deal Brexit fears. inews. https://inews.co.uk/news /politics/brexit/suspend-parliament-prorogue-boris-johnson-brexit-talks-mps-pro rogation-date-495004

Sato, T., Matsumoto, T., Kawano, H., Watanabe, T., Uematsu, Y., Sekine, K., . . . Kato, S. (2004). Brain masculinization requires androgen receptor function. *Proceedings of the National Academy of Sciences, 101*(6), 1673–1678. https://doi.org/10.1073/pnas .0305303101

Satterlie, R. A. (2011). Do jellyfish have central nervous systems? *Journal of Experimental Biology, 214*(8), 1215–1223. https://doi.org/10.1242/jeb.043687

The Sceptics. (2010, May 31). *Insight.* SBS. https://www.sbs.com.au/news/sites/sbs .com.au.news/files/transcripts/363503_insight_thesceptics_transcript.html

Schechter, D., Howard, S. M., & Gandelman, R. (1981). Dihydrotestosterone promotes fighting behavior of female mice. *Hormones and Behavior, 15*(3), 233–237. https://doi.org/10.1016/0018-506X(81)90012-X

Scheibner, V. (1993). *Vaccination: 100 years of orthodox research shows that vaccines represent a medical assault on the immune system.* Blackheath. https://catalog.nlm.nih .gov/discovery/search?vid=01NLM_INST:01NLM_INST&query=lds04,exact,9421665

Schiffman, S. S., & Graham, B. G. (2000). Taste and smell perception affect appetite and immunity in the elderly. *European Journal of Clinical Nutrition, 54*(3), S54–S63. https://doi.org/10.1038/sj.ejcn.1601026

Schlichter, K. (2019, September 9). "Climate change" is a hoax. Townhall. https:// townhall.com/columnists/kurtschlichter/2019/09/09/climate-change-is-a-hoax -n2552748

Scottish Legal News. (2019, July 23). MPs and peers to launch Scottish court challenge to block Brexit prorogation of Parliament. *Scottish Legal News.* https:// scottishlegal.com/article/mps-and-peers-to-launch-scottish-court-challenge-to-block -brexit-prorogation-of-parliament

Searle, J. R. (1980). Minds, brains, and programs. *Behavioral and Brain Sciences, 3*(3), 417–424. https://doi.org/10.1017/S0140525X00005756

Searle, J. R. (1983). *Intentionality: An essay in the philosophy of mind.* Cambridge University Press.

Searle, J. R. (1992). *Rediscovering the mind.* MIT Press.

Selemon, L. D., & Zecevic, N. (2015). Schizophrenia: A tale of two critical periods for prefrontal cortical development. *Translational Psychiatry, 5*(8), e623–e623. https:// doi.org/10.1038/tp.2015.115

Semendeferi, K., Damasio, H., Frank, R., & Van Hoesen, G. W. (1997). The evolution of the frontal lobes: A volumetric analysis based on three-dimensional reconstructions of magnetic resonance scans of human and ape brains. *Journal of Human Evolution, 32*(4), 375–388. https://doi.org/10.1006/jhev.1996.0099

Semendeferi, K., Lu, A., Schenker, N., & Damasio, H. (2002). Humans and great apes share a large frontal cortex. *Nature Neuroscience, 5*(3), 272–276. https://doi.org/10 .1038/nn814

Seth, A. K. (2013). Interoceptive inference, emotion, and the embodied self. *Trends in Cognitive Sciences, 17*(11), 565–573. https://doi.org/10.1016/j.tics.2013.09.007

Shanahan, M. (2016, Spring). The frame problem. In E. N. Zalta (Ed.), *The Stanford encyclopedia of philosophy.* Metaphysics Research Lab, Stanford University. https:// plato.stanford.edu/archives/spr2016/entries/frame-problem/

Sharma, K. N. (1990). Varna and Jati in Indian traditional perspective. *Sociological Bulletin, 39*(1–2), 15–31. https://doi.org/10.1177/0038022919900102

Shave, R. E., Lieberman, D. E., Drane, A. L., Brown, M. G., Batterham, A. M., Worthington, S., . . . Baggish, A. L. (2019). Selection of endurance capabilities and the trade-off between pressure and volume in the evolution of the human heart. *Proceedings of the National Academy of Sciences, 116*(40), 19905–19910. https://doi.org/10.1073/pnas.1906902116

Shen, Y., Tan, S., Sordoni, A., & Courville, A. (2019). Ordered neurons: Integrating tree structures into recurrent neural networks. *ICLR 2019.* http://arxiv.org/abs/1810.09536

Sherif, M. (1988). *The robbers cave experiment: Intergroup conflict and cooperation.* Wesleyan University Press. [Originally published as *Intergroup conflict and group relations.*]

Sherman, M. (2020, January 30). Dershowitz says his impeachment argument was misinterpreted. *PBS NewsHour.* https://www.pbs.org/newshour/nation/dershowitz-says-his-impeachment-argument-was-misinterpreted

Sherman, P. W. (1985). Alarm calls of Belding's ground squirrels to aerial predators: Nepotism or self-preservation? *Behavioral Ecology and Sociobiology, 17*(4), 313–323. https://doi.org/10.1007/BF00293209

Sherrington, C. (1952). *The integrative action of the nervous system.* CUP Archive.

Sherwood, L., & Kell, R. (2009). *Human physiology: From cells to systems.* Nelson College Indigenous.

Shettleworth, S. J. (1973). Food reinforcement and the organization of behaviour in golden hamsters. In R. A. Hinde & J. S. Hinde (Eds.), *Constraints on learning: Limitations and predispositions* (pp. 243–263). Academic Press.

Shettleworth, S. J. (2010). Clever animals and killjoy explanations in comparative psychology. *Trends in Cognitive Sciences, 14*(11), 477–481. https://doi.org/10.1016/j.tics.2010.07.002

Shewmon, D. A., Holmes, G. L., & Byrne, P. A. (1999). Consciousness in congenitally decorticate children: Developmental vegetative state as self-fulfilling prophecy. *Developmental Medicine & Child Neurology, 41*(6), 364–374. https://doi.org/10.1111/j.1469-8749.1999.tb00621.x

Silbereis, J. C., Pochareddy, S., Zhu, Y., Li, M., & Sestan, N. (2016). The cellular and molecular landscapes of the developing human central nervous system. *Neuron, 89*(2), 248–268. https://doi.org/10.1016/j.neuron.2015.12.008

Silk, J. B., Brosnan, S. F., Vonk, J., Henrich, J., Povinelli, D. J., Richardson, A. S., . . . Schapiro, S. J. (2005). Chimpanzees are indifferent to the welfare of unrelated group members. *Nature, 437*(7063), 1357–1359. https://doi.org/10.1038/nature04243

Simon, H. A. (1955). A behavioral model of rational choice. *The Quarterly Journal of Economics, 69*(February), 99–118.

Simon, H. A. (1997). *Models of bounded rationality: Empirically grounded economic reason.* MIT Press.

Simpson, H. B., & Vicario, D. S. (1991). Early estrogen treatment of female zebra finches masculinizes the brain pathway for learned vocalizations. *Journal of Neurobiology, 22*(7), 777–793. https://doi.org/10.1002/neu.480220711

Singer, B., & Toates, F. M. (1987). Sexual motivation. *The Journal of Sex Research, 23*(4), 481–501. https://doi.org/10.1080/00224498709551386

Skinner, B. F. (1953). *Science and human behavior.* Macmillan.

Skinner, B. F. (1957). *Verbal behavior.* Appleton-Century-Crofts.

Skinner, B. F. (1984). Behaviorism at fifty. *Behavioral and Brain Sciences, 7*(4), 615–667. https://doi.org/10.1017/S0140525X00027618

Slimp, J. C., Hart, B. L., & Goy, R. W. (1978). Heterosexual, autosexual and social behavior of adult male rhesus monkeys with medial preoptic-anterior hypothalamic lesions. *Brain Research, 142*(1), 105–122. https://doi.org/10.1016/0006-8993(78)90180-4

Sloane, S., Baillargeon, R., & Premack, D. (2012). Do infants have a sense of fairness? *Psychological Science, 23*(2), 196–204. https://doi.org/10.1177/0956797611422072

Sloman, S. A. (1996). The empirical case for two systems of reasoning. *Psychological Bulletin, 119*(1), 3–22.

Sloman, S. A., & Lagnado, D. A. (2005). The problem of induction. In K. J. Holyoak & R. G. Morrison (Eds.), *The Cambridge handbook of thinking and reasoning* (pp. 95–116). Cambridge University Press.

Smolensky, P. (1988). On the proper treatment of connectionism. *Behavioral and Brain Sciences, 11*(1), 1–74.

Smulders, T. V., Sasson, A. D., & DeVoogd, T. J. (1995). Seasonal variation in hippocampal volume in a food-storing bird, the black-capped chickadee. *Journal of Neurobiology, 27*(1), 15–25. https://doi.org/10.1002/neu.480270103

Smyth, C. M., & Bremner, W. J. (1998). Klinefelter syndrome. *Archives of Internal Medicine, 158*(12), 1309–1314. https://doi.org/10.1001/archinte.158.12.1309

Socher, R., Lin, C. C.-Y., Ng, A. Y., & Manning, C. D. (2011). Parsing natural scenes and natural language with recursive neural networks. *Proceedings of the 28th International Conference on Machine Learning,* 129–136.

Spence, C., Okajima, K., Cheok, A. D., Petit, O., & Michel, C. (2016). Eating with our eyes: From visual hunger to digital satiation. *Brain and Cognition, 110*, 53–63. https://doi.org/10.1016/j.bandc.2015.08.006

Spencer, H. (1882). *The principles of biology* (Vol. 2). D. Appleton.

Sperry, R. W. (1944). Optic nerve regeneration with return of vision in anurans. *Journal of Neurophysiology, 7*(1), 57–69. https://doi.org/10.1152/jn.1944.7.1.57

Stanovich, K. E. (2004). *The robot's rebellion: Finding meaning in the age of Darwin.* University of Chicago Press.

Stanovich, K. E., & West, R. F. (2000). Individual differences in reasoning: Implications for the rationality debate. *Behavioral and Brain Sciences, 23*(5), 645–665.

Stevens, J. R., & Hauser, M. D. (2004). Why be nice? Psychological constraints on the evolution of cooperation. *Trends in Cognitive Sciences, 8*(2), 60–65. https://doi.org/10.1016/j.tics.2003.12.003

Stiles, J., & Jernigan, T. L. (2010). The basics of brain development. *Neuropsychology Review, 20*(4), 327–348. https://doi.org/10.1007/s11065-010-9148-4

Stollstorff, M., Vartanian, O., & Goel, V. (2012). Levels of conflict in reasoning modulate right lateral prefrontal cortex. *Brain Research, 1428*, 24–32. https://doi.org/10.1016/j.brainres.2011.05.045

Stuart, K. (2017). *Sugar.* Skyhorse. https://www.goodreads.com/work/quotes/49789773-sugar

Sturdy, C. B., & Nicoladis, E. (2017). How much of language acquisition does operant conditioning explain? *Frontiers in Psychology, 8.* https://doi.org/10.3389/fpsyg.2017.01918

Sun, Y.-H., Chen, S.-P., Wang, Y.-P., Hu, W., & Zhu, Z.-Y. (2005). Cytoplasmic impact on cross-genus cloned fish derived from transgenic common carp (*Cyprinus carpio*) nuclei and goldfish (*Carassius auratus*) enucleated eggs. *Biology of Reproduction, 72*(3), 510–515. https://doi.org/10.1095/biolreprod.104.031302

Sundaram, V., & Jackson, C. (2018). "Monstrous men" and "sex scandals": The myth of exceptional deviance in sexual harassment and violence in education. *Palgrave Communications, 4*(1), 147. https://doi.org/10.1057/s41599-018-0202-9

Sur, M., Garraghty, P. E., & Roe, A. W. (1988). Experimentally induced visual projections into auditory thalamus and cortex. *Science, 242*(4884), 1437–1441. https://doi.org/10.1126/science.2462279

Sur, M., & Rubenstein, J. L. R. (2005). Patterning and plasticity of the cerebral cortex. *Science, 310*(5749), 805–810. https://doi.org/10.1126/science.1112070

Swaab, D. F. (2007). Sexual differentiation of the brain and behavior. *Best Practice & Research Clinical Endocrinology & Metabolism, 21*(3), 431–444. https://doi.org/10.1016/j.beem.2007.04.003

Swift, J. ([1726] 2012). *Gulliver's travels.* Broadview Press.

Tajfel, H. (1970). Experiments in intergroup discrimination. *Scientific American*, *223*(5), 96–103. JSTOR. https://www.jstor.org/stable/e24927650

Talmy, L. (1983). How language structures space. In H. L. Pick Jr. & L. P. Acredolo (Eds.), *Spatial orientation* (pp. 225–282). Springer. http://link.springer.com/chapter /10.1007/978-1-4615-9325-6_11

Tang, Y., Nyengaard, J. R., Groot, D. M. G. D., & Gundersen, H. J. G. (2001). Total regional and global number of synapses in the human brain neocortex. *Synapse*, *41*(3), 258–273. https://doi.org/10.1002/syn.1083

Taylor, M. G., Rhodes, M., & Gelman, S. A. (2009). Boys will be boys; cows will be cows: Children's essentialist reasoning about gender categories and animal species. *Child Development*, *80*(2), 461–481. https://doi.org/10.1111/j.1467-8624.2009 .01272.x

ten Cate, C. (2009). Niko Tinbergen and the red patch on the herring gull's beak. *Animal Behaviour*, *77*(4), 785–794. https://doi.org/10.1016/j.anbehav.2008.12.021

Thompson, R. A., & Nelson, C. A. (2001). Developmental science and the media: Early brain development. *American Psychologist*, *56*(1), 5–15. https://doi.org/10.1037/00 03-066X.56.1.5

Thorndike, E. L. (1927). The law of effect. *The American Journal of Psychology*, *39*(1/4), 212–222. https://doi.org/10.2307/1415413

Tillyard, E. M. W. (2011). *The Elizabethan world picture*. Transaction.

Tinbergen, N. (1951). *The study of instinct*. Clarendon Press.

Tinbergen, N. (1953). *Social behaviour in animals: With special reference to vertebrates*. Chapman and Hall.

Tobet, S. A., & Baum, M. J. (1987). Role for prenatal estrogen in the development of masculine sexual behavior in the male ferret. *Hormones and Behavior*, *21*(4), 419–429. https://doi.org/10.1016/0018-506X(87)90001-8

Todd, P., & Gigerenzer, G. (2000). Précis of simple heuristics that make us smart. *Behavioral and Brain Sciences*, *23*(5), 727–741.

Tomasello, M. (1995). Language is not an instinct. *Cognitive Development*, *10*(1), 131–156. https://doi.org/10.1016/0885-2014(95)90021-7

Tooby, J., & Cosmides, L. (1995). The psychological foundations of culture. In J. H. Barkow, L. Cosmides, & J. Tooby (Eds.), *The adapted mind: Evolutionary psychology and the generation of culture* (pp. 19–136). Oxford University Press.

Toronchuk, J. A., & Ellis, G. F. R. (2013). Affective neuronal selection: The nature of the primordial emotion systems. *Frontiers in Psychology*, *3*. https://doi.org/10.3389 /fpsyg.2012.00589

Trayhurn, P., & Bing, C. (2006). Appetite and energy balance signals from adipocytes. *Philosophical Transactions of the Royal Society B: Biological Sciences, 361*(1471), 1237–1249. https://doi.org/10.1098/rstb.2006.1859

Trippas, D., Thompson, V. A., & Handley, S. J. (2017). When fast logic meets slow belief: Evidence for a parallel-processing model of belief bias. *Memory & Cognition, 45*, 539–552. https://doi.org/10.3758/s13421-016-0680-1

Trivers, R. L. (1971). The evolution of reciprocal altruism. *The Quarterly Review of Biology, 46*(1), 35–57. https://doi.org/10.1086/406755

Troyer, J. (2003). Introduction. In *The Classical Utilitarians: Bentham and Mill.* Hackett.

Trump, Donald. (2019, September 28). Twitter tweet. @realDonaldTrump. https://deadline.com/2019/09/president-donald-trump-tweetstorm-the-saturday-edition-41-1202747231/

Tschöp, M., Smiley, D. L., & Heiman, M. L. (2000). Ghrelin induces adiposity in rodents. *Nature, 407*(6806), 908–913. https://doi.org/10.1038/35038090

Tsujii, T., Masuda, S., Akiyama, T., & Watanabe, S. (2010). The role of inferior frontal cortex in belief-bias reasoning: An rTMS study. *Neuropsychologia, 48*(7), 2005–2008. https://doi.org/10.1016/j.neuropsychologia.2010.03.021

Tsujii, T., Sakatani, K., Masuda, S., Akiyama, T., & Watanabe, S. (2011). Evaluating the roles of the inferior frontal gyrus and superior parietal lobule in deductive reasoning: An rTMS study. *NeuroImage, 58*(2), 640–646. https://doi.org/10.1016/j.neuroimage.2011.06.076

Tucker, R., & Collins, M. (2012). What makes champions? A review of the relative contribution of genes and training to sporting success. *British Journal of Sports Medicine, 46*(8), 555–561. https://doi.org/10.1136/bjsports-2011-090548

Turton, M. D., O'Shea, D., Gunn, I., Beak, S. A., Edwards, C. M. B., Meeran, K., . . . Bloom, S. R. (1996). A role for glucagon-like peptide-1 in the central regulation of feeding. *Nature, 379*(6560), 69–72. https://doi.org/10.1038/379069a0

Tversky, A., & Kahneman, D. (1974). Judgment under uncertainty: Heuristics and biases. *Science, 185*(4157), 1124–1131. JSTOR. https://www.jstor.org/stable/20159081

Tversky, A., & Kahneman, D. (1982). Evidential impact of base rates. In D. Kahneman, P. Slovic, & A. Tversky (Eds.), *Judgment under uncertainty: Heuristics and biases* (pp. 153–160). Cambridge University Press. https://doi.org/10.1017/CBO9780511809477.011

Tversky, A., & Kahneman, D. (1983). Extensional versus intuitive reasoning: The conjunction fallacy in probability judgment. *Psychological Review, 90*(4), 293–315. https://doi.org/10.1037/0033-295X.90.4.293

Ulber, J., Hamann, K., & Tomasello, M. (2017). Young children, but not chimpanzees, are averse to disadvantageous and advantageous inequities. *Journal of Experimental Child Psychology, 155*, 48–66. https://doi.org/10.1016/j.jecp.2016.10.013

Vaish, A., Missana, M., & Tomasello, M. (2011). Three-year-old children intervene in third-party moral transgressions. *British Journal of Developmental Psychology, 29*(1), 124–130. https://doi.org/10.1348/026151010X532888

Varki, N., Anderson, D., Herndon, J. G., Pham, T., Gregg, C. J., Cheriyan, M., . . . Varki, A. (2009). Heart disease is common in humans and chimpanzees, but is caused by different pathological processes. *Evolutionary Applications, 2*(1), 101–112. https://doi.org/10.1111/j.1752-4571.2008.00064.x

Venkatraman, A., Edlow, B. L., & Immordino-Yang, M. H. (2017). The brainstem in emotion: A review. *Frontiers in Neuroanatomy, 11*. https://doi.org/10.3389/fnana.2017.00015

Ventureyra, V. A. G., Pallier, C., & Yoo, H.-Y. (2004). The loss of first language phonetic perception in adopted Koreans. *Journal of Neurolinguistics, 17*(1), 79–91. https://doi.org/10.1016/S0911-6044(03)00053-8

Vigen, T. (n.d.). US spending on science, space, and technology correlates with suicides by hanging, strangulation and suffocation. Tylervigen.com. Retrieved May 17, 2020, from http://tylervigen.com/spurious-correlations

Voland, E., & Schiefenhövel, W. (Eds.). (2009). *The biological evolution of religious mind and behavior*. Springer. https://doi.org/10.1007/978-3-642-00128-4

vom Saal, F. S. (1979). Prenatal exposure to androgen influences morphology and aggressive behavior of male and female mice. *Hormones and Behavior, 12*(1), 1–11. https://doi.org/10.1016/0018-506X(79)90021-7

Wade, M. J., & Breden, F. (1980). The evolution of cheating and selfish behavior. *Behavioral Ecology and Sociobiology, 7*(3), 167–172. https://doi.org/10.1007/BF002 99360

Wade, P. (2020, July 16). White House: "Science should not stand in the way" of school openings. *Rolling Stone*. https://www.rollingstone.com/politics/politics-news/mcenany-science-should-not-stand-in-the-way-of-school-openings-1030018/

Wakefield, A. J., Murch, S. H., Anthony, A., Linnell, J., Casson, D. M., Malik, M., . . . Walker-Smith, J. A. (1998). RETRACTED: Ileal-lymphoid-nodular hyperplasia, non-specific colitis, and pervasive developmental disorder in children. *The Lancet, 351*(9103), 637–641. https://doi.org/10.1016/S0140-6736(97)11096-0

Wallace, A. R. ([1871] 2009). *Contributions to the theory of natural selection: A series of essays* (2nd ed.). Cambridge University Press. https://doi.org/10.1017/CBO97805 11693106

Ward, I. L., & Renz, F. J. (1972). Consequences of perinatal hormone manipulation on the adult sexual behavior of female rats. *Journal of Comparative and Physiological Psychology, 78*(3), 349–355. https://doi.org/10.1037/h0032375

Waterson, R. H., Lander, E. S., Wilson, R. K., & the Chimpanzee Sequencing and Analysis Consortium. (2005). Initial sequence of the chimpanzee genome and comparison with the human genome. *Nature, 437*(7055), 69–87. https://doi.org/10.1038/nature04072

Welsh, M., Suzuki, H., & Yamada, G. (2014). The masculinization programming window. *Understanding differences and disorders of sex development (DSD), 27*, 17–27. https://doi.org/10.1159/000363609

West, M. J., & King, A. P. (1988). Female visual displays affect the development of male song in the cowbird. *Nature, 334*(6179), 244–246. https://doi.org/10.1038/334244a0

Wettergren, A., Wøjdemann, M., & Holst, J. J. (1998). Glucagon-like peptide-1 inhibits gastropancreatic function by inhibiting central parasympathetic outflow. *American Journal of Physiology—Gastrointestinal and Liver Physiology, 275*(5), G984–G992. https://doi.org/10.1152/ajpgi.1998.275.5.G984

Wheeler, B. C. (2009). Monkeys crying wolf? Tufted capuchin monkeys use antipredator calls to usurp resources from conspecifics. *Proceedings of the Royal Society B: Biological Sciences, 276*(1669), 3013–3018. https://doi.org/10.1098/rspb.2009.0544

White, N. M., & Milner, P. M. (1992). The psychobiology of reinforcers. *Annual Review of Psychology, 43*(1), 443–471. https://doi.org/10.1146/annurev.ps.43.020192.002303

The White House. (2020, February 26). Remarks by President Trump, Vice President Pence, and members of the Coronavirus Task Force in press conference. The White House. https://trumpwhitehouse.archives.gov/briefings-statements/remarks-president-trump-vice-president-pence-members-coronavirus-task-force-press-briefing-10/

Wilczynski, W. (2009). Evolution of the brain in amphibians. In M. D. Binder, N. Hirokawa, & U. Windhorst (Eds.), *Encyclopedia of neuroscience* (pp. 1301–1305). Springer. https://doi.org/10.1007/978-3-540-29678-2_3148

Wilkins, M. C. (1928). The effect of changed material on the ability to do formal syllogistic reasoning. *Archives of Psychology, 16*(102), 5–83.

Wilkinson, G. S. (1984). Reciprocal food sharing in the vampire bat. *Nature, 308*(5955), 181–184. https://doi.org/10.1038/308181a0

Willerman, L., Schultz, R., Neal Rutledge, J., & Bigler, E. D. (1991). In vivo brain size and intelligence. *Intelligence, 15*(2), 223–228. https://doi.org/10.1016/0160-2896(91)90031-8

Wilson, D. S., & Wilson, E. O. (2007). Rethinking the theoretical foundation of sociobiology. *The Quarterly Review of Biology, 82*(4), 327–348. https://doi.org/10.1086/522809

Wilson, E. O. (1975). *Sociobiology: The new synthesis*. Harvard University Press. https://www.jstor.org/stable/j.ctvjnrttd

Wilson, M. L., Boesch, C., Fruth, B., Furuichi, T., Gilby, I. C., Hashimoto, C., . . . Wrangham, R. W. (2014). Lethal aggression in *Pan* is better explained by adaptive strategies than human impacts. *Nature, 513*(7518), 414–417. https://doi.org/10.1038/nature13727

Wirth, S., Yanike, M., Frank, L. M., Smith, A. C., Brown, E. N., & Suzuki, W. A. (2003). Single neurons in the monkey hippocampus and learning of new associations. *Science, 300*(5625), 1578–1581. https://doi.org/10.1126/science.1084324

Wrangham, R. W. (1999). Evolution of coalitionary killing. *American Journal of Physical Anthropology, 110*(S29), 1–30. https://doi.org/10.1002/(SICI)1096-8644(1999)110:29+<1::AID-AJPA2>3.0.CO;2-E

Wrangham, R. W., & Glowacki, L. (2012). Intergroup aggression in chimpanzees and war in nomadic hunter-gatherers. *Human Nature, 23*(1), 5–29. https://doi.org/10.1007/s12110-012-9132-1

Wren, A. M., Seal, L. J., Cohen, M. A., Brynes, A. E., Frost, G. S., Murphy, K. G., . . . Bloom, S. R. (2001). Ghrelin enhances appetite and increases food intake in humans. *The Journal of Clinical Endocrinology & Metabolism, 86*(12), 5992–5995. https://doi.org/10.1210/jcem.86.12.8111

Wu, J., Balliet, D., & Van Lange, P. A. M. (2016). Reputation management: Why and how gossip enhances generosity. *Evolution and Human Behavior, 37*(3), 193–201. https://doi.org/10.1016/j.evolhumbehav.2015.11.001

Wu, M. V., & Shah, N. M. (2011). Control of masculinization of the brain and behavior. *Current Opinion in Neurobiology, 21*(1), 116–123. https://doi.org/10.1016/j.conb.2010.09.014

Yeo, G. S. H., & Heisler, L. K. (2012). Unraveling the brain regulation of appetite: Lessons from genetics. *Nature Neuroscience, 15*(10), 1343–1349. https://doi.org/10.1038/nn.3211

Yeomans, M. R. (1996). Palatability and the micro-structure of feeding in humans: The appetizer effect. *Appetite, 27*(2), 119–133. https://doi.org/10.1006/appe.1996.0040

Yeomans, M. R., Blundell, J. E., & Leshem, M. (2004). Palatability: Response to nutritional need or need-free stimulation of appetite? *British Journal of Nutrition, 92*(S1), S3–S14. https://doi.org/10.1079/BJN20041134

Yeomans, M. R., & Gray, R. W. (1996). Selective effects of naltrexone on food pleasantness and intake. *Physiology & Behavior, 60*(2), 439–446. https://doi.org/10.1016/S0031-9384(96)80017-5

Young, D. G. (2019). *Irony and outrage: The polarized landscape of rage, fear, and laughter in the United States*. Oxford University Press.

Young, P. T. (1947). Studies of food preference, appetite and dietary habit. VII. Palatability in relation to learning and performance. *Journal of Comparative and Physiological Psychology, 40*(2), 37–72. https://doi.org/10.1037/h0061360

Young, P. T. (1959). The role of affective processes in learning and motivation. *Psychological Review, 66*(2), 104–125. https://doi.org/10.1037/h0045997

Young, P. T., & Asdourian, D. (1957). Relative acceptability of sodium chloride and sucrose solutions. *Journal of Comparative and Physiological Psychology, 50*(5), 499–503. https://doi.org/10.1037/h0047303

Young, P. T., & Falk, J. L. (1956). The relative acceptability of sodium chloride solutions as a function of concentration and water need. *Journal of Comparative and Physiological Psychology, 49*(6), 569–575. https://doi.org/10.1037/h0047744

Young, P. T., & Greene, J. T. (1953). Quantity of food ingested as a measure of relative acceptability. *Journal of Comparative and Physiological Psychology, 46*(4), 288–294. https://doi.org/10.1037/h0061302

Young, P. T., & Shuford, E. H., Jr. (1954). Intensity, duration, and repetition of hedonic processes as related to acquisition of motives. *Journal of Comparative and Physiological Psychology, 47*(4), 298–305. https://doi.org/10.1037/h0055982

Zabel, C. J., Glickman, S. E., Frank, L. G., Woodmansee, K. B., & Keppel, G. (1992). Coalition formation in a colony of prepubertal spotted hyenas. In A. Harcourt & F. B. M. de Waal (Eds.), *Coalitions and alliances in humans and other animals* (pp. 113–134). Oxford University Press.

Zha, X., & Xu, X. (2015). Dissecting the hypothalamic pathways that underlie innate behaviors. *Neuroscience Bulletin, 31*(6), 629–648. https://doi.org/10.1007/s12264-015-1564-2

Zucker, K. J., Bradley, S. J., Oliver, G., Blake, J., Fleming, S., & Hood, J. (1996). Psychosexual development of women with congenital adrenal hyperplasia. *Hormones and Behavior, 30*(4), 300–318. https://doi.org/10.1006/hbeh.1996.0038

Zuloaga, D. G., Puts, D. A., Jordan, C. L., & Breedlove, S. M. (2008). The role of androgen receptors in the masculinization of brain and behavior: What we've learned from the testicular feminization mutation. *Hormones and Behavior, 53*(5), 613–626. https://doi.org/10.1016/j.yhbeh.2008.01.013

Index

Locators followed by *f, t,* or *b* indicate a figure, table, or box.

Press
et Rossi
Main Street, 9th floor
02142

edu
tt@mit.edu
253-2882

authorized representative in the EU for product safety and compliance is

Access System Europe Oü, 16879218
tamäe tee 50,
10621

requests@easproject.com
56 968 939

9780262045476
ase ID: 152941459

www.ingramcontent.com/pod-product-compliance
Lightning Source LLC
Chambersburg PA
CBHW032302280326
41932CB00009B/667